PATHWAYS TO SOLUTIONS, FIXED POINTS, AND EQUILIBRIA

W. I. Zangwill
University of Chicago
Chicago, Illinois

C. B. Garcia
University of Chicago
Chicago, Illinois

Prentice-Hall, Inc. Englewood Cliffs, N.J. 07632

Prentice-Hall Series in Computational Mathematics

Cleve Moler, Advisor

©1981 by Prentice-Hall, Inc., *Englewood Cliffs, N.J. 07632*

Editorial/production supervision by *Aliza Greenblatt*
Interior design by *Karen J. Clemments*
Manufacturing buyers: *Gordon Osbourne* and *Joyce Levatino*

Printed in the United States of America

10 9 8 7 6 5 4 3 2 1

PRENTICE-HALL INTERNATIONAL, INC., *London*
PRENTICE-HALL OF AUSTRALIA PTY. LIMITED, *Sydney*
PRENTICE-HALL OF CANADA, LTD., *Toronto*
PRENTICE-HALL OF INDIA PRIVATE LIMITED, *New Delhi*
PRENTICE-HALL OF JAPAN, INC., *Tokyo*
PRENTICE-HALL OF SOUTHEAST ASIA PTE. LTD., *Singapore*
WHITEHALL BOOKS LIMITED, *Wellington, New Zealand*

Contents

Foreword

Numerical methods for the solution of systems of nonlinear equations have long been available. Until quite recently, however, they were local in character in the sense of requiring an accurate estimate of the correct solution in order to guarantee convergence of the algorithm. Global methods were developed approximately two decades ago in an attempt to solve a class of problems which were central to economics, though surely peripheral to the concerns of most numerical analysts.

The economic problem was to find a vector of prices which would equate supply and demand in a system of interrelated markets. We are, of course, all familiar with the elementary economic considerations which suggest the way in which prices move toward equilibrium. If, at a particular price, the demand for a specific commodity exceeds its supply, then an increase in the price of that commodity will presumably decrease its demand, call forth an increased supply, and narrow the gap between the two sides of the market.

If such a price adjustment mechanism is to have general economic validity, it should be capable of providing an algorithm for the solution of the equilibrium equations which arise when the economic problem is given mathematical form. The most natural translation of this market adjustment mechanism is to a system of nonlinear, first-order differential equations in which the rate of change of price is proportional to the discrepancy between supply and demand. But economic intuition and mathematical technique come to an abrupt conflict at this point. There is no reason for these differential equations to converge and, from a mathematical point of view, the complete disregard of information conveyed by first-order derivatives would seem to lead to particularly inefficient algorithms.

The difficulties of numerical calculation were avoided, for many years, by alternative developments in mathematical economics. The central concern became that of demonstrating the existence of equilibrium prices. This involved making use of non-constructive methods, such as the Brouwer and Kakutani fixed point theorems. The shift in point of view which allowed fixed point techniques to be used as effective computational algorithms took place in the 1960s. Since then the field has developed substantially. Not only has there been an enormous improvement in the numerical methods themselves, and an extension of the range of applications to problem areas far from mathematical economics, but there has also been a substantial increase in our understanding of how fixed point methods relate to other branches of mathematics.

In these pages, Garcia and Zangwill bring this material together in an elegant and lucid presentation, which makes these important developments available to the general reader, provides a wealth of fascinating examples, and bears their unique intellectual signatures.

Herbert E. Scarf
New Haven, Connecticut, 1981

Preface

The problem underlying the finding of a solution, an equilibrium, or a fixed point is virtually the same in each case and is of immense practical importance. Yet that problem remained unsolved until only recently when an exciting field emerged, a field we call path following. In not much over a decade, this field has witnessed brilliant work and today its practical and theoretical aspects are being developed at an ever-increasing rate. Fundamental issues are attacked. Solutions to nonlinear equations are needed in virtually every aspect of life. Equilibrium means balance and most systems operate in some sort of equilibrium, be they physical, economic, social, or biological. Fixed points are a basic mathematical notion and a prime means of calculating equilibria.

We have published a number of papers on path following, and now in this book attempt to demonstrate the field's broad implications. The underlying idea is truly simple: from where we are now, we follow a path to the point desired, that is, to the solution, fixed point, or equilibrium. It is like following a road on a map, although since our "roads" are in n dimensions, our trip will be sometimes perilous, frequently fun, and nearly always fascinating.

Historical Perspective

The historical antecedents to path following are several. One that is over a century old is the homotopy approach, which verifies when it is theoretically possible to reach the point desired. It does not specify any means to reach that point, however. In other words, it does not state how to acutally calculate the solution, fixed point, or equilibrium, but merely that such a point exists. Another major step was achieved by Davidenko [1953a, b], who added certain path-following aspects to the homotopy idea. The differential equations he conceived provided impetus to the field and were applied to a variety of problems, including integral equations, matrix eigenvalue problems, and determinant evaluations.

Something was still missing and, as is widely documented, there was no general approach for generating a solution, fixed point, or equilibrium. For example, according to McKenzie [1976], the economist Frank Graham submitted the problem to mathematicians of great repute, who were unable to furnish the formula "perhaps because they were not sufficiently interested to devote it the necessary time." The verbal tradition at Princeton was that the mathematician was John von Neumann, who was at the Institute for Advanced Study at the time.

The breakthrough in actually determining the solution, fixed point, or equilibrium was due to Scarf [1967], who, curiously, did not use either a homotopy or a path-following approach, but an entirely novel technique based upon what he called primitive sets. Related to this work is an earlier work by Cohen [1967]. The theoretical principle Scarf employed was that of complementarity, developed by Lemke and Howson [1964].

As Scarf's method became better understood, Hansen and Scarf [1969] and Kuhn [1968] replaced the primitive sets with simplices, and Eaves [1972] pointed out the relation to homotopies. Eaves [1972], Eaves and Saigal [1972], Kuhn and MacKinnon [1975], and Merrill [1971] conceived some of the early algorithms, and numerous others have since built and continue to build on these ideas.

Working from differential equations, Kellogg, Li, and Yorke [1976] also proved the Brouwer fixed-point theorem. Their method built on the earlier efforts of Hirsch [1963].

As it evolved, the Scarf ideas led to algorithms which generate piecewise linear paths. The Davidenko ideas led to algorithms that produce differentiable paths. Today, the research stemming from Scarf's and Davidenko's ideas have largely intertwined. In both instances a path is to be followed; it is merely the particular form of the path that differs. Which form is superior, a piecewise-linear or a differentiable path, usually depends upon the problem under consideration. Nevertheless, the principle underlying both approaches is now known to be essentially the same: follow a path to the desired point.

This Book

The foundation concept of this book is path following, and, as mentioned, there are two common forms, differentiable paths and piecewise-linear paths. This book commences with differentiable paths, as that seems superior pedagogically and produces elegant mathematical proofs. Piecewise-linear paths require the paraphernalia of simplices, triangulations, and pivoting, which, when presented too early, often confuse the student. Differentiable paths also permit the important notion of orientation to be elucidated nicely, and that is done in Chapter 2 via the basic differential equations. Orientation for a piecewise-linear path is cumbersome, and requires an unintuitive definition using the sign of a deter-minant. By contrast, the basic differential equations define orientation naturally and more powerfully as the direction of movement along the path.

Only later in the book, after the key ideas are well established, are piece-wise-linear paths introduced, and at that point they are relatively easy to grasp. After completing the text the reader should be conversant with both approaches and be able to apply whichever is needed to solve a particular problem.

Applications

Applications are emphasized and a large section of the book is devoted solely to the practical uses of the field. Equilibrium provides a fundamental means to represent the behavior of many systems, and to help apply it we introduce equilibrium programming. The equilibrium programming model is quite general, and by changing the constraints and objective functions, uses in economics, game theory, networks, transportation, engineering, and many other fields are easily obtained.

A different application area stresses the dynamic behavior of a system, and that is explored by presenting the new and interesting field of catastrophe theory.

Overall, we have tried to include a wide range of applied material, and we hope that engineers, economists, social scientists, chemists, physicists, managers, philosophers, political scientists, computer scientists, and others will find the book relevant to their work. Path following is a fundamental technique whose applications should grow enormously in the years ahead. One goal of this book is to provide the reader with the insight and understanding necessary to partake in these continuing and very exciting developments.

Research Results

For the more theoretically minded reader, the book contains considerable new research, including many new theorems and proofs. It is important to note that we develop the field using fairly elementary mathematics. In contrast, other presentations may lose the reader in a welter of simplices, triangulations, and other preliminaries or require an extensive background in differential topology. The mathematical simplifications were possible because our research revealed how to dispense with much of the abstruse apparatus previously required and thereby clarify the central concepts. We hope that the new results presented herein not only aid the researcher and the practitioner, but better highlight the fundamental ideas underlying path following, fixed points, equilibrium, catas-trophes, applications, algorithms, and related areas.

Prerequisites

The text assumes some knowledge of nonlinear programming, advanced calculus, and linear algebra. The nonlinear programming prerequisites include linear programming pivot theory, convexity, and Kuhn–Tucker theory. In advanced calculus, the text assumes the usual topics of limits and sequences, continuity, partial differentiation, and Taylor's expansion. As for linear algebra, the standard knowledge of matrix and vector manipulations and linear independence are necessary. Appendices are provided at the end of the book to refresh the reader's understanding of these topics.

Book Organization

The book is subdivided into four parts. The first deals with the basic theory of path following. Chapter 1 introduces the homotopy idea and demonstrates the existence of a path. Chapter 2 discusses the basic differential equations that both enable us to follow the path in n dimensions and provide information on the path direction. The third chapter unites, using the path-following approach, certain powerful mathematical theorems, such as the Brouwer theorem, the Leray–Schauder theorem, an existence-of-solutions theorem, and theorems on degree and homotopy invariance.

Part II of the book deals with applications. In Chapter 4, we revise the Kuhn–Tucker conditions of nonlinear programming, discuss the path-following approach for it, and extend nonlinear programming into the important area of dynamics. Chapter 5 explains the idea of equilibrium, defines the equilibrium programming problem, and provides a path-following algorithm for calculating the equilibrium point. The next four chapters emphasize the importance of equilibrium programming in real-world uses. Chapters 6 and 7 analyze the economic equilibrium problem and state a direct and novel approach for proving competitive equilibrium without resorting to a fixed-point theorem. Chapter 8 explores game theory, and Chapter 9 extends the approach to networks, traffic equilibria, and elasticity. Chapter 10 concludes the applications section by connecting path following with the very important field of catastrophe theory (Thom [1972]).

The third part of the book develops the algorithms and techniques needed for solving specific problems. Chapters 11 through 14 articulate algorithms that are based on triangulations, simplices, and simplicial pivoting, whereas Chapters 15 and 16 describe algorithms based on differential equations. These algorithms are global in nature and can follow the path all the way to a solution. Newton methods, which are more local in nature but are often very rapid, are also analyzed and given a path-following interpretation. If the functions are of a special kind, such as separable or contractive, then special-purpose algorithms can be devised, as shown in Chapter 17.

The fourth and last part of the book discusses certain fundamental concepts. Chapter 18 demonstrates how all solutions to a given system of equations are obtained. Chapters 19 and 20 define the linear complementarity problem and describe the Lemke algorithm. These two chapters are noteworthy, since much of the literature on path following was inspired by Lemke's algorithm. Point-to-set mappings and the Kakutani fixed-point theorem are discussed in Chapter 21.

Such mappings occur frequently in economic theory and are vital to understanding multivalued systems. Finally, Chapter 22 discusses Sard's theorem and Weierstrass's theorem in order to illustrate the "stability," "genericity," and "typicalness" of the approach. These two theorems are important and demonstrate that the assumption of regularity and differentiability assumed throughout the book are extremely weak and can be eliminated by perturbation arguments.

Course Outline

The text contains sufficient material to serve as a first-year graduate-level course in such areas as mathematical programming, numerical analysis, or economic equilibrium theory. The student in mathematical programming can cover the basic chapters—1, 2, and 3—then the topics on nonlinear programming, equilibrium programming, networks—Chapters 4 and 5 and Sections 9.1 through 9.6—and finally, the linear complementarity problem—Chapters 19 and 20. A course in numerical analysis might consist of Chapters 1, 2, 3, 11, 12, 13, 15, 16, 18, and 22. A course in economic equilibrium theory could consist of Chapters 1, 2, 3, 4, 5, 6, 7, 8, and 21.

A reader interested in other topics, such as algorithms, mathematics, management, engineering, or social science, can easily find the subjects needed.

Appreciation

Over the years many people have stimulated our thoughts and prompted us to probe ever deeper in our efforts to comprehend this vast and fascinating field of research. Herbert Scarf, whose ideas founded much of the field, has continually been an inspiration. Carlton Lemke also contributed much of the field's early concepts and has always been supportive. B. C. Eaves, Harold Kuhn, Werner Rheinboldt, and F. J. Gould have been helpful in many ways. To Dean Rosett, Associate Dean Graves, and our colleagues we express thanks for making the University of Chicago such an exciting and stimulating environment. Also, our deep appreciation extends to typists Maggie Newman, Evelyn Shropshire, and Nerissa Walton and administrative supervisors Jane Hilmers and Elizabeth Kinnear for their fine efforts. Last, and perhaps unusually, we wish to express appreciation to each other. The book, as all our research, has truly been a joint venture, with each of us spurring on and helping the other. Indeed, we have expressly used our names in both orders, Garcia–Zangwill and Zangwill–Garcia, in the book to emphasize this. Thus our thanks to each other.

C. B. Garcia
University of Chicago

W. I. Zangwill
University of Chicago

Follow the **Yellow** Brick Road.*

*Adaptation from *Little Wizard Stories of Oz*, by L. Frank Baum, illustrated by John R. Neill, C. 1914 with the permission of Contemporary Books, Inc.

I

BASIC THEORY

Solutions, Homotopies, and Paths

1.1 SOLVING EQUATIONS

You as an executive, economist, scientist, butcher, baker, or even candlestick maker will undoubtedly encounter and employ mathematical representations of real situations. The mathematics may be elementary or it may be sophisticated, but at some point you will use it.

Viewing yourself as such a user, consider: What is the most important problem in mathematics? That is a simplistic, even naive, question to pose. If we ask 10 people, we would receive at least 11 different answers. For mathematics that is actually applied and used, however, we feel that the question has at least a partial answer. To achieve a result for a real problem generally requires obtaining the solution to a system of equations. Indeed, determining the solution is fundamental, because what generally provides the practical meaning to the mathematics is the solution and its interpretation.

This book explores an especially broad class of these problems, obtaining solutions to nonlinear systems of equations or their equivalent. Usually, these systems have n variables and n equations, and we want to identify the points $x = (x_1, x_2, \ldots, x_n)$ that solve the system. Systems such as these, some with hundreds of variables, arise in virtually every aspect of life. Problems that do not even look like these—for instance, game problems, equilibria, fixed points, and programming problems—can be reduced to this form. Economics, biology, engineering, statistics, sociology, business, and other fields abound in problems that require solving systems of this nature. Solving systems of nonlinear equations is indeed a basic, fascinating, and crucial problem. This book presents some of the most recent developments not only in solving these and related systems but in applying them to a panoply of problems. Later in the book we detail such applications as nonlinear programming, equilibrium in systems, dynamics and control, fixed points, economic equilibrium, organizational operation, transportation, catastrophe behavior, stress and elasticity, plus a host of others. These new concepts and applications have ignited great interest and enthusiasm among researchers throughout the world. We hope this book provides the reader not only with the tools to understand this new field but also with some of the excitement.

1.2 THE HOMOTOPY PRINCIPLE

Suppose that a system of nonlinear equations is given for you to solve. As we have discussed, the system could represent any one of a vast range of problems. Indeed, the system itself could have arisen out of almost any real problem in almost any field and, as mentioned, later in the book we discuss in detail how such systems arise. But suppose that we are faced now with the task of solving such a system.

A simple example of a system that might arise is

$$(x_1)^2 + x_1 x_2 = 3$$
$$-7 + 2 \sin x_1 + x_2 = 0.$$

Another example might be

$$x_1 + x_2 - 3x_3 = 7$$
$$x_1^2 + 2x_2 = 4$$
$$-x_1 + x_2 x_3 = 3.$$

Our immediate goal is to find a point $x = (x_1, \ldots, x_n)$ that solves such a system. How might we accomplish this?

One approach is to start with another system of equations to which we already know the solution. Usually, this is a particularly simple system that has an obvious solution. We then take this simple system and mathematically "bend" it into the original system. While bending the system we carefully watch the solution, as it also "bends" from the obvious solution into the solution we seek. This bending notion underlies a key idea that we shall develop shortly, the *homotopy* concept.

Most of us have probably used this concept at one time or another. For example, the analyst often starts with a simpler system or model, one with an easily obtained solution. He or she then "bends" or twists the simple system into the desired system. Carefully done, this process solves the initial given problem.

Intuitive and nonrigorous, this approach seems to work, although usually no general explanation is provided for why and how. Our task is to provide a solid, rigorous, and general formulation for this procedure, and that is achieved with the homotopy concept. Indeed, homotopies will permit us to solve a truly impressive range of problems, possibly more than any other competitive method known.

EXAMPLE 1.2.1 (*A Linear Example*)

To begin, let us see how this "bending" notion might operate for a linear system. Let A be an $n \times n$ matrix and b an $n \times 1$ vector, and suppose that we seek an $n \times 1$ vector x^* to solve

$$Ax = b$$

or equivalently, and as it is often written,

$$Ax - b = 0.$$

Suppose we already know that the $n \times 1$ vector x^0 solves the system

$$Ax^0 - b - d = 0$$

for some given vector d. Typically, d is expressly chosen so that the system $Ax = b + d$ is extremely easy to solve. This system will serve as our simple system and has by its construction an already known or obvious solution x^0.

Now consider the system for the scalar t, where $0 \leq t \leq 1$:

$$Ax - b - d + td = 0$$

or

$$Ax = b + d - td.$$

Observe that this is a system of n equations in $n + 1$ variables $x_1, \ldots, x_n$ and t. The additional variable t is called the *homotopy parameter*. It will serve to "bend" the system.

By setting $t = 0$, we obtain

$$Ax = b + d,$$

which we know has the solution x^0. If $t = 1$, we obtain

$$Ax = b,$$

which has as its solution x^*, the point we are seeking. For an intermediate t, $0 < t < 1$, the system is partially "bent." If, say, $t = \frac{1}{2}$, then

$$Ax = b + \tfrac{1}{2}d$$

with a solution that, no doubt, is somewhere between x^0 and x^*.

To explicitly show how x depends upon t, let us write $x(t) = (x_1(t), x_2(t),$

$\ldots, x_n(t))$. Then at any t, if A is an invertible matrix,

$$x(t) = A^{-1}[b + d - td]$$
$$= x^0 - tA^{-1}d,$$

where A^{-1} denotes the inverse of A. Obviously,

$$x^0 = x(0) \qquad \text{and} \qquad x^* = x(1).$$

Moreover, if we increase t from 0 to 1, a path is obtained from x^0 to x^*. Presumably, we could actually follow this path from x^0 to x^* and thereby solve the original system.

Rarely would one use this means for solving a system of linear equations, as other means are generally available. But Example 1.2.1 does serve as an excellent introduction to the homotopy method for nonlinear systems. In particular, for the homotopy approach:

1. We start with a simple system that is specially chosen to have an obvious solution. In this example the simple system is $Ax = b + d$ with known solution x^0.

2. We set up a system of equations with an additional variable t that yields the simple system at $t = 0$, while at $t = 1$ it yields the original system.

3. We follow the path $x(t)$ of the system's solution from $t = 0$ to $t = 1$, thereby solving the original problem.

Example 1.2.1 followed this approach, and t served to "bend" the simple system into the original one. Note particularly that as t changed, we obtained a path from the obvious solution, x^0, to the solution we sought, x^*.

EXAMPLE 1.2.2 (*A Nonlinear Example*)

To further illustrate the homotopy notion, let us apply it to solving the nonlinear system

$$(x_1)^3 - 3(x_1)^2 + 8x_1 + 3x_2 - 36 = 0$$
$$(x_1)^2 \qquad + x_2 + 4 = 0.$$

Take as our initial system of equations

$$(x_1)^3 + 8x_1 + 3x_2 = 0$$
$$x_2 = 0.$$

The only real solution to this is obviously $(x_1^0, x_2^0) = (0, 0)$.

Next, utilize our homotopy parameter t to construct a system that yields our initial system at $t = 0$ and the originally given system at $t = 1$. That system is

$$(x_1)^3 + 8x_1 + 3x_2 - t(3(x_1)^2 + 36) = 0$$
$$x_2 + t((x_1)^2 + 4) = 0.$$

Letting $(x_1(t), x_2(t))$ be its solution, we see that $(x_1(0), x_2(0)) = (0, 0)$, while $(x_1(1), x_2(1))$ is the point we wish to obtain.

If we eliminate x_2, the system factors into

$$(x_1 - 6t)((x_1)^2 + 8) = 0.$$

Thus

$$x_1(t) = 6t$$
$$x_2(t) = -36t^3 - 4t.$$

The point $(x_1(t), x_2(t))$ describes a path from $(x_1(0), x_2(0))$ as t increases from 0 to 1 (see Figure 1.2.1). Following this path leads us directly to the solution, $(x_1(1), x_2(1)) = (6, -40)$.

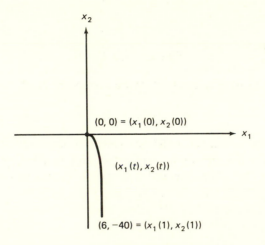

The path $(x_1(t), x_2(t))$ as t increases from 0 to 1.

Figure 1.2.1

SPECIFICS

With the homotopy idea fairly clear, we must now state it more precisely.

Let R^n denote *Euclidean n space*. A function $F: R^n \rightarrow R^n$ means that $F(x)$ has n components, $F(x) = (F_1(x), \ldots, F_n(x))$, and that x has n components, $x = (x_1, \ldots, x_n)$, so that

$$F_i(x) = F_i(x_1, \ldots, x_n), \qquad i = 1, \ldots, n.$$

We are desirous of solving the $n \times n$ system of nonlinear equations

$$F(x) = 0$$

or, in detail, solving

$$F_1(x_1, \ldots, x_n) = 0$$
$$F_2(x_1, \ldots, x_n) = 0$$
$$\cdot$$
$$\cdot$$
$$\cdot$$
$$F_n(x_1, \ldots, x_n) = 0$$

for a solution $x^* = (x_1^*, \ldots, x_n^*)$.

The homotopy process to obtain x^* is as follows. First set up a simple system

$$E_1(x_1, \ldots, x_n) = 0$$
$$\cdot$$
$$\cdot$$
$$\cdot$$
$$E_n(x_1, \ldots, x_n) = 0,$$

which we select to have solution x^0. More succinctly, $E: R^n \longrightarrow R^n$ and x^0 solves

$$E(x) = 0.$$

Then define a special function, called a *homotopy function*, $H(x, t)$: $R^{n+1} \longrightarrow R^n$, which has the original n variables plus an extra one, t. Here $(x, t) = (x_1, \ldots, x_n, t) \in R^{n+1}$. The homotopy function H must be constructed so that

$$H(x, 0) = E(x)$$
$$H(x, 1) = F(x).$$

It follows that at $t = 0$,

$$H(x, 0) = 0$$

has a solution x^0, which we already know, and at $t = 1$,

$$H(x, 1) = 0$$

has solution x^*, which we seek. In general for arbitrary t, $x(t)$ solves

$$H(x(t), t) = 0.$$

The idea is to start at $x(0) = x^0$ and then increase t until we reach $x(1) = x^*$. Generally, $x(t)$ will generate a path that we can follow from $t = 0$ to $t = 1$, thereby solving the original system.

Our previous examples followed this approach. For instance, in Example 1.2.1, $E(x) = Ax - b - d$, $F(x) = Ax - b$, and $H(x) = Ax - b - (1 - t)d$. In

Example 1.2.2,

$$H_1(x, t) = (x_1)^3 + 8x_1 + 3x_2 - t(3(x_1)^2 + 36) = 0$$
$$H_2(x, t) = \qquad\qquad x_2 + t(\ (x_1)^2 + \ 4) = 0.$$

In sum, the homotopy approach specifies a homotopy function $H(x, t)$: $R^{n+1} \longrightarrow R^n$, where $H(x(t), t) = 0$, $H(x, 0) = E(x)$, and $H(x, 1) = F(x)$, so that:

1. $H(x, 0) = 0$ is easily solved for $x(0)$.

2. $x(1)$, the solution to $H(x, 1) = 0$, is the solution we want.

3. The path $x(t)$ leads from $x(0)$ to $x(1)$ as t increases.

Then by following the path from $x(0)$, we find the solution $x^* = x(1)$. In a nutshell this specifies the homotopy approach of following a path to the solution. This approach, as we will see as the book unfolds, is remarkably powerful.

EXAMPLE 1.2.3   (A Piecewise-Nonlinear Example)

When the problems become more and more complicated, the idea of following the path $x(t)$ becomes more and more useful. Even in the following rather straightforward one-variable example, it provides insight. Let

$$F(x) = \begin{cases} 2x - 5, & x \le 1 \\ x^2 - 4, & x \ge 1 \end{cases}$$

and suppose that we wish to solve

$$F(x) = 0$$

via the homotopy approach.

Here F is a *piecewise* function. By "piecewise" is meant that the entire domain is divided into subsets (pieces), and on each subset the function is defined differently. In this example the domain is divided into two parts, $x \le 1$ and $x \ge 1$, on which F is different. However, the function values are required to match up on the boundaries of the pieces, so that the function is continuous on the whole domain. In the example the boundary is $x = 1$ and $F(x) = -3$ there.

In general, there could be many, even an infinite number of subsets dividing the domain. Almost all practical applications use nicely shaped subsets, say polyhedra. For example, in two dimensions the domain might be divided into rectangles or triangles (Figure 1.2.2). In three dimensions the subsets might be like boxes or tetrahedra. Also, to simplify, we require that any two intersecting subsets must intersect only on their common boundary. Figure 1.2.2 depicts that.

Although this will be discussed more in later chapters, in short a piecewise function F has its domain divided into subsets such that

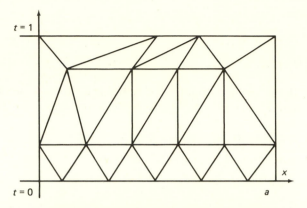

The region $0 \le x \le a$ and $0 \le t \le 1$ is subdivided into triangles. The function $F(x, t)$ may be defined differently on each triangle, but must match up on the boundaries.

A Piecewise Function

Figure 1.2.2

1. Two intersecting subsets intersect only on their common boundary.

2. F is different on each subset.

3. F must have the same value on the boundary where two subsets intersect.

Our example is piecewise, but also note that it is piecewise nonlinear, because at least one part is nonlinear. A *piecewise-linear* function would be linear on each subset of the domain. A *piecewise-differentiable* function would be differentiable on each subset of the domain.

Let us now apply the homotopy approach to the example. Define

$$H(x, t) = (1 - t)x + tF(x) = 0.$$

We have selected a particularly straightforward, simple function E. In fact,

$$H(x, 0) = x,$$

so that

$$x(0) = 0.$$

By definition of F,

$$H(x, t) = \begin{cases} (1 - t)x + t(2x - 5), & x \le 1 \\ (1 - t)x + t(x^2 - 4), & x \ge 1. \end{cases}$$

At $t = 0$, $x(0) = 0$; thus if we increase t from 0,

$$H(x, t) = (1 - t)x + t(2x - 5) = 0$$

and

$$x(t) = \frac{5t}{1 + t} \qquad \text{if } 0 \le t \le \tfrac{1}{4}.$$

At $t = \tfrac{1}{4}$, $x(\tfrac{1}{4}) = 1$.

To increase t above $\frac{1}{4}$, x increases above 1, so by definition of F,

$$H(x, t) = (1 - t)x + t(x^2 - 4) = 0 \quad \text{if } \tfrac{1}{4} \le t \le 1.$$

This yields, after some algebra,

$$x(t) = \frac{(t - 1) + \sqrt{17t^2 - 2t + 1}}{2t}$$

for $\frac{1}{4} \le t \le 1$.

Increasing t from $\frac{1}{4}$ to 1, we obtain the solution

$$x(1) = +2.$$

In this example we started with $x(0) < 1$, increased t until $x(t) = 1$, and then had to switch to the other form of F for $x(t) \ge 1$ to increase t further. Particularly in higher dimensions, where F may switch forms many times, following the path $x(t)$ can be very useful. Then $x(t)$ pinpoints the piece of the space we are on, and hence which form of F must be used.

Note that $x^* = +2$ is the only solution to $F(x) = 0$. This is true even though at $x = -2$, $F(x) = x^2 - 4 = 0$. The point $x = -2$ is not a solution because there $F(x) = 2x - 5$. Thus path following led us directly to the "right" solution.

Piecewise functions will arise often in the applications we consider later and the path will lead us to the solution.

There are instances, of course, when an equation system has more than one solution, as the next example shows.

EXAMPLE 1.2.4   (A System with Multiple Solutions)

The homotopy method can help us obtain more than one solution to a system of equations, assuming that there is more than one solution. The idea here is that instead of commencing with only one start point $x(0)$, more than one start point is utilized. Then we generate several paths, one from each of the different start points. Following these paths will then often lead us to more than one, if not all of the solutions.

To illustrate, suppose that we wish to obtain both solutions to

$$x^2 + x - 6 = 0.$$

Utilize the homotopy

$$H(x, t) = (1 - t)(x^2 - 1) + t(x^2 + x - 6) = 0.$$

At $t = 0$,

$$H(x, 0) = x^2 - 1 = 0,$$

which is selected because it has two easily obtained solutions and thus two easily obtained start points, $x^1(0) = +1$ and $x^2(0) = -1$. Solving $H(x, t) = 0$ produces two paths:

$$x^1(t) = \frac{-t + \sqrt{t^2 + 20t + 4}}{2}$$

and

$$x^2(t) = \frac{-t - \sqrt{t^2 + 20t + 4}}{2}$$

The path $x^1(t)$ goes from $+1$ to $+2$ while the path $x^2(t)$ goes from -1 to -3 (see Figure 1.2.3). Multiple starts thus provide multiple solutions, as is further explained in Chapter 18.

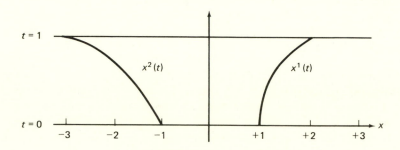

Multiple starts provide multiple solutions.

Figure 1.2.3

1.3 VARIETIES OF HOMOTOPIES

At this stage the notion of commencing at a given known point $x(0)$ and following the path $x(t)$ until t hits one should be clear. Because of the simplicity of the examples, one could solve them almost by inspection, and the path following seems superfluous. In n dimensions, however, such easy visual perceptions rarely exist, and we are forced to inch along the path. There it truly is like a dark jungle and the path is the way through the thick underbrush to the solution.

The exact path that occurs will depend directly upon the homotopy function $H(x, t)$ selected. So far we have specified that $E: R^n \longrightarrow R^n$, $F: R^n \longrightarrow R^n$, and $H: R^{n+1} \longrightarrow R^n$, where

$$H(x, 0) = E(x)$$
$$H(x, 1) = F(x),$$

but we have not elaborated upon the form of $H(x, t)$. Numerous forms exist and many will be discussed later, but three forms are the most common. Let us now specify them more precisely.

NEWTON HOMOTOPY

Example 1.2.1 utilized the homotopy

$$H(x, t) = F(x) - (1 - t)F(x^0). \qquad (1.3.1)$$

This form of homotopy is termed the *Newton homotopy* because some of the ideas behind it (as discussed in Chapter 15) come from the work of Sir Isaac Newton himself. It is a particularly simple method to start. Pick an arbitrary point x^0. Next calculate $F(x^0)$, and then let

$$E(x) = F(x) - F(x^0).$$

The function E, by construction, has solution x^0. From (1.3.1), the beginning of the path $x(0) = x^0$ is obvious and immediate. Since x^0 can be selected arbitrarily, the Newton homotopy permits us to start the path from wherever we choose, a very nice feature.

FIXED-POINT HOMOTOPY

In Example 1.2.3 a different form of homotopy was utilized,

$$H(x, t) = (1 - t)(x - x^0) + tF(x). \qquad (1.3.2)$$

This homotopy, called the *fixed-point homotopy*, will be extremely valuable later in proving fixed-point theorems as well as other results. Observe that

$$E(x) = x - x^0 = 0,$$

so that $x(0) = x^0$. The fixed-point homotopy is thus also easy to start since x^0 can be chosen arbitrarily.

Both the fixed-point and Newton homotopies can be started from any point x^0, which is one of the reasons that both are so widely employed.

LINEAR HOMOTOPY

The third homotopy form, called the *linear homotopy*, is a linear combination of the two functions E and F:

$$H(x, t) = tF(x) + (1 - t)E(x)$$
$$= E(x) + t[F(x) - E(x)]. \qquad (1.3.3)$$

Examples 1.2.2 and 1.2.4 illustrated this form. In Example 1.2.2, E was chosen to produce a convenient factorization. In Example 1.2.4, E was selected to have two start points, $x^1(0) = +1$ and $x^2(0) = -1$. When we want our starting function E to have some special properties, such as in Examples 1.2.2 and 1.2.4, the linear form is quite useful.

It might be observed that by selecting

$$E(x) = x - x^0,$$

the linear form reduces to the fixed-point form of the homotopy function. Also, if $E(x) = F(x) - F(x^0)$, then the linear form becomes the Newton form. The linear form thus subsumes the other two, but we want to highlight the other two forms because they are so useful and common.

1.4 PATH EXISTENCE

The idea of path following to a solution underlies this entire book. In our previous examples a path existed, where by "path" is meant a piecewise differentiable curve in space. However, such is not always the case. We want a path to exist so that we can follow it, and this section presents conditions that guarantee its existence.

Given a homotopy function $H: R^{n+1} \rightarrow R^n$, we must now be more explicit about solutions to

$$H(x, t) = 0.$$

In particular, define

$$H^{-1} = \{(x, t) \,|\, H(x, t) = 0\}$$

as the set of all solutions $(x, t) \in R^{n+1}$ to the system $H(x, t) = 0$. With our previous examples this set consisted of one or more paths. In Figure 1.2.3, for instance, this set consisted of two paths for $0 \le t \le 1$. The general situation, however, is much more complex, as H^{-1} could be rather arbitrary. The points (x, t) that satisfy $H(x, t) = 0$ could be all over the place and in no particular configuration. Still, H^{-1}, by definition, includes all of them.

Understand well that in H^{-1} both x and t vary. However, we also must denote the solutions for t fixed. Let

$$H^{-1}(t) = \{x \,|\, H(x, t) = 0\}.$$

Now $H^{-1}(0)$ consists of all the start points $x(0)$, or, equivalently, all solutions to

$$H(x, 0) = E(x) = 0.$$

Similarly, $H^{-1}(1)$ consists of all the points $x^* = x(1)$ which solve

$$H(x, 1) = F(x) = 0.$$

Carefully observe that H^{-1} has points $(x, t) \in R^{n+1}$ while $H^{-1}(t)$ has points x in R^n. For instance, in Example 1.2.4, for $0 \le t \le 1$,

$$H^{-1} = \left\{ \left(\frac{-t + \sqrt{t^2 + 20t + 4}}{2}, t \right), \left(\frac{-t - \sqrt{t^2 + 20t + 4}}{2}, t \right) \right\}$$

$$H^{-1}(t) = \left\{ \frac{-t + \sqrt{t^2 + 20t + 4}}{2}, \frac{-t - \sqrt{t^2 + 20t + 4}}{2} \right\}$$

$$H^{-1}(0) = \{1, -1\}$$

$$H^{-1}(1) = \{2, -3\}.$$

Intuitively, H^{-1} describes all solutions (x, t) where t is allowed to change. However, $H^{-1}(t)$ specifies only the x solutions for t fixed.

Using this notation, our task of determining when there are paths for solutions is the same as determining when H^{-1} consists of paths. In Figure 1 4.1, H^{-1} consists of differentiable curves only and hence consists of paths and only paths. However, Figure 1.4.2 depicts several situations that are not paths,

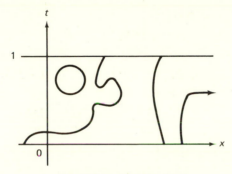

Here H^{-1} consists only of paths.

Figure 1.4.1

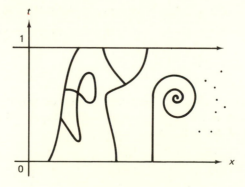

An H^{-1} consisting of isolated points, bifurcations, infinite spirals, and so on.

Figure 1.4.2

such as bifurcations, crossings, and infinite spirals. Situations such as that shown in Figure 1.4.2 make our approach extremely difficult; hence we now proceed to eliminate them and to ensure that H^{-1} consists only of paths.

IMPLICIT FUNCTION THEOREM

The key to ensuring that H^{-1} consists solely of paths is the implicit function theorem. As that theorem utilizes derivatives and Jacobians, we examine them first.

Let $L: R^m \longrightarrow R^k$ be a continuously differentiable function. The function L has k components $L = (L_1, \ldots, L_k)$ and $y = (y_1, \ldots, y_m)$ has m components. The Jacobian matrix of L at y, denoted $L'(y)$, is the $k \times m$ matrix of partial derivatives.

$$L'(y) = \begin{pmatrix} \dfrac{\partial L_1}{\partial y_1} & \cdots & \dfrac{\partial L_1}{\partial y_m} \\ \vdots & & \vdots \\ \dfrac{\partial L_k}{\partial y_1} & \cdots & \dfrac{\partial L_k}{\partial y_m} \end{pmatrix}. \tag{1.4.1}$$

For a homotopy function $H(x, t): R^{n+1} \longrightarrow R^n$, its Jacobian is an $n \times (n + 1)$ matrix which can be written as

$$H'(x, t) = \begin{pmatrix} \dfrac{\partial H_1}{\partial x_1} & \cdots & \dfrac{\partial H_1}{\partial x_n} & \dfrac{\partial H_1}{\partial t} \\ \vdots & & \vdots & \vdots \\ \dfrac{\partial H_n}{\partial x_1} & \cdots & \dfrac{\partial H_n}{\partial x_n} & \dfrac{\partial H_n}{\partial t} \end{pmatrix}. \tag{1.4.2}$$

For instance, in Example 1.2.2,

$$H_1(x, t) = (x_1)^3 + 8x_1 + 3x_2 - t(3(x_1)^2 + 36)$$
$$H_2(x, t) = \qquad\qquad\qquad x_2 + t(\ (x_1)^2 + \ 4)$$
$$H'(x, t) = \begin{pmatrix} 3(x_1)^2 - 6tx_1 + 8 & 3 & -3(x_1)^2 - 36 \\ 2tx_1 & 1 & (x_1)^2 + 4 \end{pmatrix}.$$

The Jacobian H' divides conveniently into two parts. Let $H'_x(x, t)$ be the $n \times n$ matrix which is the first n columns of $H'(x, t)$. Also let $\partial H/\partial t$ be the $(n + 1)$st column of $H'(x, t)$. Thus

$$H'_x(x, t) = \begin{pmatrix} \dfrac{\partial H_1}{\partial x_1} & \cdots & \dfrac{\partial H_1}{\partial x_n} \\ \vdots & & \vdots \\ \dfrac{\partial H_n}{\partial x_1} & \cdots & \dfrac{\partial H_n}{\partial x_n} \end{pmatrix} \tag{1.4.3}$$

$$\frac{\partial H}{\partial t} = \begin{pmatrix} \frac{\partial H_1}{\partial t} \\ \cdot \\ \cdot \\ \frac{\partial H_n}{\partial t} \end{pmatrix}. \tag{1.4.4}$$

In Example 1.2.2,

$$H'_x(x, t) = \begin{pmatrix} 3(x_1)^2 - 6tx_1 + 8 & 3 \\ 2tx_1 & 1 \end{pmatrix} \quad \text{and} \quad \frac{\partial H}{\partial t} = \begin{pmatrix} -3(x_1)^2 - 36 \\ (x_1)^2 + 4 \end{pmatrix}$$

Using this notation, H' can then be compartmentalized as

$$H'(x, t) = \left(H'_x(x, t), \frac{\partial H}{\partial t} \right). \tag{1.4.5}$$

PATHS OF SOLUTIONS—THE LINEAR CASE

We are now able to examine the solutions in H^{-1} more precisely.
Take a point $(\bar{x}, \bar{t}) \in H^{-1}$, so that

$$H(\bar{x}, \bar{t}) = 0. \tag{1.4.6}$$

If H is continuously differentiable, we can make a linear approximation of it near $(\bar{x}, \bar{t})$ using its Jacobian:

$$H_i(x, t) \doteq H_i(\bar{x}, \bar{t}) + \sum_{j=1}^{n} \frac{\partial H_i(\bar{x}, \bar{t})}{\partial x_j}(x_j - \bar{x}_j) + \frac{\partial H(\bar{x}, \bar{t})}{\partial t}(t - \bar{t}),$$

$$i = 1, \ldots, n$$

or in vector form,

$$H(x, t) \doteq H(\bar{x}, \bar{t}) + H'(\bar{x}, \bar{t}) \begin{pmatrix} x - \bar{x} \\ t - \bar{t} \end{pmatrix}, \tag{1.4.7}$$

where, via (1.4.5) and (1.4.6),

$$= H'_x(\bar{x}, \bar{t})(x - \bar{x}) + \frac{\partial H(\bar{x}, \bar{t})}{\partial t}(t - \bar{t}).$$

(The symbol $\doteq$ means approximately equal.)

Study points (x, t) in H^{-1} that are near $(\bar{x}, \bar{t})$, as we want to show that they are on a path through $(\bar{x}, \bar{t})$. By definition, all such points must satisfy

$$H(x, t) = 0. \tag{1.4.8}$$

Moreover, if we assume that H is actually linear near $(\bar{x}, \bar{t})$, then by (1.4.7) and (1.4.8), we get

$$0 = H'_x(\bar{x}, \bar{t})(x - \bar{x}) + \frac{\partial H(\bar{x}, \bar{t})}{\partial t}(t - \bar{t}).$$

In other words, if H were actually linear near $(\bar{x}, \bar{t})$, then any point $(x, t) \in H^{-1}$ near $(\bar{x}, \bar{t})$ would have to satisfy this equation. Assuming that H'_x is invertible at $(\bar{x}, \bar{t})$,

$$(x - \bar{x}) = -[H'_x]^{-1}\frac{\partial H}{\partial t}(t - \bar{t}). \tag{1.4.9}$$

This is a system of n linear equations in $n + 1$ variables $(x_1, \ldots, x_n, t)$. Its solutions must therefore form a straight line. In sum, if H is linear, the solutions to (1.4.9) form a straight line through $(\bar{x}, \bar{t})$.

Assuming that H is linear at $(\bar{x}, \bar{t})$, we obtain a very nice path for our solutions—a straight line through $(\bar{x}, \bar{t})$. In general, H will not be linear, but near $(\bar{x}, \bar{t})$ it will be close to linear. At least it will be well approximated by its linear approximation. Then, although there may not be a straight-line solution through $(\bar{x}, \bar{t})$, we get a curve or path. Indeed, as long as H is continuously differentiable and H'_x is invertible, we get a path through $(\bar{x}, \bar{t})$. And this path will not have kinks in it but will be continuously differentiable. All of these observations are summarized in the famous implicit function theorem:

IMPLICIT FUNCTION THEOREM 1.4.1   Let $H: R^{n+1} \longrightarrow R^n$ be continuously differentiable, $(\bar{x}, \bar{t}) \in H^{-1}$ and $H'_x(\bar{x}, \bar{t})$ be invertible. Then in a neighborhood of $(\bar{x}, \bar{t})$ all points (x, t) that satisfy $H(x, t) = 0$ are on a single continuously differentiable path through $(\bar{x}, \bar{t})$.

We will not reprove the theorem here (see, e.g., Ortega and Rheinboldt [1970]). But its proof is similar to our discussions. That is, if H were linear, the points $(x, t) \in H^{-1}$ would be on a single straight line. The function H is not necessarily linear but is close to linear in a neighborhood. Thus the solution points $(x, t) \in H^{-1}$ near $(\bar{x}, \bar{t})$ are also close to linear; that is, they lie on a unique, continuously differentiable path.

The implicit function theorem hypothesis immediately excludes the disasters that occur in Figure 1.4.2. All points $(x, t) \in H^{-1}$ near $(\bar{x}, \bar{t})$ are on a unique continuously differentiable path. This means that no splittings, bifurcations, forks, crossings, infinite endless spirals, and so on, can occur (Exercises 2 and 3). We are guaranteed that our path is extremely well behaved, indeed.

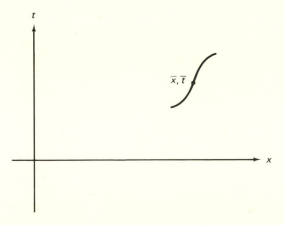

The path of points (x, t) satisfying $H(x, t) = 0$ near the point $(\bar{x}, \bar{t})$

Figure 1.4.3

PATHS FOR H^{-1}

We now commence to build on the implicit function theorem. First a more succinct notation is needed. Let

$$y = (x, t),$$

so that

$$y_i = x_i, \qquad i = 1, \ldots, n \quad \text{and} \quad y_{n+1} = t.$$

Here $x \in R^n$ and $y \in R^{n+1}$. Moreover, $(\bar{x}, \bar{t}) = \bar{y}$, $H'(x, t) = H'(y)$, and so on.

Next, define the partial Jacobian H'_{-i} to be the $n \times n$ matrix formed from H' by deleting the ith column. (The notation $-i$ means minus the ith column.) Explicitly,

$$H'_{-i} = \begin{pmatrix} \dfrac{\partial H_1}{\partial y_1} & \cdots & \dfrac{\partial H_1}{\partial y_{i-1}} & \dfrac{\partial H_1}{\partial y_{i+1}} & \cdots & \dfrac{\partial H_1}{\partial y_{n+1}} \\ \vdots & & \vdots & \vdots & & \vdots \\ \dfrac{\partial H_n}{\partial y_1} & \cdots & \dfrac{\partial H_n}{\partial y_{i-1}} & \dfrac{\partial H_n}{\partial y_{i+1}} & \cdots & \dfrac{\partial H_n}{\partial y_{n+1}} \end{pmatrix}$$

Observe that $n + 1$ matrices H'_{-i} exist. Notationally,

$$H'_x = H'_{-t} = H'_{-(n+1)}.$$

The implicit function theorem as previously stated required that H'_{-i}, for $i = n + 1$, be invertible. But actually the theorem holds as long as any H'_{-i} is invertible. (The theorem is independent of the variable chosen, for if we let y_i

for any i play the role of t, we get the same result.) In other words, if at least one H'_{-i} is invertible, a single continuously differentiable path exists for $(x, t) \in H^{-1}$ in the neighborhood of $(\bar{x}, \bar{t})$. Moreover, it does not matter if one or several of the different H_{-i}' are invertible; we always get the same path. The implicit function theorem guarantees that just one solution path exists through $\bar{y} = (\bar{x}, \bar{t})$. Several different H'_{-i} may be invertible, yet there can be but one path. In short, if any H'_{-i} is invertible, we get a unique path.

THE PATH RESULT

Suppose now that $H'(\bar{y})$, which is an $n \times (n + 1)$ Jacobian matrix, is of full rank [i.e., $H'(\bar{y})$ has rank n]. Some partial Jacobian matrix $H'_{-i}(\bar{y})$ must then be invertible. But then, as just discussed, a unique path results, Stated precisely:

> Let H be continuously differentiable and suppose that $H'(\bar{y})$ is of full rank at $\bar{y} \in H^{-1}$. Then a single continuously differentiable path describes all points $y \in H^{-1}$ in a neighborhood of $\bar{y}$.

Taking this one step further we can achieve the very important path theorem.

PATH THEOREM 1.4.2   Let $H: R^{n+1} \longrightarrow R^n$ be continuously differentiable and suppose that for every $y \in H^{-1}$, the Jacobian $H'(y)$ is of full rank. Then H^{-1} consists only of continuously differentiable paths.

Proof.   At $\bar{y} \in H^{-1}$, by hypothesis some H'_{-i} is invertible. The implicit function theorem ensures that the $y \in H^{-1}$ in a neighborhood of $\bar{y}$ are on a continuously differentiable path through $\bar{y}$. As this holds for any point $\bar{y} \in H^{-1}$, it holds for all points in H^{-1}. Thus H^{-1} consists of paths and only paths.   □

The theorem provides hypotheses ensuring that a path exists. The theorem, of course, is crucial because to follow a path we must first make sure that it exists. Chapter 2 provides the additional information on paths that we need to actually follow them and solve problems. It develops a differential equation that will enable us to easily follow a path.

SUMMARY

This chapter introduced an idea fundamental to the entire book—path following to a solution. First we showed how several sample problems could be solved by following a path. The implicit function theorem then verified when H^{-1} consists only of paths, thereby guaranteeing us that there was a path to follow. Chapter

2 analyzes these paths much more precisely and reveals both how to follow them and where they lead. That will then enable us to attack and solve wide classes of problems, nonlinear equation systems being only one of many examples.

EXERCISES/CHAPTER 1

1. Let $H(x, t) = (1 - t)x + t(x^2 - 4)$, $0 \leq t \leq 1$, $x \in R^1$. Determine the set H^{-1} and draw this set on the plane.

2. Let $H: R^2 \longrightarrow R^1$ be continuously differentiable. Suppose that for every $(x, t) \in H^{-1}$, rank $H'(x, t) = 1$. Show that a path H^{-1} cannot be of the type shown in Figure E.1.1.

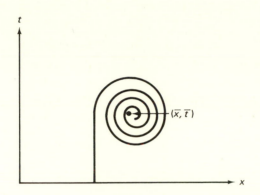

The path spirals around $(\bar{x}, \bar{t})$, and slowly converges toward it.

Figure E.1.1

3. Let $H: R^{n+1} \longrightarrow R^n$ be continuously differentiable and suppose that for every $(x, t) \in H^{-1}$, rank $H'(x, t) = n$. Show that H^{-1} cannot contain isolated points, spirals or crossings. (Hint: Consider the limit point of a spiral. Show how it must be on a continuously differentiable path, and hence could not be a limit point.)

4. Construct a continuously differentiable $H: R^2 \longrightarrow R^1$ such that H^{-1} contains a loop and such that rank $H'(x, t) = 1$ for each $(x, t) \in H^{-1}$. For each $(x, t) \in H^{-1}$, determine an index i for which $H'_{-i}(x, t)$ is nonsingular. (A loop is a path that comes back on itself, thereby forming the loop.)

5. Let H' be of less than full rank. Show by example that H^{-1} can consist of solutions such as in Figure 1.4.2: that is, (a) crossings, (b) spirals, and (c) isolated points.

6. Define $F: R^2 \longrightarrow R^2$ by

$$F_1(x) = e^{2x_1} - x_2^2 + 3 = 0$$

$$F_2(x) = 4x_2 e^{2x_1} - x_2^3 = 0$$

and let $H(x, t) = F(x) - (1 - t)F(0)$.

How many paths are there in H^{-1}? Draw a sketch of the paths on the plane with axes x_1 and x_2.

7. Let

$$E(x) = x^2 - 1 \qquad F(x) = x^2 - 5x + 6$$

and suppose that

$$H(x, t) = (1 - t)E(x) + tF(x).$$

Show that no path of H^{-1} connects a solution of $E(x) = 0$ to a solution of $F(x) = 0$.

8. Consider the equation

$$F_1(x) = x_1 - 3 = 0$$
$$F_2(x) = (x_1^2 + x_2^2)^2 - 10(x_1^2 + x_2^2) - 3 = 0$$

and let

$$H(x, t) = F(x) - (1 - t)F(0, 1).$$

Show that the path in H^{-1} starting from $x = (0, 1)$ does not contain a solution of $F(x) = 0$.

9. Define the equation system that describes Figure E.1.2.

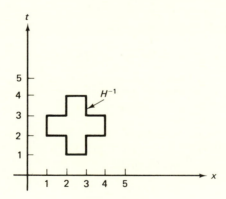

Figure E.1.2

10. Define the equation system that describes the outline of a happy face. Be creative!

11. Consider the system

$$H(x, t) = x^2 - t^2 - \epsilon = 0.$$

a. Show that H' is not of full rank everywhere in H^{-1} when $\epsilon = 0$.
b. Show that H' is of full rank everywhere in H^{-1} when $\epsilon = \pm 0.01$.
   Draw H^{-1} in each case.

12. Let $H: R^{n+1} \longrightarrow R^n$ be continuously differentiable, and suppose that $H'(x, t)$ has rank n for all $(x, t) \in H^{-1}$. Show that a path in H^{-1} is either a loop or is not contained in any compact set.

NOTES

The homotopy approach has been known to scientists at least since the nineteenth century. It is a standard tool in the theory of ordinary differential equations (Ficken [1951]). The first application to nonlinear equation systems seems to have been made by Lahaye [1948]. It has been rediscovered several times since; see, for example, Freudenstein and Roth [1963], Deist and Sefor [1967], and Sidlovskaya [1958]. Davidenko [1953a, b; 1965a, b] has applied this idea to a wide variety of problems, including integral equations, matrix eigenvalue problems, and determinant evaluations. Applications of the method to boundary-value problems are given by Bosarge [1968] and Roberts and Shipman [1967].

The first fail-safe general procedure for fixed points and related equation systems is due to Scarf [1967]. In actuality, Scarf did not use a homotopy approach but an entirely novel approach which he called primitive sets. An earlier work related to Scarf's fundamental work is Cohen [1967]. The underlying principle used by Scarf to prove convergence of his procedure is the "complementarity principle" of Lemke and Howson [1964].

2

The Basic Differential Equations

Chapter 1 introduced the concept of following a path from an easily known start point to a solution. Under quite moderate assumptions the path was shown to exist and be well behaved, so that it could be easily followed.

The path following itself was almost trivial in the previous examples, since a precise closed-form description of each path was obtained. Closed-form results seem to appear only in examples and texts, however, not in real life. Real problems are n-dimensional and far more complicated.

This chapter presents the basic differential equations that will enable us to follow the path whatever the dimension and solve real problems. Also, they provide immediate information on the path direction in space. Simply knowing the path's direction is enough to obtain certain powerful results. Indeed, the basic differential equations will shortly be seen easily to live up to their name, and let us proceed to develop them.

2.1 MOVEMENT ALONG A PATH

REGULARITY

Given a homotopy function $H: R^{n+1} \rightarrow R^n$, under a reasonable assumption we know that the points $y = (x, t) \in H^{-1}$ form nice paths. That was the final and most important conclusion of Chapter 1. Indeed, whenever H satisfies that assumption it is termed regular. Specifically, H is *regular* if $H'(y)$ is of full rank for all $y \in H^{-1}$. Regularity is an extremely mild condition which as discussed in Chapter 22, holds almost always, and will often be assumed in this and later chapters. We reiterate, regularity ensures that H^{-1} consists of continuously differentiable paths.

THE PATH PARAMETER p

Recall also that in Chapter 1 we sometimes used the notation $x(t)$ to describe a path. However, that notation has limitations, as we may not be able to write x in terms of t. Let us now adopt a totally new and more powerful notation. Specify $y = (x, t)$ as a function of p, where p is the distance moved along the path. Thus

$$y(p) = (x(p), t(p)) = (x_1(p), \ldots, x_n(p), t(p))$$

for

$$y(p) \in H^{-1}.$$

Note that x and t now both depend upon p.

By highlighting the path distance p, this new notation immediately reflects our objective, which is to move along a path. The point $y(p)$ tells us where we are after traveling a distance p along the path. For example, consider a comet traversing the solar system, its tail vividly marking its path through space. Then $y(p)$ would denote the coordinates in space of the comet and p the distance traveled along its path. Most of our other examples are more earthbound, but whether in this world or extraterrestrial, our new notation will be highly useful, because it directly reflects movement along the path.

PROCEEDING ALONG A PATH

Let us examine how to determine the path $y(p) \in H^{-1}$. Specify $\dot{y}_i \equiv \dot{y}_i(p)$ as

$$\dot{y}_i = \frac{dy_i}{dp}, \qquad i = 1, \ldots, n + 1,$$

where dy_i/dp are the derivatives of the $y_i(p)$, $i = 1, \ldots, n + 1$.

Since as p varies, $y = y(p)$ describes a path in H^{-1},

$$H(y(p)) = 0.$$

Differentiating both sides yields, by the chain rule,

$$\sum_{i=1}^{n+1} \frac{\partial H}{\partial y_i} \dot{y}_i = 0,$$

where $\partial H / \partial y_i$ is a column of the Jacobian H' or, equivalently,

$$H'(y)\dot{y} = 0. \qquad (2.1.1)$$

Observe that this is a system of n linear equations in $n + 1$ unknowns $(\dot{y}_1, \ldots, \dot{y}_{n+1})$. Here "det" means determinant and recall that H'_{-i} is the Jacobian of H with the ith column deleted. The $\dot{y}$ that satisfies equation (2.1.1) will shortly be seen to yield a fundamental result, the *basic differential equations* (BDE):

$$\dot{y}_i = (-1)^i \det H'_{-i}(y), \qquad i = 1, \ldots, n + 1. \qquad (2.1.2)$$

These are the BDE and when solved will determine a path $y(p) \in H^{-1}$. In other words, suppose that an initial starting point $y^0 = (x^0, t^0) \in H^{-1}$ is given, and let $y(p)$ solve the BDE where $y(p^0) = y^0$. Then, as we will show, $y(p)$ will be in H^{-1}. To emphasize, solving the BDE given an initial point in H^{-1} will produce a path in H^{-1} as p varies. This is a result so important that it will maintain our attention for the entire chapter.

PRELIMINARIES

The demonstration of the BDE theorem relies on two well-known results which it is helpful to review. The first is Cramer's rule for solving linear equations, which we review first. Suppose that

$$Ay = b,$$

where A is an $m \times m$ matrix, y is an $m \times 1$ vector, and b is an $m \times 1$ vector. Then Cramer's rule states:

$$y_i = \frac{\det [A^1, A^2, \ldots, A^{i-1}, b, A^{i+1}, \ldots, A^m]}{\det A}. \qquad (2.1.3)$$

Here the numerator is the determinant of A, but with its ith column replaced by b. The notation A^i indicates the ith column of A. Of course, A is square, whereas our system equation (2.1.1) is $n \times (n + 1)$, and it is precisely this difference that will help us in the BDE theorem.

The other result we need is from elementary differential equations (Birkhoff and Rota [1962]; Hirsch and Smale [1974]). Specifically, suppose that a continuously differentiable function $L: R^{n+2} \rightarrow R^{n+1}$ is given and consider the differential equations

$$\dot{y} = L(y, p), \qquad y(p^0) = y^0. \qquad (2.1.4)$$

[Here $y \in R^{n+1}$, so $(y, p) \in R^{n+2}$.] Then differential equation theory ensures that this differential equations system has a solution $y(p)$. Furthermore, given any $y^0 = y(p^0)$, this solution is unique (see Exercises 1 to 4).

Identify $L(y, p)$ with the right-hand side of the BDE (2.1.2). The differential equations result then states that the BDE will have a solution $y(p)$ and it will be unique (Exercise 5).

UNIQUENESS OF THE FORM y(p)

Let us elaborate further on the uniqueness aspect. The actual path generated will be unique, despite the fact that there are an infinite number of forms $y(p)$. For example, for $y = (y_1, y_2)$, let $y_1 = 2p$, $y_2 = 4p$, or $y_1 = p^2$, $y_2 = 2p^2$, or $y_1 = 10p^4 + 3$, $y_2 = 20p^4 + 6$. In all three cases $2y_1 = y_2$, which means as p varies all three describe the same straight-line path in y space. The same path always results in y space despite the infinite number of forms of $y(p)$. As another example, let $f: R^1 \longrightarrow R^1$ be an arbitrary function. Then $y_1 = f(p)$ and $y_2 = (f(p))^3$ become $y_2 = (y_1)^3$ in y space. In short, even though an infinite number of forms of $y(p)$ satisfy (2.1.1). as p varies they all must generate the same unique path in y space.

Indeed, the fact that so many forms of $y(p)$ generate the identical path is what permits us to create the BDE. Out of the infinite number of possible forms of $y(p)$, we select one special form, the one that satisfies the BDE. This special form, the solution to the BDE, provides an extraordinary amount of information about the path and its behavior.

THE BDE THEOREM

We will now employ these results to develop our theorem. The standard notation $H \in C^k$ is used to mean that the function H has continuous derivatives of order k. For example, $H \in C^0$ means that H is continuous, $H \in C^2$ means that H is twice continuously differentiable, and so on.

THE BDE THEOREM 2.1.1   Let $H: R^{n+1} \longrightarrow R^n$, $H \in C^2$, be regular. Given a starting point y^0 in H^{-1}, the solution of the basic differential equations (2.1.2), starting from $y(p^0) = y^0$, is unique and determines a path in H^{-1}.

Proof.   Since H is C^2, det H'_{-i} is C^1, $i = 1, \ldots, n + 1$. Therefore, as with (2.1.4), a solution to the BDE is unique.

Let $y(p)$ be a path in H^{-1} so that

$$H(y(p)) = 0.$$

Differentiating by the chain rule, we obtain

$$\sum_{i=1}^{n+1} \frac{\partial H}{\partial y_i} \dot{y}_i(p) = 0. \tag{2.1.5}$$

An infinite number of forms of $\dot{y}(p)$ satisfy (2.1.5). As mentioned, all of these forms starting from $y^0 = y(p^0)$ describe the same path, $y(p) \in H^{-1}$ as p varies (Exercise 6). Consequently, it remains only to demonstrate that the BDE are one of these forms.

Substituting the form of the BDE (2.1.2) into (2.1.5), we must verify that

$$\sum_{i=1}^{n+1} \frac{\partial H}{\partial y_i}[(-1)^i \det H'_{-i}] = 0. \tag{2.1.6}$$

[In other words, we need only prove that (2.1.6) is an identity, as then the BDE will satisfy (2.1.5) and therefore will produce a path in H^{-1}.]

By regularity for some i, say $i = 1$, H'_{-1} is nonsingular. So consider

$$\sum_{i=2}^{n+1} \frac{\partial H}{\partial y_i}[(-1)^i \det H'_{-i}] = -\frac{\partial H}{\partial y_1}[(-1)^1 \det H'_{-1}].$$

Via Cramer's rule for $i = 2, \ldots, n+1$,

$$(-1)^i \det H'_{-i}$$

$$= \frac{1}{\det H'_{-1}} \det \left(\frac{\partial H}{\partial y_2}, \frac{\partial H}{\partial y_3}, \ldots, \frac{\partial H}{\partial y_{i-1}}, \frac{\partial H}{\partial y_1} \det H'_{-1}, \frac{\partial H}{\partial y_{i+1}}, \ldots, \frac{\partial H}{\partial y_{n+1}}\right).$$

Factoring out the $\det H'_{-1}$ from the determinant and rearranging columns yield

$$(-1)^i \det H'_{-i} = \frac{\det H'_{-1}(-1)^i \det H'_{-i}}{\det H'_{-1}} = (-1)^i \det H'_{-i}.$$

We see that (2.1.6) is an identity.

The BDE thus satisfy (2.1.5), and consequently solving the BDE, given $y(p^0) = y^0$, yields a path $y(p) \in H^{-1}$. □

The theorem states that solving the BDE for $y(p)$ as p varies will permit us to follow a path in H^{-1}. Let us illustrate this with an example.

EXAMPLE 2.1.1

Let us use the BDE to solve the nonlinear system:

$$F_1(x_1, x_2) = x_1 - 2x_2 + (x_2)^2 + (x_2)^3 - 4 = 0$$
$$F_2(x_1, x_2) = -x_1 - x_2 + 2(x_2)^2 \qquad - 1 = 0.$$

Consider the Newton homotopy,

$$H(x, t) = F(x) - (1 - t)F(x^0),$$

where the initial point $x^0 = (0, 0)$ at $t = 0$. Thus we must solve

$$H_1(x, t) = x_1 - 2x_2 + (x_2)^2 + (x_2)^3 - 4 + (1 - t)4 = 0$$
$$H_2(x, t) = -x_1 - x_2 + 2(x_2)^2 \qquad - 1 + (1 - t)1 = 0.$$

The Jacobian is

$$H'(x, t) = \begin{pmatrix} 1 & -2 + 2x_2 + 3(x_2)^2 & -4 \\ -1 & -1 + 4x_2 & -1 \end{pmatrix}$$

Consequently, where $y = (x, t)$,

$$\det H'_{-1} = -2 + 14x_2 - 3(x_2)^2$$
$$\det H'_{-2} = -5$$
$$\det H'_{-3} = -3 + 6x_2 + 3(x_2)^2.$$

The BDE become

$$\dot{x}_1 = +2 - 14x_2 + 3(x_2)^2$$
$$\dot{x}_2 = -5$$
$$\dot{t} = +3 - 6x_2 - 3(x_2)^2.$$

Integrating $\dot{x}_2$ yields

$$x_2 = -5p + \text{constant}.$$

But at $p = 0$, $x_2 = 0$, so the constant is zero and we get

$$x_2 = -5p.$$

Plugging this result into $\dot{t}$, we obtain

$$\dot{t} = +3 - 6(-5p) - 3(-5p)^2,$$

so that

$$t = +3p + 15p^2 - 25p^3,$$

where again the constant is zero because $t = 0$ at $p = 0$. Similarly,

$$\dot{x}_1 = +2 - 14(-5p) + 3(-5p)^2$$

and

$$x_1 = +2p + 35p^2 + 25p^3.$$

Note that we have obtained an explicit representation for $y(p) = (x(p), t(p))$. In words, y is now expressed in terms of the path parameter p.

We must move along the path from the initial point

$$x_1 = 0, \quad x_2 = 0, \quad t = 0$$

by increasing p until $t = 1$. Notice at $p = \frac{1}{5}$, $t = 1$, so plugging $p = \frac{1}{5}$ into x_1 and x_2 yields

$$x_1^* = 2, \quad x_2^* = -1.$$

In sum, solving the BDE provides a path $(x(p), t(p))$ which by changing p we can follow from $t = 0$ to $t = 1$, thus solving the original system.

COMMENT

The BDE provide a general means to follow a path in H^{-1} and thereby obtain a solution point $x^1 \in H^{-1}(1)$. Later in the book we provide a full variety of algorithms to follow a path and obtain a solution. However, the BDE yield far more than merely a means to solve an equation system. They provide immediate and important information on the direction of the path. This direction, or orientation as it is called, is extremely useful in simplifying problems, calculating solutions, and proving certain mathematical theorems. In short, the BDE provide not only a means to follow a path but the key to numerous theoretical and practical results. We proceed with this forthwith by discussing orientation.

2.2 ORIENTATION

The BDE provide explicit expressions for $\dot{y}(p) = (\dot{x}_1(p), \ldots, \dot{x}_n(p), \dot{t}(p))$, and thus indicate how any variable changes as p changes. As we move along the path, the signs of the different $\dot{y}_i$ tell which variables are increasing, decreasing, or constant. For instance, if $\dot{x}_1(p) > 0$, then increasing p increases the variable x_1; similarly, if $\dot{t} < 0$, then increasing p decreases t.

Even further, suppose that $\dot{t} < 0$ and we know that the path looks as shown in Figure 2.2.1. Then we know that we must be between points b and c.

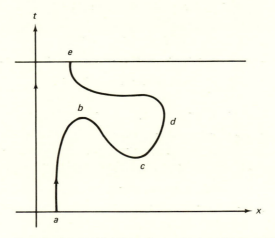

The derivatives provide information on the change in the variables as p increased along the path.

$\dot{x}(p) > 0$ from point a to point d.

$\dot{x}(p) = 0$ at point d.

$\dot{x}(p) < 0$ from point d to point e.

$\dot{t} > 0$ from a to b and from c to e.

$\dot{t} = 0$ at b and c.

$\dot{t} < 0$ from b to c.

Figure 2.2.1

Thus by examining $\dot{y}$ the BDE tell us immediately which components of $y(p)$ are increasing or decreasing, and when.

DERIVATIVE AND TANGENTS

So far, we have considered each $\dot{y}_i$ component singly, but taken together they provide even more information. In particular, the vector $\dot{y} = (\dot{y}_1, \ldots, \dot{y}_{n+1})$ is the tangent to the path at any point. This follows from the definition of derivative, because for Δ small,

$$y(p + \Delta) - y(p) \doteq \dot{y}(p)\Delta.$$

But $y(p + \Delta) - y(p)$ is a vector with both endpoints lying along the path. Thus letting $\Delta \longrightarrow 0$, we see that $\dot{y}(p)$ is the tangent at p (see Figure 2.2.2).

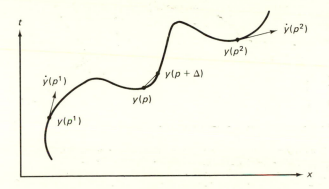

The vector $\dot{y}(p)$ is tangent to the path at any point $y(p)$

Figure 2.2.2

The tangent vector $\dot{y}(p)$ is like the friendly native guide who always points the direction along the jungle path. For example, suppose that we are given an arbitrary vector $a \in R^{n+1}$, so that a points in some direction in our space. Often we are interested in knowing if at some point $y(p)$ the path moves in a direction similar to a. The direction a could be considered a signpost pointing to our destination, and we might wish to know if the path's direction is taking us closer to or away from the destination. Mathematically, the inner product $a\dot{y}(p)$ provides that information. For instance, if $a\dot{y}(p) > 0$, then at the point $y(p)$ the path is moving at an acute angle to a.

A directly related and important situation occurs for a hyperplane

$$by = c$$

where $b \in R^{n+1}$, recall, is the normal to the hyperplane. Suppose that $b\dot{y}(p) > 0$.

Then by increasing p we move across the hyperplane toward the positive side of the hyperplane, that is, into the region $by \geq c$. Decrease p and we move to the negative side. All this follows because b is the normal and points to the positive side. Similarly, if at some point $b\dot{y}(p) = 0$, then the path is tangent to the hyperplane. Figure 2.2.3 depicts two instances of this (Exercise 13).

Thus the tangent $\dot{y}$ tells us not only the path's direction but also how we are moving relative to arbitrary vectors and hyperplanes. Such information will turn out to be quite useful, and moreover, easily obtained since $\dot{y}$ is given immediately by the BDE.

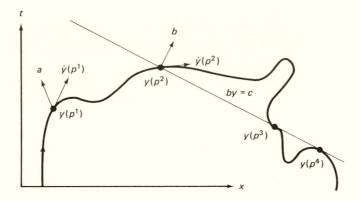

The tangent to the path $\dot{y}(p^1)$ and the direction a form acute angles at p^1. At p^2, increasing p moves us into the region $by \geq c$ because $b\dot{y}(p^2) > 0$. At p^3 and p^4, $b\dot{y}(p^3) = 0$ and $b\dot{y}(p^4) = 0$, so the path is tangent to the hyperplane. Note how the path crosses the hyperplane at p^3 but not at p^4.

Figure 2.2.3

ARE WE COMING OR GOING?

In solving the BDE it is important to be careful about certain aspects. Do not assume that increasing p will increase t. Thus suppose that we are at $t = 0$ and desire to increase t. If $i > 0$, then increasing p will increase t. However, the opposite is true if $i < 0$, because then we must decrease p to increase t. The sign of i must be checked first.

A similar phenomenon occurs if we are moving from a point a to b along the path. It might be necessary to increase p to move from a to b, or it might be necessary to decrease p. Which way to change p is not clear beforehand and must be checked when the BDE are solved. In effect, the path is a *two-way* street.

The BDE provide a path, yet do not tell us beforehand in which direction p will change along that path. After writing the BDE down, however, this can be obtained easily. If we wish to increase t, we must change p in the direction of the sign of i. More generally, suppose that a direction b is given and we want to move in a direction acute to b. Then if $b\dot{y} > 0$, we increase p; if $b\dot{y} < 0$, we decrease p. Which way we are heading is thus easily determined.

2.3   THE PIECEWISE CASE

We know that the path is a two-way street and that p can move in either direction, depending on the solution to the differential equations. An important application of this is the piecewise case introduced in Example 1.2.3.

Under the piecewise case, recall the domain is divided into pieces (sets). We assume that each piece is a polyhedral set. Recall also that two intersecting polyhedra can meet only on their common boundary. Further, the function can be defined differently on each piece, but must match up on any common boundary so that it is continuous.

Assume that a path $y(p) \in H^{-1}$ crosses the hyperplane that separates two polyhedral pieces of the space Q^1 and Q^2. On piece Q^1 let $H = H^1$; on piece Q^2 let $H = H^2$. Also, specify $by = c$ as the hyperplane where Q^1 and Q^2 intersect. For all y on that hyperplane, since H must match up,

$$H(y) = H^1(y) = H^2(y).$$

Suppose that H^1 is C^2 on piece Q^1 and H^2 is C^2 on piece Q^2. (Note that H itself need not be C^2 even though it is C^2 on Q^1 and C^2 on Q^2.)

To determine the path we would solve the BDE twice, once on Q^1 using H^1 and a second time on Q^2 using H^2. More exactly, let $y(p) \in H^{-1}$ be the path and assume that it crosses the hyperplane $by = c$ at one point $y(\bar{p})$. Let $y^k(p)$ be the value of $y(p)$ on piece Q^k, $k = 1, 2$. Thus from the BDE,

$$\dot{y}_i^k = (-1)^i \det H_{-i}^{k'},$$

where $H^{k'}$ is the Jacobian of H^k. To reiterate, the path is in two parts. The part on Q^1 is obtained by solving the BDE using H^1; the part on Q^2 is obtained by solving the BDE using H^2. Both parts meet at $y(\bar{p})$.

The question is: Will the directions for p match up properly? Since the differential equations for H^1 and for H^2 are solved separately, they might not match up, as in Figure 2.3.1. In that figure p increases along the path from e to f, but decreases from f to g. Essentially, since the two differential equations are solved separately and since each path is a two-way street, it is conceivable that the p's on the two paths head in opposite directions.

We want p to behave consistently. That is, suppose that by increasing p, $y^1(p)$ intersects the hyperplane. Then on path $y^2(p)$ we want to move away from and into the other side of the hyperplane by continuing to increase p. Conversely, suppose that by decreasing p we run into the hyperplane. Then by continuing to decrease p on the other side of the hyperplane, we want to move away from the

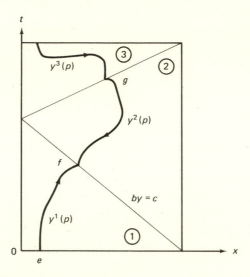

The path $y(p) \in H^{-1}$ traverses three polyhedra (triangles) 1, 2, and 3 as t increases from 0 to 1. Here the hyperplane $by = c$ is the boundary between triangles 1 and 2. The function:

$$H = H^k \text{ on triangle } k$$

where H^k is twice continuously differentiable on triangle k. Also, $y(p) = y^k(p)$ on triangle k, where y^k solves the BDE for H^k. On triangle 1, p increases from e to f; on triangle 2, p decreases from f to g.

Figure 2.3.1

hyperplane. In short, p should increase along both pieces Q^1 and Q^2 or it should decrease along both. If p is inconsistent as in Figure 2.3.1, certainly it would cause a great deal of confusion.

Fortunately, p is consistent and indeed this result holds automatically for any piecewise function H. The proof is technical and is given in the appendix to this chapter. The key to the result is that the two pieces of H must be continuous on the boundary hyperplane; and the fact that H must itself match up forces p to behave consistently.

We may thus solve the BDE separately on each piece and the p direction will automatically be consistent. If we change p in a given direction on one piece, we can continue to change p in the same direction and keep crossing piece after piece consistently. The BDE therefore can be applied to the piecewise situation directly, and p will automatically be consistent.

SUMMARY

This chapter has developed the BDE and explored some of their implications and uses. Solving the BDE will permit us to follow a path $y(p) \in H^{-1}$ and will work in the piecewise situation as well. However, their main import is that they immediately yield the direction $\dot{y}$ of path movement. Merely knowing this direction will provide the basis for many of the theorems of Chapter 3.

APPENDIX

The appendix contains two interesting, although somewhat technical extensions of the BDE. The first concerns velocity and how fast we traverse a path $y(p)$. It also provides insight into the two-way-street aspect of the path that was discussed. The second concerns the piecewise case and provides the detailed proof mentioned in that section.

VELOCITY

We previously stated that although $y(p)$ could have many forms, they all generate the same path in y space. We now present a form of $\dot{y}(p)$ which differs from the BDE slightly and provides further understanding of the BDE. The proof is like Theorem 2.1.1 and is left as Exercise 14. Also, the more succint notation is used, $H(y(p)) = H(p)$.

THEOREM A.2.1   Let $H: R^{n+1} \longrightarrow R^n$, $H \in C^2$ be regular. Also let $q(p)$: $R^1 \longrightarrow R^1$ be a continuous function where $q(p) \neq 0$ all p. Then, given $y(p^0) = y^0$, a path $y(p) \in H^{-1}$ can be obtained by solving the following differential equation system:

$$\dot{y}_i(p) = q(p)(-1)^i \det H'_{-i}(p), \qquad i = 1, \ldots, n+1. \qquad \text{(A.2.1)}$$

Differential equations (A.2.1) illustrate that many forms of $y(p)$ generate the same path in H^{-1}. Clearly, the only difference between (A.2.1) and the BDE is the factor $q(p)$ in front of each term. That factor is itself quite interesting because it adjusts the speed with which we traverse the path in p, as we now elaborate.

First note that if $q(p) = +1$, the system (A.2.1) becomes the BDE. If $q(p) = -1$, then we obtain a path that moves in the opposite direction as p increases. In other words, changing the sign of $q(p)$ changes the direction in which we traverse the path. This is precisely the two-way-street phenomenon discussed previously and lets us go either way on the path.

But the path is more than a two-way street, for we can also travel faster or slower on it. Consider carefully for Δ small:

$$y_i(p + \Delta) - y_i(p) \doteq \dot{y}_i(p)\Delta = q(p)(-1)^i \det H'_{-i} \cdot \Delta.$$

Observe that $\dot{y}$ not only gives direction of movement, but the magnitude of $\dot{y}$

provides the speed. And $q(p)$ adjusts that speed. The larger the $q(p)$, the larger the $\dot{y}(p)$ and consequently the greater the distance moved along the path per unit change in p.

It is desirable in certain applications to move along the path at unit velocity. This means that we would require

$$\|\dot{y}\| = 1.$$

[Here $\|y\|^2 = \sum_1^{n+1} (y_i)^2$ is the Euclidean norm.] Unit velocity is easily obtained; simply let

$$q(p) = \left[\sum_1^{n+1} (\det H'_{-i})^2 \right]^{-1/2}.$$

We have illustrated that there can be many forms to $\dot{y}(p)$. The particular form (A.2.1) with the factor $q(p)$ permits us to adjust the speed of movement along the path. The solution process to the new differential equations (A.2.1) is similar to the BDE. The $q(p)$ factor, however, can be useful in applications where there is a physical or economic meaning to movement along the path and speed is a consideration.

CONSISTENCY OF p
FOR THE PIECEWISE CASE

To prove that p behaves consistently for the piecewise case, let us recall the situation. The homotopy $H: R^{n+1} \longrightarrow R^n$ is a piecewise function defined on two polyhedra that intersect at a hyperplane $by = c$. The homotopy $H = H^k$ on the polyhedron Q^k, $k = 1, 2$, and $H^1(y) = H^2(y)$ on the hyperplane. A piecewise path $y(p) \in H^{-1}$ is such that $y(p) = y^k(p)$ on polyhedron Q^k, $k = 1, 2$, and

$$y(\bar{p}) = y^1(\bar{p}) = y^2(\bar{p}),$$

where $y(\bar{p})$ is the point that the path crosses the hyperplane. Moreover, the y^k are the solutions of the BDE:

$$\dot{y}_i(p) = (-1)^i \det H^{k'}_{-i}, \qquad i = 1, \ldots, n+1.$$

We must verify that p changes consistently as it crosses the hyperplane. Mathematically, this means that if

$$b\dot{y}^1(\bar{p}) > 0, \qquad \text{then } b\dot{y}^2(\bar{p}) > 0$$

or if

$$b\dot{y}^1(\bar{p}) < 0, \qquad \text{then } b\dot{y}^2(\bar{p}) < 0.$$

In words, at $\bar{p}$, if increasing p moves $y^1(p)$ toward the positive side of the hyper-

plane, then increasing p should move $y^2(p)$ toward the positive side of the hyperplane also; and conversely if we are decreasing p (see Figure A.2.1).

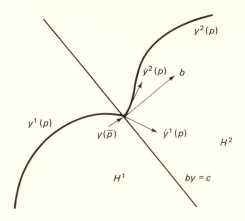

Here $H^1(y^1(p)) = 0$, $H^2(y^2(p)) = 0$, and $H(y) = H^1(y) = H^2(y)$ if y is on the hyperplane $by = c$. The vector $\dot{y}^1(\bar{p})$ is tangent to $y^1(p)$ at $\bar{p}$ and $\dot{y}^2(\bar{p})$ is tangent to $y^2(p)$ at $\bar{p}$.

Figure A.2.1

To obtain this result requires first examining some quite useful properties. Given an $n \times (n + 1)$ Jacobian H' and $b \in R^{n+1}$, expand the determinant along the first row to yield

$$\det \begin{bmatrix} -b \\ H' \end{bmatrix} = \sum_{i=1}^{n+1} b_i(-1)^i \det H'_{-i}.$$

Yet from the basic differential equations, since

$$\dot{y}_i = (-1)^i \det H'_{-i}, \qquad \text{then } \det \begin{bmatrix} -b \\ H' \end{bmatrix} = b\dot{y}(p), \qquad \text{(A.2.2)}$$

which provides a convenient representation of $b\dot{y}$ in terms of the Jacobian.

The point $y(\bar{p})$, recall, is where the path crosses the hyperplane, that is, $by(\bar{p}) = c$. Select n points u^i in the hyperplane so that the vectors

$$v^i = u^i - y(\bar{p}), \qquad i = 1, \dots, n$$

are linearly independent. As we are in R^{n+1}, this can be easily done.
Also,

$$bv^i = bu^i - by(\bar{p}) = c - c = 0, \qquad i = 1, \dots, n, \qquad \text{(A.2.3)}$$

since the u^i and $y(\bar{p})$ are in the hyperplane.

Defining the $(n + 1) \times n$ matrix,

$$V = [v^i]$$

we may express (A.2.3) as

$$bV = 0. \tag{A.2.4}$$

Clearly, since b is normal to all the v^i, the $(n + 1) \times (n + 1)$ matrix

$$[-b^\tau, V]$$

is of full rank. (Here τ indicates the transpose.) We then can calculate

$$\begin{bmatrix} -b \\ H' \end{bmatrix} [-b^\tau, V] = \begin{bmatrix} \|b\|^2 & -bV \\ -H'b^\tau & H'V \end{bmatrix}$$

$$= \begin{bmatrix} \|b\|^2 & 0 \\ -H'b^\tau & H'V \end{bmatrix}$$

using (A.2.4). Moreover, since $[-b^\tau, V]$ is full rank and thus invertible,

$$\det \begin{bmatrix} -b \\ H' \end{bmatrix} = \det [-b^\tau, V]^{-1} \det \begin{bmatrix} \|b\|^2 & 0 \\ -H'b^\tau & H'V \end{bmatrix}.$$

Consequently, from (A.2.2) and expanding the determinant,

$$b\dot{y}(\bar{p}) = \det [-b^\tau, V]^{-1} \det [H'V] \|b\|^2. \tag{A.2.5}$$

We now have a direct expression for $b\dot{y}(\bar{p})$ in easy-to-evaluate terms. In fact, this nice expression holds separately for each superscript k. That is,

$$b\dot{y}^k(\bar{p}) = \det [-b^\tau, V]^{-1} \det [H^{k'}V] \|b\|^2.$$

This is because (A.2.5) holds for either H^1 or H^2. Simply note that at any point on the hyperplane

$$H(y) = H^1(y) = H^2(y),$$

and since $y(\bar{p})$ is on the hyperplane,

$$y(\bar{p}) = y^1(\bar{p}) = y^2(\bar{p}).$$

Our goal of proving $b\dot{y}^1(\bar{p})$ and $b\dot{y}^2(\bar{p})$ of the same sign is thereby made quite easy. We need only verify that

$$\det [H^{1'}V] \quad \text{and} \quad \det [H^{2'}V] \quad \text{are of the same sign.} \tag{A.2.6}$$

THEOREM A.2.2   Given H: $R^{n+1} \rightarrow R^n$, $H \in C^2$, $y(p) \in H^{-1}$, and a hyperplane $by = c$, assume that the path $y(p)$ crosses the hyperplane at a single point $y(\bar{p})$. Suppose, if $by \leq c$, that

$$H(y) = H^1(y) \qquad \text{and} \qquad y(p) = y^1(p),$$

and if $by \geq c$, that

$$H(y) = H^2(y) \qquad \text{and} \qquad y(p) = y^2(p),$$

where H^1 and H^2 are both C^2 and regular and

$$\dot{y}_i^k = (-1)^i \det H_{-i}^{k}.$$

Then

$$b\dot{y}^1(\bar{p}) > 0 \quad \text{implies} \quad b\dot{y}^2(\bar{p}) > 0.$$

Similarly,

$$b\dot{y}^1(\bar{p}) < 0 \quad \text{implies} \quad b\dot{y}^2(\bar{p}) < 0.$$

Proof.   We need only verify (A.2.6).

Since all the u^i and $y(p)$ are in the hyperplane,

$$H(u^i) = H^1(u^i) = H^2(u^i), \qquad i = 1, \ldots, n$$

and

$$H(y(\bar{p})) = H^1(y(\bar{p})) = H^2(y(\bar{p})). \tag{A.2.7}$$

By Taylor's expansion, since $v^i = u^i - y(\bar{p})$,

$$H^k(u^i) = H^k(y(\bar{p})) + H^{k\prime}(y(\bar{p}))v^i + R^{ki}, \qquad k = 1, 2,$$

where R^{ki} is a remainder that goes to zero as u^i gets close to $y(\bar{p})$. In fact, we can arrange $\| R^{ki} \|_\infty \leq \alpha \| u^i - y(\bar{p}) \|_\infty^2$ for some constant $\alpha > 0$, where $\| \cdot \|_\infty$ is the infinity norm. Utilizing (A.2.7) yields

$$H^{1\prime}(y(\bar{p}))v^i = H^{2\prime}(y(\bar{p}))v^i + R^i \tag{A.2.8}$$

where $R^i = R^{2i} - R^{1i}$.

Carefully considering (A.2.8) as the columns of a matrix,

$$H^{1\prime}V = [H^{1\prime}v^i] = [H^{2\prime}v^i + R^i] = H^{2\prime}V + [R^i].$$

Yet R^i gets small faster than v^i as u^i approaches $y(\bar{p})$. Hence it can be shown (Exercise 15) that if

$$\det H^{1\prime}V > 0,$$

then

$$\det H^{2\prime}V > 0.$$

This provides our conclusion, as then

$$b\dot{y}^1(\bar{p}) > 0 \quad \text{implies} \quad b\dot{y}^2(\bar{p}) > 0. \qquad \square$$

Suppose that we follow the path across one piece of space to its boundary. The theorem assures us, if we cross the boundary, that we may keep changing

p in the same direction. All that is necessary is, given each H^k, to solve the BDE for y^k and everything works out. The underlying reason is that although the H^k may change on their different pieces, they are equal on their respective boundary hyperplanes. The theorem thereby tells us that p changes in one direction along the path no matter how many boundaries the piecewise path may cross.

EXERCISES/CHAPTER 2

1. Consider the differential equations (2.1.4) where $L: R^{n+2} \longrightarrow R^{n+1}$ is C^1. Prove that there is a maximal interval (α, β) containing p^0 such that the solution to the differential equations (2.1.4) exists and is unique on (α, β).

2. Prove that if $\beta < \infty$ in Exercise 1, the solution $y(p)$ diverges to infinity as $p \longrightarrow \beta$. Similarly, if $\alpha > -\infty$ in Exercise 1, $y(p)$ diverges to infinity as $p \longrightarrow \alpha$. [This means that if the solution $y(p)$ cannot be extended to a larger interval, it leaves any compact set.]

3. There is no guarantee that a solution $y(p)$ to a differential equation can be defined for all p. For example, consider
    $$\dot{y} = 1 + y^2, \qquad y(0) = 0.$$
    a. Determine the solution $y(p)$.
    b. Find the maximum open interval (α, β) containing $y(p)$.

4. Consider $\dot{x} = \frac{3}{2}x^{1/3}$, $x(0) = 0$. Show that there are two solutions to this differential equation on the set $x \geq 0$. Why does this example not contradict the statement in Exercise 1?

5. Prove that by solving the BDE (2.1.2), the entire set H^{-1} can be generated. That is, the entire length of any path in H^{-1} can be obtained. [*Hint:* Suppose that $y(p)$ is the unique solution of the BDE for $y^0 = y(p^0)$ on the maximal interval (α, β). If $\beta < \infty$, then from Exercise 2, $y(p)$ diverges to infinity as $p \longrightarrow \beta$. If $\beta = \infty$, show that the integral
    $$\int_{p^0}^{p} \left\{ \sum_{i=1}^{n+1} [\dot{y}_i(\theta)]^2 \right\}^{1/2} d\theta$$
    diverges to infinity as $p \longrightarrow \infty$. Regularity ensures that $\dot{y}(\theta) \neq 0$. Then verify that the distance traveled by $y(p)$ on the interval $[p^0, \beta)$ is infinite. Thus $y(p)$ is always moving in H^{-1} and does not get stuck near any point in H^{-1}.]

6. Let $y(p)$ be a differentiable function satisfying the differential equations (2.1.5) and $y(p^0) = y^0 \in H^{-1}$. Show that $y(p) \in H^{-1}$. [*Hint:* Define $g: R^1 \longrightarrow R^n$ by $g(p) = H(y(p))$. Then show that $g(p)$ is identically zero.]

7. Consider the BDE $dx/dp = -t$, $dt/dp = x$, $(x(0), t(0)) = (1, 0)$.
    a. Find the solution to the BDE.
    b. Show that the distance traveled on the path solution is infinite even though the path itself is finite.

8. Let $H(x, t) = x^2 - t^2 = 0$.
   a. Formulate the BDE for $H = 0$.
   b. Given $(x^0, t^0) \in H^{-1}$, determine the solution to the BDE with initial condition $(x(0), t(0)) = (x^0, t^0)$.
   c. Observe that the function L defined by (2.1.4) is C^1 for this BDE. Yet note that H^{-1} is not a path if (x^0, t^0) is the origin. Explain why Theorem 2.1.1 does not apply in this case.

9. Consider the system
$$H(x, t) = x^2 + t^2 - 1 = 0.$$
   Determine how $\dot{x}$ and $\dot{t}$ behave in each of the four quadrants.

10. Let
$$F_1(x_1, x_2) = x_1$$
$$F_2(x_1, x_2) = (x_1^2 + x_2^2)^2 - 10(x_1^2 + x_2^2) + 9$$
   and
$$H(x, t) = F(x) - (1 - t)F(x^0),$$
   where
$$F(x^0) = (-4, 9109).$$
   Use the BDE to graph H^{-1}.

11. Let
$$F_1(x_1, x_2) = -13 + x_1 - 2x_2 + 5x_2^2 - x_2^3 = 0$$
$$F_2(x_1, x_2) = -29 + x_1 - 14x_2 + x_2^2 + x_2^3 = 0$$
   and
$$H(x, t) = F(x) - (1 - t)F(15, -2).$$
   a. Will the set H^{-1} contain a solution to $F(x) = 0$? Explain.
   b. Find the two points on H^{-1} where H'_{-t} is singular.

12. Describe in words the path traced by the BDE
$$\dot{x}_2 = \frac{-2x_1}{1 + 4x_1^2}, \qquad \dot{x}_1 = \frac{1}{1 + 4x_1^2}, \qquad x(0) = 0.$$

13. Let $by = c$, $b \in R^{n+1}$, c a scalar, denote a hyperplane in R^{n+1}, and let $y(p) \in R^{n+1}$ be a point lying on some path. Show that:
   a. If $b\dot{y}(p) > 0$, then by increasing p, we move across the hyperplane into the region $by \geq c$.
   b. If $b\dot{y}(p) = 0$, then the path is tangent to the hyperplane at the point $y(p)$.

14. Prove Theorem A.2.1.

15. In Theorem A.2.2, we showed that
$$H^{1\prime}V = H^{2\prime}V + [R^i].$$
   Verify that, as $u^i \longrightarrow y(\bar{p})$, the matrix $[R^i]$ will approach zero such that $\det H^{1\prime}V$ and $\det H^{2\prime}V$ will be of the same sign.

16. Consider the piecewise system that describes the unit square on the plane. Write the BDE for this system and determine which direction the square is traversed as p increases.

NOTES

2.1. Davidenko [1953a, b] introduced the differential equation as a means of solving $H(x, t) = 0$ for $0 \le t \le 1$. However, these differential equations used t as a parameter, required H'_{-t} nonsingular, and were of the form $\dot{x}(t) = -H'_{-t}(x, t)^{-1} H_t(x, t)$. Davis [1966] discussed coordinate transformations for overcoming problems when H'_{-t} becomes singular and Meyer [1968] attacks the same problem by modifying the homotopy H. Garcia and Zangwill [1979d] developed the parameter p and the BDE (2.1.2). The BDE require no coordinate transformation and permit H'_{-t} to be singular.

2.2. The BDE immediately provide the path orientation. The traditional argument (Milnor [1965]) is more cumbersome and requires complicated matrix manipulations. Also, the traditional argument tells us only whether the path is increasing or decreasing in a variable, not the exact direction.

2.3. The piecewise case adapts reasoning developed by Eaves and Scarf [1976].

3

Fixed Points, Equations, and Degree Theory

This chapter takes us on a fascinating adventure as we establish certain truly powerful mathematical theorems: the Brouwer fixed-point theorem, an existence of solutions theorem, and after introducing the concept of degree, the homotopy invariance theorem. Although these theorems may seem disparate, they are all united by a similar concept. Given a homotopy H, we prove that the path $y(p) \in H^{-1}$ goes to the right place. In particular, suppose we ensure that the solution path $y(p)$ cannot ever pierce the boundary of a set. If $y(p)$ is already inside the set, it cannot escape from that set. With the path trapped, we are then able to force it to go to the appropriate place. This technique of forcing the path to go where we want is very powerful and is the underlying concept of this chapter.

3.1 PRELIMINARIES

We first present some observations and definitions that are quite straightforward and make our proofs fairly easy. Given a homotopy function H, recall what happens if at some point $\bar{y} = (\bar{x}, \bar{t})$,

$$\det H'_x(\bar{y}) \neq 0.$$

By the BDE and since $H'_x = H'_{-t} = H'_{-(n+1)}$,

$$\frac{dt}{dp} = (-1)^{(n+1)} \det H'_x(y) \neq 0.$$

This means that by changing p appropriately, we can get t to strictly increase (at least a little) from $\bar{t}$. Similarly, we can decrease t at least a little from $\bar{t}$. In other words, the path cannot look like Figure 3.1.1(a) but must look like Figure 3.1.1(b). To summarize, if $\det H'_x(\bar{x}, \bar{t}) \neq 0$, we can definitely move up or down at least a little from $\bar{t}$.

We previously defined H to be regular if the $n \times (n + 1)$ matrix $H'(y)$ was of full rank for all $y \in H^{-1}$. Given $\bar{t}$, define H to be *regular at* $\bar{t}$ if the $n \times n$ matrix $H'_x(x, \bar{t})$ is of full rank for all $x \in H^{-1}(\bar{t})$. Note that the full-rank condition must hold for each and every $x \in H^{-1}(\bar{t})$ [or, equivalently, at every $(x, \bar{t}) \in H^{-1}$, for $\bar{t}$ fixed].

The condition that H be regular at $\bar{t}$ means that at all solutions $(x, \bar{t}) \in H^{-1}$, $\det H'_x(x, \bar{t}) \neq 0$. But then just as discussed, at any point $(x, \bar{t}) \in H^{-1}$ the solution path must look like Figure 3.1.1(b). Regularity of H at $\bar{t}$ thereby guarantees that we can move along the path to a t higher than $\bar{t}$ and to a t lower than $\bar{t}$.

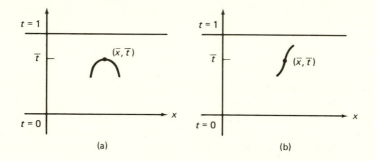

(a)   (b)

If $\det H'_x(\bar{x}, \bar{t}) \neq 0$, then the path must look like (b) and not (a).

Figure 3.1.1

Note, although we will often make regularity assumptions now on H or H at $\bar{t}$, later in the book these assumptions are totally eliminated and the theorems are displayed in their full generality. Indeed, to call regularity an assumption is almost a misnomer, because as proved in Chapter 22, it holds virtually all the time. We stress that regularity is an extremely weak assumption which later is eliminated, yet making it now helps us obtain quick, easy, and picturesque proofs.

RESTRICTED DOMAINS

We previously permitted our functions to be defined on the entire space. Often, they cannot be allowed to roam free like that but are restricted to inhabit a certain domain. For example, specify that $F: D \longrightarrow R^n$ where $D \subset R^n$. Then it is required that $x \in D$. Similarly, the variable t was defined on the entire real line, but generally we want it only on the unit interval. Call the unit interval

$$T = \{t \mid 0 \leq t \leq 1\}. \tag{3.1.1}$$

Using D and T, the domain of the homotopy function can easily be revised. Let $\times$ denote the *Cartesian product*. If $u \in A \subset R^n$ and $v \in B \subset R^m$, recall by the definition of Cartesian product that

$$A \times B = \{(u, v) \mid u \in A, \quad v \in B\} \subset R^{n+m}.$$

Via this notation,

$$D \times T = \{(x, t) \in R^{n+1} \mid x \in D, \quad 0 \leq t \leq 1\} \subset R^{n+1}. \tag{3.1.2}$$

We are now able to redefine our homotopy as $H: D \times T \to R^n$. Clearly, x is now restricted to D and t to T.

Finally, since H^{-1} and $H^{-1}(t)$ now must depend upon D and T, denote

$$H^{-1} = \{y = (x, t) \mid H(x, t) = 0, \quad x \in D, \quad t \in T\}. \tag{3.1.3}$$

And given $t \in T$,

$$H^{-1}(t) = \{x \mid H(x, t) = 0, \quad x \in D\}. \tag{3.1.4}$$

Obviously, all the definitions are as before except $x \in D$ and $t \in T$.

BOUNDARY-FREE

We begin the theorems of this chapter by introducing the idea that underlies all their proofs. It has to do with the boundary of a set. Given $D \subset R^n$, define ∂D to be the *boundary* of the set D. A point $x \in D$ is a *boundary point* if all open neighborhoods of x include points that are not in D. Thus given any boundary point, there must be points outside D that are arbitrarily close to it.

Recall that $x \in D$ is an *interior point* if there exists an open neighborhood of x entirely within D. It then follows, denoting D^0 as the interior of D, that

$$D = \partial D \cup D^0. \tag{3.1.5}$$

In words, any point in D is either a boundary point or an interior point. For instance, if D is the closure of an open set, then D^0 is the open set and ∂D is the "edge" of D^0.

We now define the fundamental concept from which we develop the theorems of this chapter. Given $H: D \times T \to R^n$, we say that H is *boundary-free* at $\bar{t}$ if

$$x \notin \partial D \quad \text{for any} \quad x \in H^{-1}(\bar{t}).$$

This means that at $\bar{t}$ and given any $y = (x, \bar{t}) \in H^{-1}$, x cannot be in the boundary of D. The concern here is where x is. Observe by (3.1.5) that if H is boundary-free at $\bar{t}$ and $y = (x, \bar{t}) \in H^{-1}$, then x must be an interior point of D. (See Figure 3.1.2.)

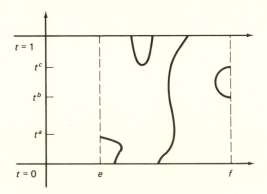

D is the interval $e \leq x \leq f$. H is not boundary-free at t^a, t^b, and t^c because at these t values the corresponding x on the solution path is on the boundary of D. The function H is boundary-free for all other t.

Figure 3.1.2

We say that H is boundary-free for t in a set if H is boundary-free for all t in that set. For instance, if H is boundary-free for $0 \leq t < 1$, then the only place where $y = (x, t) \in H^{-1}$ could possibly have $x \in \partial D$ is at $t = 1$.

The boundary-free idea provides the proof for all the theorems of this chapter. It prevents a path from encountering a boundary of D, thereby keeping it inside D. But that will force the path to go where we want it to go. We illustrate this process first with fixed-point theorems.

3.2 FIXED-POINT THEOREMS

The celebrated fixed-point theorem of Brouwer was developed very early in this century. For $D \subset R^n$, consider a function $F: D \rightarrow R^n$. Notice that F takes point $x \in D \subset R^n$ into points $F(x) \in R^n$. We may consider that F takes points in R^n and moves them to other points in R^n. The Brouwer theorem states conditions under which there exists a point $x^* \in D$ such that

$$F(x^*) = x^*. \tag{3.2.1}$$

Intuitively, the function F does not move the point x^*; rather, x^* is fixed. In practical applications this means that x^* is a point of stability or equilibrium.

We may operate on it by F, but it does not change. This concept of a stable system or equilibrium is extremely important in economics, biology, ecology, medicine, physics, chemistry, and other fields. Even philosophically, one has the notion that much of life should be fairly stable, with only a few severe discontinuities or disruptions. It is truly impressive that a concept which is so basic to mathematics has so many applications and implications.

The first fixed-point theorem we demonstrate is presented in Spartan form to highlight its mathematical essentials. The more convenient and well-known forms are presented in the corollary.

As might be anticipated, the fixed-point form of the homotopy function is utilized:

$$H(x, t) = (1 - t)(x - x^0) + t(x - F(x)). \tag{3.2.2}$$

At $t = 0$, note there is a unique point, $x^0 \in H^{-1}(0)$. Also letting I be the $n \times n$ identity matrix

$$H'_x(x^0, 0) = I,$$

so

$$\det H'_x(x^0, 0) = 1.$$

Our task is to prove that a point $x^1 \in H^{-1}(1)$ exists because then by (3.2.2), x^1 is a fixed point. That is, at $t = 1$, $H(x^1, 1) = x^1 - F(x^1) = 0$, so x^1 is a fixed point.

THEOREM 3.2.1   Given $F \in C^2$, $F: D \longrightarrow R^n$, let $D \subset R^n$ be compact and $D^0 \neq \varnothing$. For some $x^0 \in D^0$, suppose that $H: D \times T \longrightarrow R^n$ is regular and boundary-free for $0 \leq t < 1$, where

$$H(x, t) = (1 - t)(x - x^0) + t(x - F(x)).$$

Then F has a fixed point.

Proof.   As H is regular and continuously differentiable, the paths $y(p) \in H^{-1}$ exist and are continuously differentiable by Theorem 1.4.2. Thus, bifurcations, infinite spirals, and the like are automatically eliminated and the paths are well-behaved. Also, starting from $x^0 \in H^{-1}(0)$ and since $\det H'_x(x^0, 0) = +1$, we can increase t at least a little from zero.

The question is: What happens to the path after it rises at least a little from $t = 0$? Figure 3.2.1(a) cannot occur, for then there would be two solutions x^0 and $\hat{x} \in H^{-1}(0)$, which is impossible since $x^0 \in H^{-1}(0)$ is unique. Figure 3.2.1(b) cannot occur, for then at some $\bar{t} < 1$, $x \in \partial D$, and that is prohibited by the assumption that H is boundary-free. Thus the path must go up from x^0, cannot turn around, and cannot plow into a

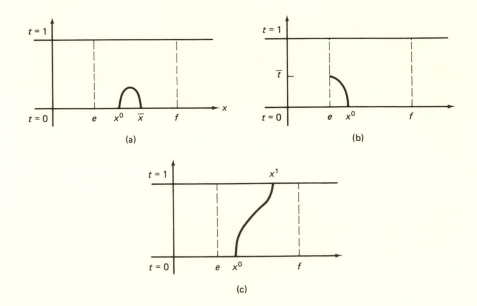

Here D is the interval $e \leq x \leq f$.

Figure 3.2.1

boundary if $t < 1$. Since D is compact and thus bounded, the path must reach $t = 1$ as in Figure 3.2.1(c). A point $x^1 \in H^{-1}(1)$ therefore exists and is the fixed point. (Note that x^1 could be in the boundary of D.) □

The proof of the theorem also provides a means to calculate the fixed point. Simply start at $y^0 = (x^0, 0)$ and follow the path $y(p)$ until $t = 1$. The path might dilly-dally and wander for a while, but ultimately it will go to a fixed point x^1.

PRACTICAL CONDITIONS

Applications to implement the theorem usefully require some practical conditions which ensure H is boundary-free for $0 \leq t < 1$. Here is a very important condition of that nature. Suppose that for some $x^0 \in D^0$,

$$x \neq sF(x) + (1 - s)x^0 \qquad \text{for } 0 < s < 1 \qquad (3.2.3)$$

whenever $x \in \partial D$. Geometrically, this means that given any $x \in \partial D$, x does not lie interior to the line segment joining x^0 and $F(x)$. We will now verify that this condition ensures that H is boundary-free.

Let

$$H(x, t) = (1 - t)(x - x^0) + t(x - F(x)).$$

Since $x^0 \in D^0$ and $x^0 \in H^{-1}(0)$ is unique, H is boundary-free at $t = 0$. Yet $H(x, t) = 0$ on the path, and solving this we see that any point x on the path must satisfy

$$x = tF(x) + (1 - t)x^0 \tag{3.2.4}$$

Via (3.2.3) and setting $s = t$, if $0 < t < 1$, any $x \in \partial D$ cannot satisfy (3.2.4), and thus cannot be on the path. Therefore, H is boundary-free for $0 \le t < 1$.

Condition (3.2.3) has a direct and immediate application to convex sets. A set D is *convex* if given $x^a \in D$ and $x^b \in D$,

$$sx^a + (1 - s)x^b \in D \qquad \text{for } 0 \le s \le 1.$$

In words, the line segment between x^a and x^b must be entirely within D. Convex sets are very nice sets, and include balls, polyhedra, and indeed any set that does not have holes or dents in it (see Exercise 1 of this chapter and also the book appendix). Also, note if $\bar{x} = sx^a + (1 - s)x^0$, where $x^0 \in D^0$ and $0 < s < 1$, then $\bar{x}$ must be strictly interior to D, that is, $\bar{x} \in D^0$.

Now suppose that we require

$$F: D \longrightarrow D, \qquad \begin{array}{l} \text{where } D \text{ is convex, compact, and has} \\ \text{a nonempty interior } D^0. \end{array} \tag{3.2.5}$$

Let us show H is then boundary-free for $0 \le t < 1$. Since $F: D \to D$, $F(x) \in D$. Given $x^0 \in D^0$, observe that the point

$$\bar{x} = sF(x) + (1 - s)x^0, \qquad 0 < s < 1$$

is in D^0. This means (3.2.3) holds because obviously if $x \in \partial D$,

$$x \ne sF(x) + (1 - s)x^0, \qquad 0 < s < 1.$$

H must be boundary-free for $0 \le t < 1$.

We see that conditions (3.2.3) or (3.2.5) ensure that H is boundary-free for $0 \le t < 1$. Let us now summarize these observations as a corollary to Theorem 3.2.1. Condition (3.2.3) is often called the Leray–Schauder condition and condition (3.2.5) is the condition for the Brouwer fixed-point theorem.

COROLLARY 3.2.2   Let $F: D \to R^n$, $F \in C^2$, and $D \subset R^n$ be compact with a nonempty interior D^0. Also let H be regular where

$$H(x, t) = (1 - t)(x - x^0) + t(x - F(x))$$

given $x^0 \in D^0$. Suppose that either:

1. (Leray–Schauder): $x \ne sF(x) + (1 - s)x^0$ for $0 < s < 1$ and $x \in \partial D$; or

2. (Brouwer): $F: D \to D$ where D is convex.

Then F has a fixed point.

Note

Our results were obtained under $F \in C^2$ and regularity in order to illustrate the elegant geometry of the path-following proof. These assumptions are weak, however, and will be eliminated in Chapter 22, where F is required only to be continuous.

Condition 1 in Corollary 3.2.2, the Leray–Schauder condition, can be varied in different ways and yields a number of useful results. As their proofs are straightforward, they are left as exercises.

AN ODD THEOREM

There are times when it is helpful to know if the number of fixed points is even or odd. For instance, if the number is odd, we are assured that at least one fixed point exists. The following theorem provides a simple condition to verify when an odd number exists and is also a nice illustration of our regularity concepts.

THEOREM 3.2.3   Given $F: D \subset R^n \to R^n$, $F \in C^2$, let D be compact and $D^0 \neq \emptyset$. Suppose that $H: D \times T \to R^n$ is regular and boundary-free for $0 \leq t \leq 1$ where

$$H(x, t) = (1 - t)(x - x^0) + t(x - F(x))$$

for some $x^0 \in D^0$. Also suppose that H is regular at $t = 1$. Then F has an odd number of fixed points.

Proof.   As in the proof of the fixed-point theorem, only one path goes from $t = 0$ to $t = 1$. We must explore what happens to other paths that leave $t = 1$.

In Figure 3.2.2, a path like A or F cannot occur because H is boundary-free for $0 \leq t \leq 1$. There cannot be two paths from $t = 0$ because $x^0 \in H^{-1}(0)$ is unique. Also, path C would be a bifurcation which is prohibited by a regular H.

Since H is regular at $t = 1$, $H'_x(x, 1)$ is nonsingular for all $x^1 \in H^{-1}(1)$. This means that we must be able to decrease t slightly. But that prohibits any paths that might be tangent at $t = 1$, such as D. Thus the only paths that can start from $t = 1$ are similar to E. And each path like that gives two points $x^1 \in H^{-1}(1)$.

We see that only paths such as B or E can occur. Counting points $x^1 \in H^{-1}(1)$, there is one given by the path from x^0, while all other possi-

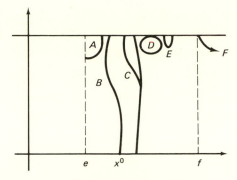

Only paths like *B* or *E* can occur.

Figure 3.2.2

ble paths yield two, so in total there must be an odd number of such points. ☐

In Chapter 22 we prove the existence of fixed points under even weaker assumptions. However, the underlying process of using boundary-freeness to ensure that the path gets to $t = 1$ should be clear. This same process will now be utilized to verify the existence of a solution to a system of nonlinear equations under an important condition.

3.3 NONLINEAR EQUATIONS

Given a system of nonlinear equations $F: R^n \longrightarrow R^n$, we would like to be sure that a solution to $F(x) = 0$ does exist. A condition which guarantees existence is that there be a compact set D and a point $x^0 \in D^0$ such that if $x \in \partial D$, then

$$F(x)^{\tau}(x - x^0) > 0 \qquad (3.3.1)$$

[or, equivalently, $F(x)^{\tau}(x - x^0) < 0$]. The set D can be selected quite arbitrarily, so this condition is very flexible. Also, recall τ denotes transpose. As an example, let $F: R^1 \longrightarrow R^1$ be a polynomial with highest power odd and its corresponding coefficient positive. Let $x^0 = 0$. Then for $x = +a$ and $x = -a$, where a is sufficiently large,

$$F(x)^{\tau}x > 0$$

because the highest power will dominate. Clearly, condition (3.3.1) then holds.

THEOREM 3.3.1   Let $F: R^n \longrightarrow R^n$, $F \in C^2$, and suppose that there is a compact set $D \subset R^n$ and point $x^0 \in D^0$ such that if $x \in \partial D$,

$$F(x)^{\tau}(x - x^0) > 0.$$

Moreover, suppose that $H: R^n \times T \longrightarrow R^n$, where

$$H(x, t) = (1 - t)(x - x^0) + tF(x) \qquad (3.3.2)$$

is regular. Then F has a solution $x^1 \in D^0$; that is,

$$F(x^1) = 0.$$

Proof.   The homotopy H has a unique point $x^0 \in H^{-1}(0)$. Also observe that by hypothesis if $x \in \partial D$, then

$$H(x, t)^{\tau}(x - x^0) = (1 - t)\|x - x^0\|^2 + tF(x)^{\tau}(x - x^0) > 0.$$

Thus H^{-1} cannot have a point on the boundary of D. In addition, at $t = 0$,

$$\det H'_x(x^0, 0) = \det I = 1.$$

The proof is now the same as in Theorem 3.2.1. The path must increase in t from $t = 0$. It cannot return to $t = 0$ because $x^0 \in H^{-1}(0)$ is unique. Moreover, the path cannot go to the boundary of D. Thus the path must go to $t = 1$. Consequently, a point $x^1 \in H^{-1}(1)$ exists and satisfies

$$F(x^1) = 0, x^1 \in D^0. \qquad \square$$

Just as with the fixed-point theorem, this theorem not only ensures that x^1 exists but provides a procedure to calculate it. Simply follow the path from x^0 using the homotopy (3.3.2) (Exercises 4 and 15).

Overall, by exploiting the boundary-free property and the structure of the homotopy H, we force the path to go where we want. The cowboys of the Old West would encircle the herd and force it to the corral. Similarly, we encircle the path and force it to the proper place, but we do this mathematically.

3.4   DEGREE THEORY

We now extend the path-following and boundary-freeness ideas to prove an index theorem, present the concept of degree, and establish the very powerful homotopy invariance theorem.

First, let us use the BDE to examine the paths in Figure 3.4.1. Path A is tangent at $t = 1$, so that there, by the BDE,

$$i(p) = (-1)^{n+1} \det H'_x(x^a, 1) = 0.$$

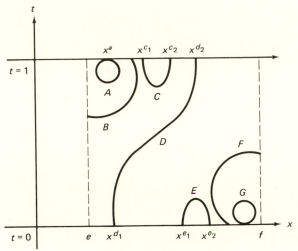

The various paths for the index theorem.

Figure 3.4.1

If H is regular at $t = 1$, path A cannot occur. Similarly, if H is regular at $t = 0$, path G cannot occur. Paths B and F can occur but, as they run into the boundary of D, are of little interest. Later we eliminate them entirely with an appropriate boundary-free assumption.

Paths C, D, and E are the important ones and we must understand them well. First consider path E. At both x^{e_1} and x^{e_2}, note $i(p) \neq 0$. In particular, if

$$i(p) > 0 \qquad \text{at } x^{e_1},$$

then $(3.4.1)$

$$i(p) < 0 \qquad \text{at } x^{e_2}.$$

In other words, suppose that increasing p increases t at x^{e_1}; then follow the path around by increasing p. It must be that at x^{e_2}, increasing p decreases t. Similarly, if

$$i(p) < 0 \qquad \text{at } x^{e_1},$$

then $(3.4.2)$

$$i(p) > 0 \qquad \text{at } x^{e_2}.$$

That is, suppose that decreasing p increases t at x^{e_1}. Then just follow the path around from x^{e_1} to x^{e_2} by decreasing p, and at x^{e_2}, $i(p) > 0$. Clearly, identical conditions hold for path C at x^{c_1} and x^{c_2}.

53

Now examine path D and observe that if

$$i(p) > 0 \qquad \text{at } x^{d_1},$$

then   (3.4.3)

$$i(p) > 0 \qquad \text{at } x^{d_2}.$$

Thus at x^{d_1} if increasing p increases t, then at x^{d_2} increasing p must also increase t. Similarly, if

$$i(p) < 0 \qquad \text{at } x^{d_1},$$

then   (3.4.4)

$$i(p) < 0 \qquad \text{at } x^{d_2}.$$

The concept here is quite simple. By following the path and observing how t is changing versus p, we can determine $i(p)$ at different points on the path. A direct application of the BDE result that $i(p) = (-1)^{n+1} \det H'_x(x, t)$ then yields certain patterns. Indeed, these patterns are very important; hence we summarize them in the next theorem, the proof of which is immediate. For it, define

$$\text{sgn } a = \begin{cases} +1 & \text{if } a > 0 \\ 0 & \text{if } a = 0 \\ -1 & \text{if } a < 0. \end{cases} \qquad (3.4.5)$$

THEOREM 3.4.1   Suppose that a regular $H: D \times T \to R^n$, $H \in C^2$, $D \subset R^n$ is also regular at $t = 0$ and $t = 1$. Suppose that a path in H^{-1} starts at x^a and ends at x^b.

If both x^a and x^b are in $H^{-1}(0)$, then

$$\text{sgn } \det H'_x(x^a, 0) = -\text{sgn } \det H'_x(x^b, 0)$$

and similarly for both x^a and x^b in $H^{-1}(1)$.

If $x^a \in H^{-1}(0)$ and $x^b \in H^{-1}(1)$, then

$$\text{sgn } \det H'_x(x^a, 0) = \text{sgn } \det H'_x(x^b, 1).$$

Implications

Theorem 3.4.1 is called an index theorem and has valuable implications. Suppose that $H(x, 0) = E(x)$ and $H(x, 1) = F(x)$. Then

$$H'_x(x, 0) = E'(x) \qquad \text{and} \qquad H'_x(x, 1) = F'(x).$$

The index theorem thereby yields:

COROLLARY 3.4.2   Let $H(x, 0) = E(x)$ and $H(x, 1) = F(x)$. If a path starts at a solution x^a to $E(x) = 0$ and returns to another solution x^b to $E(x) = 0$, then

$$\text{sgn det } E'(x^a) = -\text{sgn det } E'(x^b),$$

and analogously for F.

   If a path goes from a solution x^a of $E(x) = 0$ to a solution x^b of $F(x) = 0$, then

$$\text{sgn det } E'(x^a) = \text{sgn det } F'(x^b).$$

Note that it does not matter whether we move from x^a to x^b or from x^b to x^a; the result still holds.

DEGREE AND HOMOTOPY INVARIANCE

The index theorem permits us to demonstrate the basic theorem of degree theory, the homotopy invariance theorem.

   Consider an arbitrary continuously differentiable function $F: D \longrightarrow R^n$, where $D \subset R^n$. By analogy to our previous concepts, we define

$$F^{-1} = \{x \in D \,|\, F(x) = 0\}.$$

Thus F^{-1} is the set of all solutions $x \in D$ to the system $F(x) = 0$. Also by analogy, call F regular if at all points $x \in F^{-1}$ its $n \times n$ Jacobian F' is nonsingular. For example, suppose that $H(x, 0) = E(x)$ and $H(x, 1) = F(x)$. If H is regular at $t = 0$ and at $t = 1$, observe that both E and F are regular.

DEGREE

The degree of a function provides some insight into the number of solutions that a system of equations has. Let $F: D \longrightarrow R^n$ be regular and C^2, where $F^{-1} \cap \partial D = \varnothing$, $D \subset R^n$ is compact and with nonempty interior. We define the *degree* of F to be the integer

$$\deg(F) = \sum_{x \in F^{-1}} \text{sgn det } F'(x).$$

Thus at each solution to $F(x) = 0$, calculate the sign of the Jacobian determinant. The algebraic sum over all solutions is the degree.

   As examples, suppose that $F: R^n \longrightarrow R^n$ is $F(x) = x - x^0$. Then there is one solution $x = x^0$, $\det F'(x) = \det I = +1$, so $\deg (F) = +1$. Now suppose that $F: R^1 \longrightarrow R^1$ is $F(x) = x^3 + 2x^2 - x - 2$. This has three solutions: -2,

-1, and $+1$. Also, $F'(-2) = +3$, $F'(-1) = -2$, and $F'(+1) = +6$. Thus $\deg(F) = +1 - 1 + 1 = +1$. In fact, any regular polynomial function of one variable has degree equal to -1, 0, or 1 (Exercise 5).

It quickly follows that the degree is always finite. First observe that F regular ensures that each solution $x \in F^{-1}$ is separate and distinct. (The Jacobian F' nonsingular ensures that there is an open set around each solution containing no other solution.) But D is compact, so there can be only a finite number of such solutions (Exercise 6). Consequently, the degree is finite.

We can now establish the homotopy invariance theorem. It provides the powerful result that if we can relate two functions F and E by a homotopy, then $\deg(F) = \deg(E)$. Specifically, knowing something about the number of solutions of F informs us something about the number for E.

The theorem requires that paths such as B or F in Figure 3.4.1 cannot occur, and the following definition will suffice for that. We say that H is boundary-free if it is boundary-free for all $t \in T$. Thus H is boundary-free if for any

$$(x, t) \in H^{-1}, \qquad x \text{ is not in } \partial D.$$

Clearly, a boundary-free H eliminates B and F in Figure 3.4.1.

HOMOTOPY INVARIANCE

THEOREM 3.4.3   For $D \subset R^n$ compact and $D^0 \neq \varnothing$, let a regular H: $D \times T \rightarrow R^n$, $H \in C^2$, also be regular at $t = 0$ and $t = 1$. Suppose that H is boundary-free. Then where $H(x, 0) = E(x)$ and $H(x, 1) = F(x)$,

$$\deg(F) = \deg(E).$$

Proof.   The result follows immediately from Corollary 3.4.2. Simply examine Figure 3.4.2, which has eliminated from Figure 3.4.1 the paths

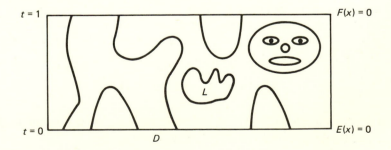

The kinds of paths possible under the assumptions of Theorem 3.4.3

Figure 3.4.2

that cannot occur. (Paths A and G in Figure 3.4.1 cannot occur due to regularity at 0 and 1, and paths B and F cannot occur due to the boundary-free condition.)

If a path in Figure 3.4.2 links a solution of $F(x) = 0$ to one of $E(x) = 0$, by Corollary 3.4.2 the signs of the determinants are the same. So the same term appears in deg (F) and deg (E). If a path starts from and returns to the same level $t = 0$, the signs of the determinants are opposite, so they do not contribute to the degree sum, and similarly for a path that starts and returns to $t = 1$. Loops such as L do not enter into the sum, so they are irrelevant.   ☐

DISCUSSION

We call two functions $E: D \longrightarrow R^n$ and $F: D \longrightarrow R^n$ *degree-equivalent* if

$$\deg (F) = \deg (E).$$

Theorem 3.4.3 provides conditions under which E and F are degree-equivalent.

The degree-equivalent concept has many applications. For instance, for $x^0 \in D$, let $E(x) = x - x^0$; notice that this has degree $+1$. Also let $\tilde{F}(x) = x - F(x)$. Now suppose that E and $\tilde{F}$ are degree-equivalent, so that

$$\deg E = \deg \tilde{F}.$$

Since the degree of E is $+1$, the degree of $\tilde{F}$ is $+1$. But then by definition of degree, $\tilde{F}(x) = 0$ must have at least one solution x^1. Immediately we see F has a fixed point

$$F(x^1) = x^1.$$

We used the concepts of degree and two functions being degree-equivalent to prove the existence of a fixed point. The degree concept thus subsumes and is more general than that of the fixed point. More succinctly, suppose that we wish to find out something about the solutions to F. If we can find an E degree-equivalent to F, then the degrees are the same, which provides much information about the solutions to F.

The path-following, notice, established not only the fixed-point theorem but degree concepts. Path-following is thus quite fundamental mathematically, and in future chapters we employ it to prove rather easily some other powerful and deep theorems.

EXERCISES/CHAPTER 3

1. Prove that the following sets are convex:
   a. $\{x \in R^n \mid Ax \leq b\}$, where A is an $m \times n$ given matrix and b is an $m \times 1$ given vector.

    b. $\{x \in R^n \,|\, \|x\| \le r\}$, where $\|\cdot\|$ is any norm and $r > 0$ a given real number.

    c. The set that is the intersection of two convex sets.

2. Let $F: D \subset R^n \longrightarrow R^n$ be C^2 and let D be the closure of an open bounded set. Suppose that given $x^0 \in D^0$, $F(x) \ne sx + (1 - s)x^0$ for $s < 1$ and $x \in \partial D$. Show that F has a fixed point. (Assume that the homotopy H you construct is regular.) Does the same result hold for $s > 1$?

3. a. Let $F: D \subset R^n \longrightarrow R^n$ be C^2 and let D be the closure of an open bounded set. Suppose that given $x^0 \in D^0$, $F(x) \ne \lambda(x - x^0)$ for $\lambda < 0$, $x \in \partial D$. Show that $F(x) = 0$ has a solution for some $x \in D$. (Use Exercise 2 and assume regularity.)

    b. Use the result in (a) to prove the *intermediate value theorem*: Let $F: [a, b] \longrightarrow R^1$ be C^2, where $F(a)$ and $F(b)$ are of opposite signs. Then there is a point c, $a < c < b$, such that $F(c) = 0$.

4. Let $F(x) = x^3 + 2x^2 - x - 2$. Using the homotopy $H(x, t) = (1 - t)x + tF(x)$, show the paths of H^{-1}.

5. Prove that any regular polynomial function of one variable has degree equal to $-1, 0$, or 1. When is the degree equal to -1? 0? 1?

6. Let $F: D \subset R^n \longrightarrow R^n$, $F \in C^2$, be regular, where D is compact. Prove that $F(x) = 0$, $x \in D$, holds for a finite number of points x.

7. Prove that the Cartesian product of closed sets is closed.

8. The *Borsuk–Ulam* theorem states that if $F: S^n \longrightarrow R^n$ is a C^2 function, where S^n is the unit sphere in R^{n+1}, there is a point x^1 such that $F(x^1) = F(-x^1)$.

    Prove the Borsuk–Ulam theorem by the following construction. First, define $H: S^n \times T \longrightarrow R^{n+1}$ by

$$H_i(y) = (1 - t)x_i + t(F_i(x) - F_i(-x)), \qquad i = 1, \ldots, n$$

$$H_{n+1}(y) = \sum_{i=1}^{n+1} (x_i)^2 - 1.$$

Assume that H is regular. Show that:

a. $H^{-1}(0)$ has exactly two points.

b. H is always boundary-free.

c. If $x \in H^{-1}(t)$, then $-x \in H^{-1}(t)$ also.

d. The two points of $H^{-1}(0)$ cannot be connected by a path.

e. The path starting from a point of $H^{-1}(0)$ must reach $t = 1$.

9. (*Winding Numbers*) Consider the C^2 mapping $F: D \subset R^n \longrightarrow R^n$, where D is the closure of an open bounded set.

    Let $\bar{w} \notin F(\partial D)$ and $v \ne 0$, $v \in R^n$. We say that a ray

$$R^+(\bar{w}, v) = \{w = \bar{w} + tv \,|\, t \ge 0\}$$

crosses $F(\partial D)$ distinctly if it is not tangent to $F(\partial D)$, if it crosses $F(\partial D)$ only a finite number of times, and if any cross-point has only one preimage in ∂D. Prove that if $R^+(\bar{w}, v)$ crosses $F(\partial D)$ distinctly, then

$$\# \, R^+(\bar{w}, v) \cap F(\partial D) \bmod 2 = \# \, F^{-1}(\bar{w}) \bmod 2, \qquad \text{(E.1.1)}$$

where $\#$ means the number of points in the set. [Intuitively, this says that the number of preimage points of $\bar{w}$ is even or odd if and only if the number of times $R^+(\bar{w}, v)$ crosses $F(\partial D)$ is even or odd.] The number of cross-points mod 2 is the *winding number* of $\bar{w}$. Note from (E.1.1) that the winding number depends only on the point $\bar{w}$ and not on the direction v. [*Hint:* Consider the homotopy H: $D \times R^+ \longrightarrow R^n$ defined by

$$H(x, t) = F(x) - \bar{w} - tv$$

where $R^+ = \{t \,|\, t \geq 0\}$. Assume that H is regular. Show that a. $\# R^+(\bar{w}, v) \cap F(\partial D) = \# \{(x, t) \in H^{-1} \,|\, x \in \partial D\}$; and b. if $(x, t) \in H^{-1}$, then t cannot become infinitely large.]

10. Let $F: D \longrightarrow R^2$ be defined by

$$F_1(x_1, x_2) = (x_1 + \tfrac{1}{2})^2$$
$$F_2(x_1, x_2) = x_2,$$

where $D = \{(x_1, x_2) \,|\, x_1^2 + x_2^2 \leq 1\}$.
a. Determine $F(\partial D)$.
b. Determine the winding number of $(1, \tfrac{1}{2})$ by considering the ray $R^+((1, \tfrac{1}{2}),$ $(1, \tfrac{1}{2}))$.

11. Let $F: D \subset R^n \longrightarrow R^n$ be a C^2 mapping, where D is the closure of an open bounded set. Show that

$$\deg(F) \bmod 2$$

equals the winding number at zero.

12. Let $F: D \longrightarrow R^2$ be defined by

$$F_1(x_1, x_2) = x_1 - 1, \qquad D = \{(x_1, x_2) \,|\, x_1^2 + x_2^2 \leq 1\}$$
$$F_2(x_1, x_2) = (x_2 + 1)^2.$$

Draw the graph of the sets $F(D)$ and $F(\partial D)$. Label the regions in $F(\partial D)$ that have one, two, or more preimages.

13. Let $F: D \longrightarrow R^2$ be defined by

$$F_1(x_1, x_2) = x_1 \sin \frac{\pi}{2} x_2$$

$$F_2(x_1, x_2) = x_1 \cos \frac{\pi}{2} x_2,$$

where $D = \{(x_1, x_1) \,|\, x_1^2 + x_2^2 \leq 1\}$. Show the preimage of the line segments $[(0, 0), (0, 1)]$ and $[(0, -1), (0, 0)]$.

14. Specify a set that has only interior points. Specify a set that has only boundary points. What are the boundary and interior points of the set $\{x \in R^1 \,|\, x = 1/n, n = 1, 2, \ldots\}$?

15. Suppose $F(x) = -12x^9 + 3x^4 - 5x + 1$. What homotopy H will guarantee that a path connects a trivial solution at $t = 0$ to a solution of F at $t = 1$? [*Hint:* Recall Theorem 3.3.1.]

NOTES

The Brouwer fixed-point theorem was proved by Brouwer [1912] using degree theory. The fixed-point theorem of Leray and Schauder [1934] assumes the condition similar to (3.2.3), where $x^0 = 0$, while Yamamuro [1963] considers an arbitrary x^0.

The concept of degree dates back to Kronecker [1869]. Our approach on degree is analytic and is related more to Nagumo [1951]. This pertinent theory of the homotopy approach goes back to works of Poincaré [1881] and Bohl [1904], and have long been among the most prominent tools in the study of equation solutions, integral equations, ordinary and partial differential equations, optimal control, variational inequalities, and so on. See, for example, Browder [1960], Cronin [1964], Krasnoselskii [1956], and Rabinowitz [1971]. For the homotopy approach, see Davidenko [1953a], Meyer [1968], Jakovlev [1964], and Branin [1972].

A constructive proof of the Brouwer fixed-point theorem was provided by Scarf [1967], who used a combinatorial principle of Lemke and Howson [1964]. Cohen [1967] gave a different proof via an inductive proof of Sperner's lemma. Eaves [1972], Kuhn and Mackinnon [1975], and Merrill [1971] cast the Scarf approach in a homotopy framework and Kellogg, Li, and Yorke [1976] emphasized fixed-point computations by differential equations based on an earlier idea of Hirsch [1963]. Garcia [1975, 1977a, b] proved the existence of a homotopy path, and proved constructively the homotopy invariance theorem. Alexander [1978], Alexander and Yorke [1978], Chow, Mallet-Paret, and Yorke [1978], Garcia and Gould [1978] Garcia and Zangwill [1979d], Keller [1978], and Smale [1976] used a differential topology approach. For a survey of the development of Brouwer's theorem, see Freudenthal [1976] and Siegberg [1976].

The approach of this chapter is taken from Garcia and Zangwill [1979d]. A recent result related to the odd theorem is provided by Eaves [1970].

APPLICATIONS

4

Nonlinear Programming: Dynamic, Parametric and Algorithmic

We hereby commence the first of several chapters on applications, and we will quickly discover how path following directly solves practical and useful problems. Nonlinear programming (NLP), our first applied topic, was selected to start the discussion of applications for two reasons. First, NLP deals with a fundamental problem—maximizing use of resources—and arises in such diverse fields as economics, business, government, mathematics, engineering, optimal control, and consumer behavior. Second, the NLP techniques we develop provide the foundation for our analysis later of even more complex situations. Most specifically, the path-following algorithm presented for NLP will be expanded in Chapter 5 into an algorithm for equilibrium programming.

In NLP an objective function must be optimized (maximized or minimized). For example, profit might be maximized or perhaps costs are to be minimized. However, the optimization cannot be done freely because of constraints on our actions. Possibly resources are restricted, or the labor supply cannot be used freely, or our own capabilities are limited. NLP thus optimizes an objective function subject to one or more constraints.

This chapter first reviews NLP and the Kuhn–Tucker (K–T) conditions for it. Especially noteworthy is the transformation of the K–T conditions into an equation system. A path is then created by introducing a control parameter. The path not only permits parametric or dynamic analyses but also yields our algorithm for solving the NLP.

4.1 STATEMENT OF NLP

For NLP an *objective function f* is given where the point $x \in R^n$ and $f(x) \in R^1$. Our goal is to select a point x that maximizes $f(x)$. However, the point x cannot be selected arbitrarily since there are *constraint functions* $g_j(x)$, $j = 1, \ldots, r$; and $h_j(x)$, $j = 1, \ldots, s$,

$$g_j(x) \geq 0, \quad j = 1, \ldots, r$$
$$h_j(x) = 0, \quad j = 1, \ldots, s.$$

Consequently, we are to maximize $f(x)$ over the set described by the foregoing constraints. A point x that achieves the maximum is called *optimal*.

STANDARD FORM

Stated in standard form, the NLP problem seeks an optimal point $\overset{*}{x}$ to

$$\max f(x)$$

subject to

$$g_j(x) \geq 0, \quad j = 1, \ldots, r$$
$$h_j(x) = 0, \quad j = 1, \ldots, s, \qquad (4.1.1)$$

where the functions are $f: R^n \rightarrow R^1$, $g_j: R^n \rightarrow R^1$, and $h_j: R^n \rightarrow R^1$. Specify the *feasible set D* as

$$D = \{x \mid g_j(x) \geq 0, \quad j = 1, \ldots, r, \quad h_j(x) = 0, \quad j = 1, \ldots, s\}. \qquad (4.1.2)$$

If $x \in D$, it is termed *feasible*. Otherwise, x is called *infeasible*.

EXAMPLE

As an example, consider the following problem, where $x = (x_1, x_2, x_3, x_4)$ and the superscript τ indicates transpose.

$$\min f(x) \equiv (x_1)^3 - 3x_2 e^{x_3} + x_3 \sin x_4$$

subject to

$$g_1(x) \equiv x_1 + x_2 + x_3 \quad + \frac{2}{x_4} + \frac{2}{3} \geq 0$$
$$g_2(x) \equiv 3x_1 - x_2 - (x_3)^3 + 2e^{x_4} \quad \leq 0$$
$$g_3(x) \equiv 2x_1 + x_2 - x_3 \quad + 3 = 0$$
$$g_4(x) \equiv x_1 \quad \geq 0$$
$$g_5(x) \equiv \quad x_2 \quad \geq 0$$
$$g_6(x) \equiv \quad x_3 \quad \geq 0$$
$$g_7(x) \equiv \quad x_4 \quad \geq 0.$$

The problem here is to minimize rather than to maximize as in the standard form. A minimization can always be transformed to a maximization by multiplying the objective function by -1. Note the equality for g_3 and the reversed inequality for g_2.

4.2 THE KUHN–TUCKER EQUATIONS

THE KUHN–TUCKER CONDITIONS

Suppose that for NLP (4.1.1), the functions f and $g_j, j = 1, \ldots, r$, are concave and continuously differentiable and that the $h_j, j = 1, \ldots, s$, are linear functions. (See the book appendix for a review of concave functions.) The Kuhn–Tucker (K–T) conditions are one of the most basic results in nonlinear programming and provide *necessary* and *sufficient* conditions for a point x to be optimal.

The K–T conditions hold at x if there exist $\lambda = (\lambda_1, \ldots, \lambda_r) \in R^r$, and $\mu = (\mu_1, \ldots, \mu_s) \in R^s$, such that (x, λ, μ) satisfies

$$\nabla f(x)^{\tau} + \sum_{j=1}^{r} \lambda_j \nabla g_j(x)^{\tau} + \sum_{j=1}^{s} \mu_j \nabla h_j(x)^{\tau} = 0$$

$$\lambda_j \geq 0, \quad g_j(x) \geq 0, \quad j = 1, \ldots, r$$

$$\lambda_j g_j(x) = 0, \quad j = 1, \ldots, r$$

$$h_j(x) = 0, \quad j = 1, \ldots, s. \tag{4.2.1}$$

Note that λ_j must be nonnegative, but μ_j can be positive, negative, or zero.

A *constraint qualification* (CQ) to ensure that the constraint functions are well behaved is also needed, although in practice it holds virtually always. Explicitly, the CQ is as follows:

1. There is a point $x^0 \in R^n$ such that

$$g_j(x^0) > 0, \quad j = 1, \ldots, r$$

$$h_j(x^0) = 0, \quad j = 1, \ldots, s. \tag{4.2.2}$$

2. For any $x \in D$, the vectors $\nabla h_j(x), j = 1, \ldots, s$, are linearly independent. [Here $\nabla h_j(x) = (\partial h_j / \partial x_1, \ldots, \partial h_j / \partial x_n)$ denotes the gradient row vector.]

Intuitively, part 1 assumes that the concave constraints $g_j \geq 0$, $j = 1, \ldots, r$, have an "interior" point, x^0. Part 2 merely requires that the linear constraints have linearly independent gradients.

The famous K–T theorem, a proof of which can be found in any standard nonlinear programming text, is:

THEOREM 4.2.1   Consider the NLP (4.1.1). Let the functions f and g_j, $j = 1, \ldots, r$, be concave and C^1 and the functions $h_j, j = 1, \ldots, s$, be linear. Suppose that the CQ (4.2.2) holds. If $\overset{*}{x}$ is an optimal solution of the NLP (4.1.1), then there exist $\overset{*}{\lambda} = (\overset{*}{\lambda}_1, \ldots, \overset{*}{\lambda}_r)$, $\overset{*}{\mu} = (\overset{*}{\mu}_1, \ldots, \overset{*}{\mu}_s)$ such that the K–T conditions (4.2.1) hold at $(\overset{*}{x}, \overset{*}{\lambda}, \overset{*}{\mu})$. Conversely, if $(\overset{*}{x}, \overset{*}{\lambda}, \overset{*}{\mu})$ satisfies the K–T conditions (4.2.1), then $\overset{*}{x}$ is an optimal solution of the NLP (4.1.1).

THE KUHN–TUCKER EQUATIONS

The K–T conditions (4.2.1) are mixtures of equalities and inequalities. But inequalities tend to be cumbersome and will hamper our attack on the NLP (4.1.1) via path following. Hence we will now eliminate them and recast the K–T conditions as a system consisting solely of equations.

The reformulation is quite simple. To begin, let k be a positive integer, and given $\alpha \in R^1$, define

$$\alpha^+ = [\max \{0, \alpha\}]^k \tag{4.2.3}$$

$$\alpha^- = [\max \{0, -\alpha\}]^k. \tag{4.2.4}$$

For example, let $k = 2$. Then $3^+ = 9$, $0^+ = 0$, $(-3)^+ = 0$, $3^- = 0$, $0^- = 0$, and $(-3)^- = 9$. Note that $\alpha^- = (-\alpha)^+$. If $\alpha \geq 0$, then $\alpha^+ \geq 0$ and $\alpha^- = 0$, while if $\alpha \leq 0$, then $\alpha^+ = 0$ and $\alpha^- \geq 0$. Thus we always have

$$\alpha^+ \geq 0, \qquad \alpha^- \geq 0, \qquad \alpha^+ \alpha^- = 0. \tag{4.2.5}$$

Another valuable feature of α^+ and α^- is that they are both $(k - 1)$-continuously differentiable (Exercise 1). Thus if $k = 1$, α^+ and α^- are continuous. If $k = 2$, then α^+ and α^- are C^1. [To better understand this, plot a graph of the broken line $y = \max \{0, \alpha\}$. The broken line is not differentiable at the origin and in fact has an elbow or kink there. However, note that the function $y = \alpha^+ = [\max \{0, \alpha\}]^2$ changes smoothly and hence is differentiable at the origin. Intuitively, transformations (4.2.3) and (4.2.4) smooth out the kinks.]

Using $\alpha_j^+ = [\max \{0, \alpha_j\}]^k$ and $\alpha_j^- = [\max \{0, -\alpha_j\}]^k$, we can recast (4.2.1) and create the *K–T equations* for the NLP (4.1.1)

$$\nabla f(x)^\tau + \sum_{j=1}^{r} \alpha_j^+ \nabla g_j(x)^\tau + \sum_{j=1}^{s} \mu_j \nabla h_j(x)^\tau = 0$$

$$\alpha_j^- - g_j(x) = 0, \qquad j = 1, \ldots, r$$

$$h_j(x) = 0, \qquad j = 1, \ldots, s. \tag{4.2.6}$$

Here $\alpha = (\alpha_1, \ldots, \alpha_r) \in R^r$.

The K–T equations are precisely equivalent to the K–T conditions, as is easily shown (Exercise 2). More explicitly, if $(\overset{*}{x}, \overset{*}{\alpha}, \overset{*}{\mu})$ satisfies the K–T equations, then $(\overset{*}{x}, \overset{*}{\lambda}, \overset{*}{\mu})$ satisfies the K–T conditions where

$$\overset{*}{\lambda}_j \equiv \overset{*}{\alpha}_j^+, \qquad j = 1, \ldots, r. \tag{4.2.7}$$

Conversely, if $(\overset{*}{x}, \overset{*}{\lambda}, \overset{*}{\mu})$ satisfies the K–T conditions, then $(\overset{*}{x}, \overset{*}{\alpha}, \overset{*}{\mu})$ satisfies the K–T equations, where

$$\overset{*}{\alpha}_j \equiv \begin{cases} (\overset{*}{\lambda}_j)^{1/k} & \text{if } g_j(\overset{*}{x}) = 0 \\ -(g_j(\overset{*}{x}))^{1/k} & \text{if } g_j(\overset{*}{x}) > 0, \end{cases} \qquad j = 1, \ldots, r. \tag{4.2.8}$$

Formulas (4.2.7) and (4.2.8) provide a one-to-one relationship between (4.2.1) and (4.2.6). Indeed, under the assumptions of Theorem 4.2.1, finding an optimal solution to the NLP (4.1.1) is equivalent to finding a solution to the K–T equations (4.2.6).

As mentioned, by choice of k we can make α_j^+ and α_j^- as differentiable as we like. Thus suppose the f and the g_j are C^l. By choosing $k \geq l$ the entire left hand side of (4.2.6) is then differentiable of order C^{l-1}. (One derivative of f and g_j was already taken to form (4.2.6) so only $l - 1$ are left.) This means we can make the K–T equations smooth, thereby eliminating the kinks that the K–T conditions have due to the inequalities.

To summarize, the K–T conditions and the K–T equations are equivalent and they both are necessary and sufficient for $\overset{*}{x}$ to be optimal for the NLP (4.1.1). The K–T conditions, however, possess inequalities, and the resultant kinks could make path following quite cumbersome. The K–T equations have only equalities and are smooth. Most importantly, as we will see in the next section, the K–T equations will permit us to easily path–follow.

AN EXAMPLE

A simple example might clarify both the K–T conditions and the K–T equations. Suppose that we wish to solve the NLP

$$\max x_1 - (x_2)^2 - x_3$$

subject to

$$x_1 - (x_2)^2 + 2x_2x_3 - (x_3)^2 + 3 \geq 0$$

$$x_1 \qquad\qquad\qquad\qquad\qquad \geq 0$$

$$2x_1 - \qquad\quad x_2 + x_3 \qquad = 0.$$

The K–T conditions for this problem are

$$
\begin{aligned}
1 \; +\lambda_1 + \qquad\qquad + \lambda_2 \qquad\qquad + 2\mu_3 &= 0 \\
-2x_2 - \lambda_1 2(x_2 - x_3) \qquad\qquad\qquad - \mu_3 &= 0 \\
-1 \; + \lambda_1 2(x_2 - x_3) \qquad\qquad\qquad + \mu_3 &= 0 \\
x_1 - (x_2)^2 \qquad + 2x_2 x_3 - (x_3)^2 + 3 \; &\geq 0 \\
x_1 \qquad\qquad\qquad\qquad\qquad\qquad &\geq 0 \\
2x_1 - x_2 \qquad\qquad + x_3 \qquad &= 0 \\
\lambda_1(x_1 - (x_2)^2 + 2x_2 x_3 - (x_3)^2 + 3) \; &= 0 \\
\lambda_2 x_1 \qquad\qquad\qquad\qquad\qquad &= 0 \\
\lambda_1 \qquad\qquad\qquad\qquad\qquad\qquad &\geq 0 \\
\lambda_2 \qquad\qquad\qquad\qquad\qquad\qquad &\geq 0.
\end{aligned}
$$

The K–T equations are, letting $k = 2$,

$$
\begin{aligned}
1 \; + \; (\max\{0, \alpha_1\})^2 \qquad\qquad + (\max\{0, \alpha_2\})^2 \qquad + 2\mu_3 &= 0 \\
-2x_2 - 2(\max\{0, \alpha_1\})^2(x_2 - x_3) \qquad\qquad\qquad - \mu_3 &= 0 \\
-1 \; + 2(\max\{0, \alpha_1\})^2(x_2 - x_3) \qquad\qquad\qquad + \mu_3 &= 0 \\
(\max\{0, -\alpha_1\})^2 - [x_1 - (x_2)^2 + 2x_2 x_3 - (x_3)^2 + 3] \; &= 0 \\
(\max\{0, -\alpha_2\})^2 - x_1 \qquad\qquad\qquad\qquad &= 0 \\
2x_1 \qquad - x_2 \; + x_3 \qquad &= 0.
\end{aligned}
$$

Note how the K–T equations require fewer relations (equalities or inequalities).

4.3 DYNAMIC AND PARAMETRIC NLP

The NLP (4.1.1) is static, and as formulated it assumes that the optimal point $\overset{*}{x}$ remains the same forever. More realistically, time progresses or conditions change, so the optimal point would vary over time. For example, with a new product more profit might be made in the future since there might be a delay until consumers accept it. In other circumstances weather conditions may change over time, or perhaps wage rates will go up in the future. As such changes occur, clearly the optimal point might change as well.

We are led to consider a *dynamic* or *parametric NLP* in which our functions f, g_j, and h_j depend upon a parameter, t. The optimal point will then depend upon t also. Moreover, t could represent any parameter that changes (e.g., a control parameter, pressure, temperature, interest rates, etc.). Time is only one interpretation of t, although for our needs a convenient one.

Especially, perceive as t changes and as the optimal point changes that we obtain a path. That is, as t changes, a path of optimal points is generated. If t is time, for example, this could be the path of the optimal point over time. Earlier chapters had t a homotopy parameter and we obtained a path. If t is an arbitrary parameter, clearly we also obtain a path.

Our dynamic NLP is quite important. In the first place it provides a very useful model because it describes how a process changes over time (or as a parameter changes). Futher, and most important relative to our theory, it yields a path and thus permits us to use our entire path-following analysis.

Let us now make this introductory discussion about dynamic NLP more precise.

DYNAMIC NLP IN STANDARD FORM

Let $t \in R^1$ be a variable which as mentioned could indicate time or some other parameter. Also, let $f: R^{n+1} \longrightarrow R^1$, $g_j: R^{n+1} \longrightarrow R^1$, $j = 1, \ldots, r$, and $h_j: R^{n+1} \longrightarrow R^1$, $j = 1, \ldots, s$. Here $f(x, t), g_j(x, t)$, and $h_j(x, t)$, where (x, t) $\in R^{n+1}$, so our functions now depend upon t. For any t fixed we wish to solve the NLP for the corresponding optimal point $\overset{*}{x}$. In detail, at any given t the dynamic NLP seeks a maximizing point to

$$\max_x f(x, t)$$

subject to

$$g_j(x, t) \geq 0, \qquad j = 1, \ldots, r$$
$$h_j(x, t) = 0, \qquad j = 1, \ldots, s. \qquad (4.3.1)$$

Remember t is a parameter here and for each t fixed we optimize over x only.

KUHN–TUCKER CONDITIONS
AND EQUATIONS FOR THE DYNAMIC NLP

Given the dynamic NLP (4.3.1) the K–T conditions and K–T equations for it follow immediately from (4.2.1) and (4.2.6), respectively.

K–T Conditions for the Dynamic NLP

For any t fixed,

$$\nabla_x f(x, t)^\tau + \sum_{j=1}^r \lambda_j \nabla_x g_j(x, t)^\tau + \sum_{j=1}^s \mu_j \nabla_x h_j(x, t)^\tau = 0$$
$$\lambda_j g_j(x, t) = 0, \qquad j = 1, \ldots, r$$
$$g_j(x, t) \geq 0, \qquad j = 1, \ldots, r$$
$$h_j(x, t) = 0, \qquad j = 1, \ldots, s$$
$$\lambda_j \geq 0, \qquad j = 1, \ldots, r. \qquad (4.3.2)$$

K–T Equations for the Dynamic NLP

For any t fixed,

$$\nabla_x f(x, t)^\tau + \sum_{j=1}^{r} \alpha_j^+ \nabla_x g_j(x, t)^\tau + \sum_{j=1}^{s} \mu_j \nabla_x h_j(x, t)^\tau = 0$$

$$\alpha_j^- - g_j(x, t) = 0, \qquad j = 1, \ldots, r$$

$$h_j(x, t) = 0, \qquad j = 1, \ldots, s. \tag{4.3.3}$$

Here ∇_x means differentiation with respect to x only. For example,

$$\nabla_x g_j(x, t) = \left(\frac{\partial g_j(x, t)}{\partial x_1}, \ldots, \frac{\partial g_j(x, t)}{\partial x_n} \right).$$

Theorem 4.2.1 extends easily to the dynamic case. Expressly for the dynamic case, we suppose that for any t fixed, the f and g_j are concave in x and the h_j are linear in x. Also, we generalize the CQ to the following:

1. For any t there is an x^0 such that

    $$g_j(x^0, t) > 0, \qquad j = 1, \ldots, r$$

    $$h_j(x^0, t) = 0, \qquad j = 1, \ldots, s \tag{4.3.4}$$

    (note here the x^0 can depend upon t); and

2. For any (x, t) satisfying the constraints of (4.3.1), the vectors $\{\nabla_x h_j(x, t)\}_{j=1}^{s}$ are linearly independent.

Immediately from Theorem 4.2.1 we have:

THEOREM 4.3.1   Considering the dynamic NLP (4.3.1), suppose that CQ (4.3.4) holds, and let all functions be continuously differentiable. Also let f and $g_j, j = 1, \ldots, r$, be concave in x and $h_j, j = 1, \ldots, s$, be linear in x. Then for $\overset{*}{t}$ fixed $\overset{*}{x}$ is optimal if and only if:

1. for some $\overset{*}{\lambda} \in R^r$ and $\overset{*}{\mu} \in R^s$, the point $(\overset{*}{x}, \overset{*}{\lambda}, \overset{*}{\mu}, \overset{*}{t})$ satisfies the K–T conditions (4.3.2); or, equivalently,

2. for some $\overset{*}{\alpha} \in R^r$ and $\overset{*}{\mu} \in R^s$, the point $(\overset{*}{x}, \overset{*}{\alpha}, \overset{*}{\mu}, \overset{*}{t})$ satisfies the K–T equations (4.3.3).

REMARKS

Theorem 4.3.1 states that we can solve the dynamic NLP by solving the corresponding K–T equations (or conditions) at each t. It might therefore seem that as t changes, the K–T equations must be resolved for each possible value of t. Clearly, that would be a horrendous amount of effort. Fortunately, our path-following techniques will come to the rescue; and instead of continually resolving equations all the time, we will simply follow a path.

We also remark that the K–T equations are meaningful even if the functions are not appropriately concave or linear. In this more general case the K–T equations need not yield a maximum point, but possibly a saddle point or even a local minimum. The exploration of this more general situation is quite fascinating, as it can lead to "catastrophes," and this more advanced phenomenon is discussed in Chapter 10.

4.4 PATH EXISTENCE
FOR THE DYNAMIC NLP

We must now analyze how to generate a path of optimal solutions to (4.3.3). As t changes, the optimal solutions change. By following this path of optimal points, we obtain a powerful method for solving the dynamic NLP.

To formalize this, let $H(x, \alpha, \mu, t)$ denote the left-hand side of (4.3.3).

$$H(x, \alpha, \mu, t) \equiv \begin{bmatrix} \nabla_x f(x, t)^\tau + \sum_{j=1}^{r} \alpha_j^+ \nabla_x g_j(x, t)^\tau + \sum_{j=1}^{s} \mu_j \nabla_x h_j(x, t)^\tau \\ \alpha_1^- - g_1(x, t) \\ \cdot \\ \cdot \\ \cdot \\ \alpha_r^- - g_r(x, t) \\ h_1(x, t) \\ \cdot \\ \cdot \\ \cdot \\ h_s(x, t) \end{bmatrix} \qquad (4.4.1)$$

With $H = (H_i)$, definition (4.4.1) means that we let H_i equal the ith component on the right side. Here $H: R^{n+r+s+1} \longrightarrow R^{n+r+s}$ because $x \in R^n$, $\alpha \in R^r$, $\mu \in R^s$, and $t \in R^1$, and there are $n + r + s$ components on the right side of (4.4.1).

Using H, the K–T equations (4.3.3) can be rewritten succinctly as

$$H(x, \alpha, \mu, t) = 0. \qquad (4.4.2)$$

Equation (4.4.2) is the key to the path following, and just as in the previous chapters our interest is in the set of solutions H^{-1} to (4.4.2). Notice that any point $(x, \alpha, \mu, t) \in H^{-1}$ is a K–T point for that t. This fact is quite important because it means that analyzing H^{-1} will provide us with paths of K–T points, as we now show.

Let x^0 be a known optimal solution to the dynamic NLP at $t = t^0$. It is then generally easy to calculate the $\alpha^0 \in R^r$, $\mu^0 \in R^s$ so that $(x^0, \alpha^0, \mu^0, t^0) \in H^{-1}$. In other words, given x^0 and t^0, we can obtain α^0, μ^0 so that $(x^0, \alpha^0, \mu^0, t^0)$ solves the K–T equations.

Now letting f and all g_j and h_j be C^3 and also setting $k = 3$ so that α_j^+ and α_j^- are C^2, H is C^2. Also, just as in earlier chapters, we make the following regularity assumption:

$$H'(x, \alpha, \mu, t) \quad \text{is of full rank for all} \quad (x, \alpha, \mu, t) \in H^{-1}. \qquad (4.4.3)$$

Since the Jacobian H' is an $(n + r + s) \times (n + r + s + 1)$ matrix, full rank means rank $(n + r + s)$. Recall that regularity is a very mild condition and, in fact, is eliminated in Chapter 22.

We can now state the theorem which ensures that H^{-1} consists of paths of K–T points. The theorem follows immediately from Chapter 1.

THEOREM 4.4.1 Let H as defined in (4.4.1) be C^2 and regular. Then H^{-1} consists only of continuously differentiable paths.

The theorem tells us that starting from $(x^0, \alpha^0, \mu^0, t^0) \in H^{-1}$, we can generate a path $(x(p), \alpha(p), \mu(p), t(p)) \in H^{-1}$. Moreover, by Theorem 4.3.1, $x(p)$ will be the optimal solution to the dynamic NLP at $t(p)$.

We have hereby determined a major result: that path following provides the optimal solution to the dynamic NLP as t varies. Succinctly, if (x, α, μ, t) is in the path, x maximizes the NLP at t. This path result is highly valuable in solving any parametric NLP or dynamic NLP. Moreover, in the next section we utilize this idea to create a path-following algorithm that solves the NLP itself.

4.5 SOLVING THE NLP BY PATH FOLLOWING

Consider the NLP (4.1.1), which for convenience we here restate:

$$\max f(x)$$
$$g_j(x) \geq 0, \qquad j = 1, \ldots, r$$
$$h_j(x) = 0, \qquad j = 1, \ldots, s. \qquad (4.5.1)$$

Assume that the f and $g_j, j = 1, \ldots, r$, are concave and C^3; the $h_j, j = 1, \ldots, s$, are linear; and that CQ (4.2.2) holds.

Our purpose in this section is to develop an algorithm for solving NLP (4.5.1) that is based upon path following. Certainly, there are many well-known methods for solving the NLP, and they employ a variety of techniques. Path following provides not only a novel approach, but even more crucially and, as we will see, generalizes to solve equilibrium programming in Chapter 5.

To begin let us pose an NLP which instead of f uses a simpler objective function,

$$f^0(x) \equiv -\tfrac{1}{2}\|x - x^0\|^2 = -\tfrac{1}{2}\sum_{i=1}^{n}(x_i - x_i^0)^2,$$

where x^0 is the point for the CQ (4.2.2) and is assumed known. Also $\|\cdot\|$ is the Euclidean norm. The point x^0 satisfies

$$\begin{aligned}
g_j(x^0) &> 0, & j &= 1,\ldots,r \\
h_j(x^0) &= 0, & j &= 1,\ldots,s.
\end{aligned} \tag{4.5.2}$$

Explicitly, this simpler NLP is

$$\max f^0(x) = -\tfrac{1}{2}\|x - x^0\|^2$$

subject to

$$\begin{aligned}
g_j(x) &\geq 0, & j &= 1,\ldots,r \\
h_j(x) &= 0, & j &= 1,\ldots,s.
\end{aligned} \tag{4.5.3}$$

Let

$$D = \{x \,|\, g_j(x) \geq 0, \quad j = 1,\ldots,r, \quad h_j(x) = 0, \quad j = 1,\ldots,s\}.$$

By (4.5.2), $x^0 \in D$, and we easily conclude that x^0 is the unique solution to NLP (4.5.3) (Exercise 4).

HOMOTOPY APPROACH

Let us now consider a slightly more complex NLP:

$$\max_{x} f(x, t) \equiv (1 - t)f^0(x) + tf(x)$$

$$\begin{aligned}
g_j(x) &\geq 0, & j &= 1,\ldots,r \\
h_j(x) &= 0, & j &= 1,\ldots,s
\end{aligned} \tag{4.5.4}$$

for t fixed, $0 \leq t \leq 1$. Denote x^t as an optimal point for (4.5.4) at t, $0 \leq t \leq 1$.

If $t = 0$, NLP (4.5.4) becomes NLP (4.5.3), which has known solution x^0. At $t = 1$ it becomes NLP (4.5.1) with optimal point x^1.

The idea is the same as that we have used numerous times previously. Start from $t = 0$ with known solution x^0 and then by increasing t, path–follow until at $t = 1$ we reach x^1. Further, since (4.5.4) is a dynamic NLP, via Theorem 4.4.1 the K–T equations can be used to path–follow. That is, the K–T equations to (4.5.4) provide a path of points, where if (x, α, μ, t) solves the K–T equations, then x is optimal for (4.5.4) at t. The approach we will take, therefore, is to path–follow in the K–T equations from $t = 0$ to $t = 1$ and thereby solve (4.5.1).

More exactly, at $t = 0$ we start with a point $(x^0, \alpha^0, \mu^0, t = 0)$ that satisfies the K–T equations for (4.5.4). The point x^0 will be optimal for (4.5.4) at $t = 0$. Then we will path–follow in the K–T equations until at $t = 1$ we obtain $(x^1, \alpha^1, \mu^1, 1)$. The point x^1 will then solve (4.5.1). Let us now specify this process in detail and validate that it works.

As mentioned, the path following will be done in the K–T equations for (4.5.4) which are

$$H(x, \alpha, \mu, t) \equiv \begin{cases} -(1 - t)(x - x^0) + t\nabla f(x)^{\tau} + \sum_{j=1}^{r} \alpha_j^+ \nabla g_j(x)^{\tau} \\ \qquad\qquad\qquad + \sum_{j=1}^{s} \mu_j \nabla h_j(x)^{\tau} = 0 \\ \alpha_j^- - g_j(x) = 0, \qquad j = 1, \ldots, r \\ h_j(x) = 0, \qquad j = 1, \ldots, s, \end{cases} \qquad (4.5.5)$$

where we let $\alpha_j^+ = [\max \{0, \alpha_j\}]^3$ and $\alpha_j^- = [\max \{0, -\alpha_j\}]^3$. We are assured that H is C^2 since the f, g_j and h_j are assumed C^3.

Note that a point on the path consists of the full vector (x, α, μ, t) $\in H^{-1}$, and not just x. Also notice that $-\frac{1}{2} \| x - x^0 \|^2$ is strictly concave; hence $f(x, t)$ is strictly concave in x for $0 \le t < 1$. This means that x^t, the optimal point at t, is unique for $0 \le t < 1$ (Exercise 5).

STARTING THE PATH

To start the path, observe that at $t = 0$ we obtain

$$\begin{aligned} x &= x^0 \\ \alpha_j^- &= g_j(x^0), \qquad \alpha_j^+ = 0, \quad j = 1, \ldots, r \\ \mu_j &= 0, \qquad\qquad\qquad j = 1, \ldots, s \end{aligned} \qquad (4.5.6)$$

as the starting solution to the K–T equations. We already know that x^0 is the unique optimal point to (4.5.4) at $t = 0$, so no other x value can solve (4.5.5) at $t = 0$. Moreover, at $t = 0$, since x^0 is unique, by examining (4.5.5) it is easily seen that the α_j and μ_j must also be unique. Thus the (x, α, μ, t) given by (4.5.6) is the unique point that solves (4.5.5) at $t = 0$ and hence is the unique point in H^{-1} at $t = 0$. [To show that (4.5.6) is the unique solution to (4.5.5) at $t = 0$, note that we had to prove not only the uniqueness of x, but also the uniqueness of α and μ.]

It turns out that the solution (x, α, μ, t) to (4.5.5) is also unique for t near zero. By strict concavity, as mentioned, x^t is unique for such t, so it remains to verify that the α_j and μ_j are unique. Recall from CQ that at $t = 0$,

$$\alpha_j^- = g_j(x^0) > 0, \qquad j = 1, \ldots, r.$$

For t near zero the optimal point x^t will be close to x^0, so

$$\alpha_j^- = g_j(x^t) > 0 \tag{4.5.7}$$

for t sufficiently small. This forces

$$\alpha_j^+ = 0, \qquad j = 1, \ldots, r. \tag{4.5.8}$$

Since x^t is unique, (4.5.7) and (4.5.8) ensure that

$$\alpha_j^+ \quad \text{and} \quad \alpha_j^-, \qquad j = 1, \ldots, r$$

will be unique also.

Using (4.5.7) and (4.5.8), the K–T equations for t near zero become

$$-(1 - t)(x^t - x^0) + t\nabla f(x^t)^{\tau} + \sum_{j=1}^{s} \mu_j \nabla h_j(x^t)^{\tau} = 0.$$

By CQ the $\nabla h_j(x^t), j = 1, \ldots, s$, are linearly independent. The $\mu_j, j = 1, \ldots, s$, must then be unique. Since x is unique and the α_j and the μ_j are unique, we conclude that the solution to the K–T equation (x, α, μ, t) is unique for t sufficiently near zero.

This uniqueness relates immediately to the path behavior. Examining Figure 4.5.1, perceive that the uniqueness at $t = 0$ excludes path B. The loop, path C, is unique at $t = 0$ but not unique for t near zero, so it also cannot occur. The path must therefore rise from $t = 0$ and never return to $t = 0$.

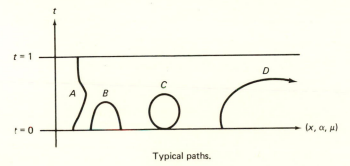

Typical paths.

Figure 4.5.1

PATH BOUNDEDNESS

With paths B and C excluded, to prove that the path reaches $t = 1$, we need only verify that it cannot run off to infinity (i.e., that path D cannot occur). Technically, we must establish that the solutions to the K–T equations are bounded. Interestingly, it is easier to prove that the solutions to the K–T conditions are bounded. Since the K–T conditions and K–T equations are equivalent, that will immediately validate boundedness for the path.

Let the feasible set D be compact. If (x, α, μ, t) is in H^{-1}, then x must be bounded since any x on the path is feasible. We will now examine the K–T conditions and prove that the λ_j and μ_j are also bounded.

The next lemma not only provides that conclusion but is a nice application of linear programming duality theory. In it

$$f(x, t) = (1 - t)f^0(x) + tf(x)$$
$$\nabla_x f(x, t) = -(1 - t)(x - x^0)^\tau + t\nabla f(x)$$

and

$$A = \max_{x,t} \{\nabla_x f(x, t)(x - x^0)\}, \tag{4.5.9}$$

where the maximum is over $x \in D$ and $0 \le t \le 1$. Because D is compact, A is a finite number. Restated, (4.5.9) becomes

$$\nabla_x f(x, t)(x - x^0) \le A \tag{4.5.10}$$

for any $x \in D$, and any t, $0 \le t \le 1$.

LEMMA 4.5.1   Let (x, λ, μ, t) be a solution to the K–T conditions, $0 \le t \le 1$. Then

$$\lambda_j \le \frac{A}{g_j(x^0)}, \qquad j = 1, \ldots, r.$$

Proof. Consider the linear programming problem in the variables λ_j and μ_j, where $x \in D$ and t are fixed:

$$\max \sum_{j=1}^{r} \lambda_j g_j(x^0)$$

$$\sum_{j=1}^{r} \lambda_j \nabla g_j(x)^\tau + \sum_{j=1}^{s} \mu_j \nabla h_j(x)^\tau = -\nabla_x f(x, t)^\tau$$
$$\lambda_j g_j(x) = 0, \qquad j = 1, \ldots, r$$
$$\lambda_j \quad\ \ge 0, \qquad j = 1, \ldots, r. \tag{4.5.11}$$

Notice that the constraints of (4.5.11) are simply taken from the K–T conditions of (4.5.4).

The dual problem to (4.5.11) is

$$\min -\nabla_x f(x, t)\beta$$

$$g_j(x)\gamma_j + \nabla g_j(x)\beta \geq g_j(x^0), \qquad j = 1, \ldots, r$$

$$\nabla h_j(x)\beta = 0, \qquad j = 1, \ldots, s, \qquad (4.5.12)$$

where $\beta = (\beta_1, \ldots, \beta_n)$.

From concavity

$$g_j(x) + \nabla g_j(x)(x^0 - x) \geq g_j(x^0), \qquad j = 1, \ldots, r,$$

and by linearity and since by feasibility $h_j(x) = 0, j = 1, \ldots, s,$

$$h_j(x) + \nabla h_j(x)(x^0 - x) = \nabla h_j(x)(x^0 - x) = 0, \qquad j = 1, \ldots, s.$$

We see that $\gamma_j = 1, j = 1, \ldots, r,$ and $\beta = (x^0 - x)$ is a feasible solution for the dual.

Employing the duality theorem of linear programming, since $\gamma_j = 1, \beta = (x^0 - x)$ is dual-feasible,

$$\sum_{j=1}^{r} \lambda_j g_j(x^0) \leq \nabla_x f(x, t)(x - x^0) \qquad (4.5.13)$$

for any primal–feasible λ.

Now calling on (4.5.10) and (4.5.13),

$$\sum_{j=1}^{r} \lambda_j g_j(x^0) \leq A. \qquad (4.5.14)$$

Notice that (4.5.14) holds no matter what $x \in D$ or t we select to form (4.5.11). Further, since $\lambda_j \geq 0, j = 1, \ldots, r,$ and by CQ, $g_j(x^0) > 0,$ $j = 1, \ldots, r,$

$$\lambda_j \leq \frac{A}{g_j(x^0)}, \qquad j = 1, \ldots, r. \qquad \square$$

Lemma 4.5.1 provides that for any (x, λ, μ, t) that solve the K–T conditions, the $\lambda_j, j = 1, \ldots, r,$ are bounded. Also, for any such $(x, \lambda, \mu, t),$

$$\nabla_x f(x, t)^\tau + \sum_{j=1}^{r} \lambda_j \nabla g_j(x)^\tau + \sum_{j=1}^{s} \mu_j \nabla h_j(x)^\tau = 0.$$

Via CQ the $\nabla h_j(x), j = 1, \ldots, s,$ are linearly independent. This means that $\mu_j, j = 1, \ldots, s,$ cannot diverge to infinity unless some $\lambda_j, j = 1, \ldots, r,$ does. Since the latter is prohibited, conclude that all $\mu_j, j = 1, \ldots, s,$ are bounded.

We have determined that any (x, λ, μ, t) which satisfies the K–T conditions must lie in a bounded set. It follows immediately from (4.2.8) that the α of the K–T equations is also bounded. Thus H^{-1}, that is, the set of points (x, α, μ, t) that solve the K–T equations, is bounded (Exercise 6). The path is bounded and must therefore reach $t = 1$.

Let us summarize the proof that the algorithm works.

THEOREM 4.5.2 For the NLP (4.5.1), suppose that the f and g_j, $j = 1, \ldots, r$, are concave and C^3; the $h_j, j = 1, \ldots, s$, are linear, CQ (4.2.2) holds where x^0 is given, the H defined by (4.5.5) is regular, and D is compact. Then by starting at

$$t = 0, \qquad x = x^0$$

$$\alpha_j^- = g_j(x^0), \quad \alpha_j^+ = 0, \qquad j = 1, \ldots, r$$

$$\mu_j = 0, \qquad\qquad\qquad j = 1, \ldots, s,$$

we can follow the path in H^{-1} until $t = 1$ and thereby solve (4.5.1).

PROOF RECAPITULATION

For problem (4.5.4), CQ (4.2.2) immediately implies CQ (4.3.4), so all hypotheses of Theorem 4.3.1 hold. We are thus assured that for any t, a solution (x, α, μ, t) to the K–T equations will yield an optimal x for (4.5.4). Moreover, regularity provides that the solutions H^{-1} to the K–T equations (4.5.5) form a path.

At $t = 0$, the starting solution to (4.5.5) is unique. Further, the solution is unique for t small. The path then cannot return to $t = 0$. Also, $x \in D$ is bounded and we verified that all α_j and μ_j are bounded. The path therefore cannot run off to infinity. The path must reach $t = 1$.

SUMMARY

This chapter applied our path-following technology to NLP. Because path following applies most easily to equations, first we eliminated the inequalities in the K–T conditions and formed the K–T equations. Next, by adding a parameter t, a path was created, that path being the optimal solutions to the dynamic NLP as t varied. Finally, the dynamic NLP was used to develop a path-following algorithm for the NLP itself.

This chapter is quite fundamental, and indeed its concepts provide the foundation for the equilibrium programming problem of Chapter 5.

EXERCISES/CHAPTER 4

1. Show that $\alpha^+ = [\max\{0, \alpha\}]^k$ is $(k - 1)$-continuously differentiable.

2. Prove that the K–T equations and the K–T conditions are equivalent.

3. Prove Theorem 4.3.1.

4. Prove that x^0 is the unique solution to (4.5.3).

5. Let f and $g_j, j = 1, \ldots, r$, be concave and $h_j, j = 1, \ldots, s$, be linear. Prove that for $0 \le t < 1$, the function $f(x, t)$ of (4.5.4) is strictly concave and that the optimal point for (4.5.4) is unique.

6. Prove that H^{-1} is bounded, given the assumptions of Theorem 4.5.2.

7. Formulate the K–T equations for the problem

$$\min f(x)$$
$$g_j(x) \le 0, \qquad j = 1, \ldots, r$$
$$h_j(x) = 0, \qquad j = 1, \ldots, s$$
$$x_j \ge 0, \qquad j = 1, \ldots, n.$$

8. Formulate the K–T equations for the problem

$$\min cx$$
$$Ax \le b$$
$$x \ge 0.$$

9. Formulate the K–T equations for the problem

$$\max 3(x_1)^2 + 4x_1 x_2 + (x_2)^2 - 3x_1 + e^{x_1 + 4x_2}$$

subject to

$$(x_1)^2 + (x_2)^2 \le 1$$
$$x_1 + 3x_2 = 0$$
$$(x_2)^2 \ge \tfrac{1}{2}$$
$$x_1 \ge 0.$$

10. By setting up (4.5.4) for the NLP

$$\min 2x^2 - 3x$$
$$5 \ge x \ge 0,$$

find the solution path for (4.5.4) given $x^0 = 1$.

NOTES

4.1. Standard texts in nonlinear programming include Zangwill [1969], Mangasarian [1969], and Avriel [1976].

4.2. The Kuhn–Tucker conditions were formulated by Karush [1939], and Kuhn and Tucker [1951]. The CQ assumption (4.2.1) is Slater's CQ [1950]. Other constraint qualifications exist. For extensive discussions on various forms of CQ, see Bazaraa, Goode, and Shetty [1972] and Gould and Tolle [1972]. Spingarn and Rockafellar [1979] proved that the CQ holds almost always.

    The transformation of the K–T conditions to K–T equations is standard in nonlinear complementarity theory. See Eaves and Scarf [1976] for the linear case; Charnes, Garcia, and Lemke [1977], Kojima [1979], and Megiddo and Kojima [1977] for the nonlinear case.

4.3. The path-following approach to dynamic NLP described here is taken from Garcia and Zangwill [1979e]. A closely related topic is sensitivity analysis and perturbation. For example, see Rockafellar [1970].

    Note that it may be advantageous to construct special numerical schemes for following the path of optimal points rather than solving the equations (4.5.5) directly. For example, one might resort to use of active constraint sets for following the path.

5

Equilibrium Programming

Equilibrium is one of the most profound concepts to explain the behavior of a system yet conceived by human intellect. Equilibrium means balance, and a system is nearly always in equilibrium, seeking an equilibrium or revising its equilibrium. The concept provides basic understanding of a complicated system, and in this chapter we present a powerful procedure for investigating the equilibrium of a system—equilibrium programming.

Equilibrium programming is applicable to almost any system—biological, physical, engineering, social, organizational, or whatever—and it enables us to formally analyze such a system. This chapter first defines the equilibrium programming problem and then illustrates its import with several examples. Finally, we utilize path following to determine the equilibrium point of the system. Later chapters extend these results and present even more applications of equilibrium programming, including the very important cases of economic equilibria, game theory, and networks.

5.1 THE EQUILIBRIUM PROGRAMMING PROBLEM

This chapter develops techniques to analyze a system. But first, what do we mean by a system? A system is an entity comprised of components and whose components interact in some manner. For example, society could be a system with the components being the individuals. A business firm could be a system with the components the various divisions of the company, say, finance, sales, marketing, production, and so on. An economy could be a system. There the components would be the various interacting economic agents, including firms, customers, banks, governments, and so on. The body could be such a system with the organs as components. The ecosystem is yet another example of such a system.

Let M be the number of components in the system. Each component i of the system has certain variables or actions it controls. Specifically, let the vector $x^i = (x^i_1, \ldots, x^i_{n_i})$ indicate the actions that component $i, i = 1, \ldots, M,$ can take. Obviously, $x^i \in R^{n_i}$, and think of component i as controlling the variables x^i. For a society x^i could be the actions that person i takes; for the body x^i might be what the heart does; for an economy x^i could be the actions of a particular economic agent. Note that n_i may differ for different i, so different components may have more or fewer variables to control. Nevertheless, we assume that component i controls the variables x^i.

Continuing, let $x = (x^1, \ldots, x^M)$ indicate the actions of all M components, where $x \in R^N$ for $N = \sum_{i=1}^{M} n_i$. Vector x denotes the state of the entire system and describes the actions of all M components.

Because the system is interdependent, the actions of component i, x^i, cannot be taken arbitrarily. In a country, for instance, suppose that the government component changes the pollution laws. That will certainly influence the action of other components of the country. Generally, the actions of component i, x^i, will depend upon the actions of the other components. To express this, let $x^{\bar{i}}$ be the vector consisting of all x except x^i, where $x^{\bar{i}} \in R^{N-n_i}$:

$$x^{\bar{i}} = (x^1, \ldots, x^{i-1}, x^{i+1}, \ldots, x^M).$$

The vector $x^{\bar{i}}$ is very important and depicts the actions of all components other than component i. Specifically, because $x^{\bar{i}}$ denotes what the others do, the actions of component i will depend upon $x^{\bar{i}}$. To highlight this we write $x = (x^i, x^{\bar{i}})$, where the components are rearranged as necessary. That is, $x = (x^1, x^{\bar{1}}) = (x^2, x^{\bar{2}}) = (x^i, x^{\bar{i}})$, and so on. This notation clearly distinguishes x^i, the actions of component i, from the other actions, $x^{\bar{i}}$.

To repeat, component i only controls x^i. Because $x^{\bar{i}}$ expresses the action of all others, x^i, the actions of component i, depends on $x^{\bar{i}}$.

We suppose further that each component of the system is attempting to maximize something. An individual might pursue happiness, the economic agent wealth or utility, and even the heart tries to do its best. Specify $f^i: R^N \to R^1$ and let $f^i(x^i, x^{\bar{i}})$ be the function that component i tries to optimize. Note that because component i only controls x^i, it can optimize f^i only over the x^i. Also, its optimization function, f^i, depends upon the actions of everyone else, $x^{\bar{i}}$. That is, $f^i(x^i, x^{\bar{i}})$ also depends upon $x^{\bar{i}}$, but component i can optimize f^i only over the x^i.

CONSTRAINTS

To make our model even more realistic, we assume that component i cannot choose x^i arbitrarily but has constraints. Also, these constraints can depend upon $x^{\bar{i}}$.

Explicitly, the constraints on component i are:

$$g^i_j(x^i, x^{\bar{i}}) \geq 0, \qquad j = 1, \ldots, r_i$$
$$h^i_j(x^i, x^{\bar{i}}) = 0, \qquad j = 1, \ldots, s_i, \qquad (5.1.1)$$

where $x^{\bar{i}}$ is given. As before, component i can select only x^i to satisfy the constraints and considers $x^{\bar{i}}$ as given.

Certainly, some of the constraints need not depend upon the other variables $x^{\bar{i}}$, although in many cases they will. For example, suppose that two divisions of a company use the same warehouse. If the second division already uses two-thirds of the warehouse, that severely limits the warehouse capacity for the first division. Or suppose that only a limited supply is available of a given labor type, and the total of that labor employed by all industries must not exceed that amount. Then a constraint would specify that the amount of labor available to industry i would depend upon the amount already used up by the other industries. We see that the constraints on component i can easily depend upon the actions of the others, namely $x^{\bar{i}}$, and (5.1.1) expresses that.

SUBPROBLEM

Overall then, component i is given the actions of others, $x^{\bar{i}}$, and selects the variables it controls, x^i, to optimize $f^i(x^i, x^{\bar{i}})$, where x^i must satisfy the constraints (5.1.1).

Mathematically, given $x^{\bar{i}}$, component i determines x^i to

$$\max_{x^i} f^i(x^i, x^{\bar{i}})$$

subject to

$$g^i_j(x^i, x^{\bar{i}}) \geq 0, \qquad j = 1, \ldots, r_i$$
$$h^i_j(x^i, x^{\bar{i}}) = 0, \qquad j = 1, \ldots, s_i. \qquad (5.1.2)$$

We call this problem for component i subproblem i.

In the system each component i has its own subproblem and is continually resolving it given the most current information on the actions of the others $x^{\bar{i}}$. That is, each component i attempts to find the optimal x^i for its own subproblem given $x^{\bar{i}}$. We call this process of all M components continually resolving their respective subproblems, *equilibrium programming* (EP).

EQUILIBRIUM PROGRAMMING

To summarize, we suppose that a system has M components and that each component i controls the variables $x^i, i = 1, \ldots, M, x^i \in R^{n_i}$. Each component i is attempting to optimize its own objective function f^i by selecting the best x^i. But the vector x^i cannot be chosen freely because the actions of the other components $x^{\bar{i}}$ limit the choice of x^i. Specifically, each component i attempts to solve its own subproblem i for the optimal x^i given $x^{\bar{i}}$. This process comprises EP.

THE EQUILIBRIUM POINT

A point $\bar{x} = (\bar{x}^i, \bar{x}^{\bar{i}}) \in R^N$ is called an *equilibrium point*, if given $\bar{x}^{\bar{i}}$, the vector $\bar{x}^i$ solves subproblem i for each $i = 1, \ldots, M$. In other words, suppose that $\bar{x}$ is an equilibrium point and component i is given that $x^{\bar{i}} = \bar{x}^{\bar{i}}$. Then $\bar{x}^i$ will optimize component i's subproblem given $\bar{x}^{\bar{i}}$. Mathematically, an equilibrium point $\bar{x} = (\bar{x}^i, \bar{x}^{\bar{i}})$ satisfies for $i = 1, \ldots, M$,

$$f^i(\bar{x}^i, \bar{x}^{\bar{i}}) = \max_{x^i} f^i(x^i, \bar{x}^{\bar{i}})$$

$$g^i_j(x^i, \bar{x}^{\bar{i}}) \geq 0, \qquad j = 1, \ldots, r_i$$

$$h^i_j(x^i, \bar{x}^{\bar{i}}) = 0, \qquad j = 1, \ldots, s_i.$$

Notice that an equilibrium point $\bar{x} = (\bar{x}^1, \ldots, \bar{x}^M)$ is *stable*. Suppose that each component $i, i = 1, \ldots, M$, selects $\bar{x}^i$ as its action. Then no one has an incentive to change. The vector $\bar{x}^i$ already optimizes subproblem i given $\bar{x}^{\bar{i}}$. Thus once the system has arrived at an equilibrium point $\bar{x}$, no component i can do better by itself. Considering the variables x^i that component i controls, component i has no incentive to deviate from $\bar{x}^i$. It cannot choose any x^i that does better (other points x^i may do as well, but no point does better). Thus an equilibrium point is stable. At an equilibrium point $\bar{x}$ and considering its own subproblem, no component i has an incentive to move, so $\bar{x}$ is stable.

5.2 NONLINEAR PROGRAMMING VERSUS EQUILIBRIUM PROGRAMMING

To place EP in perspective it may be instructive to view it as an extension of nonlinear programming (NLP). Clearly, for $M = 1$ there is only one component and EP reduces to an NLP. However, for M arbitrary let us analyze an NLP model of a system and contrast that with an EP model of the same system.

To begin, consider a system consisting of M components, $M \geq 2$; however, let us first model that system via NLP. Specifically, suppose that there is a function $b: R^N \rightarrow R^1$, which is the objective function of the entire system. The objective function b represents the profit, cost, utility, social welfare, or whatever of the system taken as a whole. Given the state of the system $x = (x^1, \ldots, x^M)$, $b(x)$ specifies the objective function value of the whole system at x.

Analyze the NLP:

$$\max_x b(x)$$

subject to

$$g_j^i(x) \geq 0, \quad j = 1, 2, \ldots, r_i \text{ and } i = 1, \ldots, M$$
$$h_j^i(x) = 0, \quad j = 1, 2, \ldots, s_i \text{ and } i = 1, \ldots, M. \quad (5.2.1)$$

Observe that the constraints of (5.2.1) include the constraints for all the M subproblems of the system. Thus suppose that $x = (x^1, \ldots, x^M)$ satisfies the constraints of (5.2.1). Then the same $x = (x^i, x^i)$ is feasible for all subproblems i for the EP (5.1.2). Consequently, if x is feasible for (5.2.1), it is also feasible for all subproblems. This means that the optimal point $\overset{*}{x}$ for (5.2.1) yields the highest system objective value to b and also is feasible for all components. Solving NLP (5.2.1) thus yields a point $\overset{*}{x}$ which is "best" for the system taken as a whole.

In the NLP formulation (5.2.1) one objective function is maximized over the entire vector $x = (x^1, \ldots, x^M)$, yielding an optimal point $\overset{*}{x}$. The EP formulation is different. EP has M separate subproblems (5.1.2) and each subproblem is being optimized only over x^i. The fact that NLP has one objective function and EP has M objective functions is the crucial difference between these two formulations. NLP assumes that there is one decision–maker who does all the optimization for the entire system. EP assumes that there are M different decision–makers, each of whom tries to optimize his or her own component of the system. We reiterate the main difference between the two. NLP has one decision–maker who optimizes the entire system. EP has M decision–makers, each of whom optimizes a portion of the system.

COUNTRY EXAMPLE

An excellent illustration of the two approaches is a country. For NLP, consider the government as the single decision–maker and that it has developed an objective function b for the country as a whole. The optimal point $\overset{*}{x}$ would then describe what each component of the country is supposed to do. That is, the government would solve the NLP for $\overset{*}{x}$ and everyone is then supposed to implement $\overset{*}{x}$.

In real countries the people are unlikely to slavishly implement $\overset{*}{x}$. Rather, many components comprise a country—the government, firms, individuals, and so on—and each component pursues its own objectives given the opportunities and limitations created by other components. Real countries thus behave like an EP, and not like an NLP. Each component makes certain decisions and optimizes its own subproblem given the actions of the other component, which is an EP.

A similar example is a corporation. In the usual NLP formulation, b is profit and the corporate head maximizes b for an optimal point $\overset{*}{x}$. Then each component of the company is to implement (using the obvious notation) what $\overset{*}{x} = ((\overset{*}{x})^1, (\overset{*}{x})^2, \ldots, (\overset{*}{x})^M)$ prescribes. Thus component 1 of the company is to do $(\overset{*}{x})^1$, component 2 is to do $(\overset{*}{x})^2$, and so on. That is the standard NLP model of a firm.

Only in the tiniest of companies does every component do exactly what the firm's head tells it to do. Generally, there are several components in a firm—the head, sales, finance, production, and so on. Although the head is one component and decides many of the variables, it does not decide all the variables. Other components decide certain variables, too. For example, marketing typically tries to optimize sales, and production tries to optimize items made. In actuality the firm operates as an EP, with each component making certain decisions and optimizing its own subproblem dependent upon what others do.

Let us stress that under EP certain components of the system can be more powerful than others, but EP permits that. For example, the government component of a country would be powerful and control actions x^i, such as taxes, laws, incentives, and so on. The other components—the firms, universities, individuals, and so on—given the government's actions, would nevertheless solve their own subproblems.

The identical situation holds in a corporation. The head component of the firm would decide certain important variables, such as budgets, staffing, and material. Yet given the head's actions, the other components of the firm would still solve their own subproblems and make decisions on the variables they control.

We perceive the contrast between NLP and EP. NLP allows only one decision–maker. EP permits more than one decision–maker, so that different people can control different variables. The government and corporate head may wish they were the sole decision–makers, so that the system were an NLP and everyone did the $\overset{*}{x}$ they desire. The government's or head's actions are the actions of just one of the components in the entire system, however, because other people make decisions as well. The real situation is an EP, and not an NLP.

OPTIMAL POINT VERSUS EQUILIBRIUM POINT

The two different ways of modeling, NLP and EP, provide much insight into the management of a system. It is fundamental to note that the optimal point $\overset{*}{x}$ to the NLP formulation (5.2.1) need not be an equilibrium point $\bar{x}$ to the EP formulation (5.1.2). The objective function b of the NLP and the f^i

functions of the EP are generally quite different. The optimal point $\overset{*}{x}$ need not be an equilibrium point to EP, and usually it will not be.

This observation is important. Since $\overset{*}{x} \neq \bar{x}$ typically, the optimal point $\overset{*}{x}$ of the NLP will not be what the individual components of the EP pursue. Using the obvious notation $\overset{*}{x} = ((\overset{*}{x})^i, (\overset{*}{x})^j)$, this means that given $(\overset{*}{x})^j$, the ith component will not do $(\overset{*}{x})^i$ because $(\overset{*}{x})^i$ does not solve his subproblem. Instead, the ith component will do something else. Thus even though the NLP arrives at $\overset{*}{x}$, because $\overset{*}{x}$ is usually not an equilibrium point, the components will do something else. The optimum point, $\overset{*}{x}$, will then not even be stable.

Relative to the country example, suppose that there is a dictator who solves an NLP. That is, the dictator decides on an objective function b for the country and solves the NLP for $\overset{*}{x}$. The country is then supposed to implement $\overset{*}{x}$. However, the country in fact consists of different decision-making components, so it is really solving an EP for an equilibrium point $\bar{x}$. As mentioned, in general $\overset{*}{x} \neq \bar{x}$; that is, what the dictator wants, $\overset{*}{x}$, and what the people want, $\bar{x}$, are different. Thus either the dictator propagandizes so that everyone embraces his or her objective or the dictator uses force to get each individual component of the country to do $\overset{*}{x}$. Furthermore, because $\overset{*}{x}$ is not stable, there is a natural tendency to leave $\overset{*}{x}$. This means the dictator cannot let up but must continue to force the country to stay at $\overset{*}{x}$ even after the country arrives at $\overset{*}{x}$ initially.

Most governments do have some objective function b which they believe (perhaps erroneously) to be in the interest of the country and solve it for $\overset{*}{x}$. But since it has many components, the country actually operates as an EP and is presumably functioning at some equilibrium $\bar{x}$ where $\bar{x} \neq \overset{*}{x}$. Although dictators may use force, most governments use milder incentives to promote movement from $\bar{x}$ toward $\overset{*}{x}$. Indeed, the long-run success of a country often depends upon how close the optimal point $\overset{*}{x}$ and the equilibrium point $\bar{x}$ are to one another. If $\overset{*}{x}$ and $\bar{x}$ are too far apart, the system does not operate well and revolution may be near because the wishes of the government, $\overset{*}{x}$, and the wishes of the people, $\bar{x}$, are quite different. In a well-functioning system, what is desirable for the system as a whole, $\overset{*}{x}$, and what the individual components want to do, $\bar{x}$, are usually fairly close.

Clearly, analogous reasoning holds for a corporate head who solves for $\overset{*}{x}$ to maximize profits whereas the true operation of the firm, because there are many decision–makers, is likely to be at some equilibrium $\bar{x}$. The head generally tries to move the actual functioning of the firm, $\bar{x}$, closer to what he or she wants, $\overset{*}{x}$. Also, discrepancies between $\overset{*}{x}$ and $\bar{x}$ can lead to worker discontent, lower profits, strikes, and so on.

In summary, EP assumes that there are many different decision–makers, each optimizing his or her own part of the system, dependent upon what the others do. NLP assumes that there is one decision–maker who makes the decision for everyone.

EP VERSUS MULTIPLE-OBJECTIVE FUNCTION

It should also be clear by this time that EP is different from a multiple-objective function problem. The multiple-objective problem has a single decision–maker trying to optimize several different objectives at once. EP, in contrast, has several different decision–makers (each perhaps with multiple objectives), and most crucially the decision–makers interact in a system. EP thus stresses the system aspect and that there are M decision–makers interacting and interrelating in a system.

5.3 EXAMPLES OF EP

MATRIX PAYOFF EXAMPLES

Our first detailed example of EP has objective functions that are specified in terms of matrices and very simple constraints. Later, in Chapter 8, on Game Theory, this example is extended. For now suppose that the system consists of two people, so that $M = 2$. Person 1 has various actions available, $x^1 \in R^{n_1}$, and person 2 has possible actions $x^2 \in R^{n_2}$. Also suppose that there are two $n_1 \times n_2$ matrices A^1 and A^2, which form the two objective functions as follows. If player 1 takes action x^1 and player 2 chooses x^2, the *payoff* (objective function value) to person 1 is

$$f^1(x) = (x^1)^{\tau} A^1 x^2$$

and to person 2 is

$$f^2(x) = (x^1)^{\tau} A^2 x^2.$$

For simplicity suppose that the only constraints are for $i = 1, 2$,

$$\sum_{j=1}^{n_i} x_j^i = 1, \quad x_j^i \geq 0, \quad j = 1, \ldots, n_i.$$

Notice how simple these constraints are.

Under these considerations subproblem 1 becomes: Given x^2, choose x^1 to

$$\max_{x^1} (x^1)^{\tau} A^1 x^2$$

subject to

$$\sum_{j=1}^{n_1} x_j^1 = 1, \quad x_j^1 \geq 0, \quad j = 1, \ldots, n_1. \tag{5.3.1}$$

Similarly, subproblem 2 becomes: Given x^1, choose x^2 to

$$\max_{x^2} (x^1)^\tau A^2 x^2$$

subject to

$$\sum_{j=1}^{n_2} x_j^2 = 1, \quad x_j^2 \geq 0, \quad j = 1, \dots, n_2. \tag{5.3.2}$$

Let matrix A^i, $i = 1, 2$, have components A_{jk}^i, $j = 1, \dots, n_1$, $k = 1, \dots, n_2$. Suppose that

$$x_j^1 = \begin{cases} 0, & j \neq q \\ 1, & j = q \end{cases} \quad \text{and} \quad x_j^2 = \begin{cases} 0, & j \neq r \\ 1, & j = r. \end{cases}$$

Then the payoff to person 1 is $A_{q,r}^1$ and to person 2 is $A_{q,r}^2$. More generally, given an arbitrary x^1 and x^2, player i, $i = 1, 2$, receives a payoff of

$$\sum_{j=1}^{n_1} \sum_{k=1}^{n_2} x_j^1 A_{jk}^i x_k^2.$$

A convenient way to describe this problem is via the $n_1 \times n_2$ payoff table P, whose entries are (A_{jk}^1, A_{jk}^2). That is,

$$P = \begin{bmatrix} (A_{11}^1, A_{11}^2) & \cdots & (A_{1n_2}^1, A_{1n_2}^2) \\ \vdots & & \vdots \\ (A_{n_1,1}^1, A_{n_1,1}^2) & \cdots & (A_{n_1,n_2}^1, A_{n_1,n_2}^2) \end{bmatrix}.$$

PARTICULAR EXAMPLES

Some special cases are quite instructive. Let $n_1 = n_2 = 2$ and suppose that

$$A^1 = \begin{bmatrix} 1 & 0 \\ 0 & -1 \end{bmatrix} \quad \text{and} \quad A^2 = \begin{bmatrix} 2 & 0 \\ 0 & -1 \end{bmatrix}.$$

Then

$$P = \begin{bmatrix} (1, 2) & (0, 0) \\ (0, 0) & (-1, -1) \end{bmatrix}.$$

The objective functions become

$$f^1(x^1, x^2) = (x^1)^\tau A^1 x^2 = x_1^1 x_1^2 - x_2^1 x_2^2$$
$$f^2(x^1, x^2) = (x^1)^\tau A^2 x^2 = 2x_1^1 x_1^2 - x_2^1 x_2^2. \tag{5.3.3}$$

To analyze the situation, suppose that person 2 initially selects action

$$x^2 = (x_1^2, x_2^2) = (0, 1)$$

and that player 1 must solve subproblem 1 given this particular x^2. By (5.3.3), subproblem 1 becomes

$$\max \; -x_2^1$$

$$x_1^1 + x_2^1 = 1, \qquad x_1^1 \geq 0, \qquad x_2^1 \geq 0.$$

Solving, if person 2 does $x^2 = (0, 1)$, person 1 responds with $x^1 = (1, 0)$.

Now how does person 2 respond if player 1 does $x^1 = (1, 0)$? The corresponding subproblem 2, since we are given $x^1 = (1, 0)$, is as follows:

$$\max \; 2x_1^2$$

$$x_1^2 + x_2^2 = 1, \qquad x_1^2 \geq 0, \qquad x_2^2 \geq 0,$$

which yields $x^2 = (1, 0)$.

At this point, if person 1 solves subproblem 1 given $x^2 = (1, 0)$, the point $x^1 = (1, 0)$ occurs again. Indeed, $x^1 = (1, 0)$, $x^2 = (1, 0)$ is an equilibrium point. Specifically, suppose that both persons are at $\bar{x} = (\bar{x}^1, \bar{x}^2) = ((1, 0), (1, 0))$. If person 1 solves subproblem 1 given $\bar{x}^2 = (1, 0)$, then he or she again arrives at $\bar{x}^1 = (1, 0)$. Similarly, if person 2 solves subproblem 2 given $\bar{x}^1 = (1, 0)$, he or she selects $\bar{x}^2 = (1, 0)$. Thus we achieve an equilibrium and it is stable; neither person has an incentive to do anything else.

A MORE SUBTLE EXAMPLE

To complicate the situation, suppose that now

$$A^1 = \begin{bmatrix} 1 & 0 \\ 0 & 2 \end{bmatrix} \quad \text{and} \quad A^2 = \begin{bmatrix} 2 & 0 \\ 0 & 4 \end{bmatrix},$$

so that

$$P = \begin{bmatrix} (1, 2) & (0, 0) \\ (0, 0) & (2, 4) \end{bmatrix}.$$

Perceive that $\bar{x}^1 = (1, 0)$, $\bar{x}^2 = (1, 0)$ is still an equilibrium point. However, it is easy to calculate and verify that $\bar{x}^1 = (0, 1)$, $\bar{x}^2 = (0, 1)$ is also an equilibrium point. (It is not unusual for an EP to have more than one equilibrium point.)

Note that the objective function values are higher to both persons 1 and 2 at $\bar{x}^1 = (0, 1)$ and $\bar{x}^2 = (0, 1)$. At $\bar{x}^1 = (1, 0)$, $\bar{x}^2 = (1, 0)$, person 1 has an objective function value of 1 and person 2 has an objective function value of 2. At $\bar{x}^1 = (0, 1)$, $\bar{x}^2 = (0, 1)$ person 1 has an objective function value of 2 and person 2 has an objective function value of 4. Thus both persons 1 and 2 would prefer equilibrium point $\bar{x}^1 = (0, 1)$, $\bar{x}^2 = (0, 1)$.

STABILITY AND COMMUNICATIONS

We have already noted that an equilibrium point is stable. That is, if persons 1 and 2 are at $\bar{x}^1 = (1, 0)$ and $\bar{x}^2 = (1, 0)$, they tend to stay there. This is despite the fact that the new equilibrium point, $\bar{x} = (\bar{x}^1, \bar{x}^2)$, is better for both of them. What is going on here?

Equilibrium is based on unilateral action. In the subproblem notice that only x^i is permitted to change. Equilibria are stable in the sense that no person has an incentive to move acting solely by himself or herself. At equilibrium there is no incentive to move unilaterally. Thus equilibrium is stable in this sense.

To move from $\bar{x}^1 = (1, 0)$, $\bar{x}^2 = (1, 0)$ to the better equilibrium $\bar{x}^1 = (0, 1)$, $\bar{x}^2 = (0, 1)$ requires joint action. Both persons must change. If one person changes without the other changing, he or she loses something, which, especially when it really counts, he or she is unwilling to do. Both people must change.

Whether both people can agree to change jointly to a better equilibrium depends upon the assumption of the model, and is a crucial stipulation. For instance, if communication is permitted between the two people, then since the two can communicate, they presumably can change their strategies jointly. However, in many circumstances no communication is available, or even if communication is available, the two parties do not trust each other. Such situations are frequently studied under the topic of game theory and arise often in war, conflict, or legal disputes. Then, owing to lack of communication or trust, the people may not be able to move to a better equilibrium. The unilateral nature of their actions dooms them to an inferior equilibrium.

Another factor is that in real situations the persons may not know that another better equilibrium exists. Because of limited information, it might not be clear what the objective function values would be in a new situation. Thus the people are again restricted to an inferior equilibrium point.

We perceive the unilateral nature of equilibrium. It is stable under unilateral action by any of the components i. In certain circumstances, because of joint action, it might be possible to move from one equilibrium point to a better one. Yet in situations of poor communication or little trust joint action may be very difficult or impossible to achieve.

CONFLICTING EQUILIBRIA

We just presented a situation where one equilibrium point was clearly superior to another in that both parties were better off at the second point. In many circumstances of multiple equilibria that is not true. Consider

$$A^1 = \begin{bmatrix} 2, & 0 \\ 0, & 1 \end{bmatrix}, \qquad A^2 = \begin{bmatrix} 1, & 0 \\ 0, & 2 \end{bmatrix},$$

so that

$$P = \begin{bmatrix} (2,1) & (0,0) \\ (0,0) & (1,2) \end{bmatrix}.$$

Again there are two equilibrium points, $\bar{x} = ((1,0),(1,0))$ and $\bar{\bar{x}} = ((0,1),(0,1))$. But now the objective values have been revised. Person 1 prefers $\bar{\bar{x}}$ because he or she receives 2 there, whereas at $\bar{x}$ he or she obtains only 1. Person 2 prefers $\bar{x}$ because he or she receives 2 there, whereas at $\bar{\bar{x}}$ he or she obtains 1. Thus neither equilibrium point is clearly better than the other because each person prefers a different one.

In this situation, even with communication, there is little we can do, as each person desires something different. This is a classic example of conflict. As an aside, note that in real life such situations sometimes promote bizarre behavior in which each person may try to harm the other and even himself or herself in order to get his or her own way.

In summary, equilibrium is stable under unilateral action. In many circumstances unilateral action is the only action possible. Moving to another equilibrium requires joint action which, depending upon the situation, may or may not be achievable.

With this introduction to the nature of equilibrium, we now present two different examples of EP, one of a decentralized firm, the other of a centralized firm.

DECENTRALIZED FIRM EXAMPLE

An international firm has two divisions: one makes TV sets, the other makes electronic digital watches. The TV set division is located in green and lush rolling hills just outside of Kyoto, Japan. The digital watch division is situated in a posh northern suburb of Chicago with a magnificent view of Lake Michigan. To simplify the situation greatly, suppose that the TV set is made of only two components, one of which is a special microchip circuit. The watch is made of two components, but one component is the same special microchip circuit. Thus both TV sets and watches use the same special circuit as one part, but the two products require different second parts.

The company purchases microchip circuits on the world market, but because of their advanced nature can obtain only a maximum total of d circuits. One problem the company faces is to allocate this total amount available, d, between the two divisions.

The company also desires to maximize total profits from its two divisions. After long consideration it was determined that the best means to achieve this is to let each division optimize profits independently of the other. That is, the Kyoto division would attempt to maximize the profits in yen from TV sets, and separately the Chicago division would maximize the electronic watch profits in U.S. dollars.

The decision to permit independent maximization was based on several very practical reasons. First, coordination of the two divisions, almost half a world and two languages apart, turned out to be almost impossible. Second, separate optimization permits each division to operate as a profit center, thereby enhancing management motivation. Finally, there was great difficulty determining a common monetary unit because of the changing value of the yen versus the U.S. dollar. Thus it was decided to let each division independently optimize its own profits in its own local currency.

Notationally, let $x^1 = (x_1^1, x_2^1, x_3^1)$ be the variables the Chicago division controls, where x_1^1 is the number of microchip circuits, x_2^1 is the amount of the other component, and x_3^1 is the number of digital watches made. Let us assume that every watch requires three circuits and two of the other part, so that

$$3x_3^1 = x_1^1$$
$$2x_3^1 = x_2^1.$$

Also, profit per watch is \$10 and an amount d_2^1 of the second watch part is available, so that

$$0 \leq x_2^1 \leq d_2^1.$$

The Kyoto division variables are $x^2 = (x_1^2, x_2^2, x_3^2)$, where x_1^2 is the number of microchip circuits used, x_2^2 is the number of the second TV part needed, and x_3^2 is the number of TV sets made. Let each TV set require one microchip circuit and two of the second part, so that

$$x_3^2 = x_1^2$$
$$2x_3^2 = x_2^2.$$

The profit on TV sets is 100 yen per set. Also,

$$0 \leq x_2^2 \leq d_2^2,$$

where d_2^2 is the amount of the second TV part available.

Since a total of d circuits is available, there is a constraint on both divisions of

$$x_1^1 + x_1^2 \leq d.$$

The EP is for each division to separately maximize its profits subject to the various constraints. Specifically, the Chicago subproblem becomes: Given x^2,

$$\max 10x_3^1$$
$$x_1^1 + x_1^2 \leq d$$
$$3x_3^1 - x_1^1 = 0$$
$$2x_3^1 - x_2^1 = 0$$
$$x_1^1 \geq 0$$
$$0 \leq x_2^1 \leq d_2^1.$$

For Kyoto the subproblem becomes: Given x^1,

$$\max 100x_3^2$$
$$x_1^1 + x_1^2 \leq d$$
$$x_3^2 - x_1^2 = 0$$
$$2x_3^2 - x_2^2 = 0$$
$$x_1^2 \geq 0$$
$$0 \leq x_2^2 \leq d_2^2.$$

These two problems would then be solved for an equilibrium point $\bar{x} = (\bar{x}^1, \bar{x}^2)$. The $\bar{x}_1^1$ value would specify how many circuits go to Chicago, and $\bar{x}_1^2$ is the number of circuits that Kyoto would get.

The equilibrium would be stable in the sense that given $\bar{x}^2$, Chicago would solve its subproblem and obtain $\bar{x}^1$. Similarly, given $\bar{x}^1$, Kyoto would solve its subproblem and obtain $\bar{x}^2$.

Note that there might be multiple equilibria. (See Exercise 11.) And because equilibria are stable, whichever equilibrium the company determines, it would tend to stay there. To move to a different equilibrium would take joint effort by Chicago and Kyoto. Although theoretically possible, in this situation a joint effort would be extremely difficult. As mentioned, there is a great cost not only in coordination but in decreased management morale. Also, with no common monetary unit it is not clear how to select among the equilibria to find the best. Thus the equilibrium found would be fairly stable. Any joint changes between Kyoto and Chicago would in all likelihood be difficult and slow.

A CENTRALIZED FIRM EXAMPLE

The previous example presented a firm which due to its international nature, was highly decentralized. Consequently, EP, because it has several different optimization problems at once, was appropriate. Let us now consider a more centralized firm:

Suppose that a firm has three components—headquarters, production, and marketing—which we index as components $i = 1, 2,$ and 3. Headquarters desires to maximize the overall profit of the firm and to do so makes monthly allocations of budget, personnel, and material to production and marketing, that is, to components 2 and 3. Component i, $i = 2, 3$, then must make day-to-day decisions based upon the monthly limits of budget, personnel, and material established by headquarters.

For $i = 2, 3$, let u_0^i be the monthly budget headquarters allots to component i, v_0^i be the personnel headquarters allots per month to component i, and w_0^i be the material allotted per month to component i.

In any month headquarters itself has a total budget available of d, total personnel available of e, and total material available of f. And headquarters must make its allotments subject to these restrictions. Thus the headquarters' constraints on the allocations to the two components are

$$u_0^2 + u_0^3 \leq d$$
$$v_0^2 + v_0^3 \leq e$$
$$w_0^2 + w_0^3 \leq f.$$

Per our previous notation, let $x^1 = (u_0^2, u_0^3, v_0^2, v_0^3, w_0^2, w_0^3)$ since these are the variables headquarters controls.

Consider now the production and marketing components. For component $i = 2, 3$, let u_k^i be the amount of budget used in day k of the month, $k = 1, \ldots, 30$. Similarly, let v_k^i be the amount of personnel used and w_k^i be the amount of material used on day k of the month. Component i, $i = 2, 3$, controls these day-to-day variables given its monthly allocation from headquarters. Considering the headquarters' monthly allocation to component i, $i = 2, 3$:

$$\sum_{k=1}^{30} u_k^i \leq u_0^i$$
$$\sum_{k=1}^{30} v_k^i \leq v_0^i$$
$$\sum_{k=1}^{30} w_k^i \leq w_0^i.$$

Notationally, specify

$$x^i = (u_1^i, \ldots, u_{30}^i, v_1^i, \ldots, v_{30}^i, w_1^i, \ldots, w_{30}^i), \qquad i = 2, 3.$$

Again, x^i are the variables that component i controls.

Given its allocations from headquarters, component i has costs $c^i(x^i)$ which it attempts to minimize by adjusting x^i, $i = 2, 3$. That is, production minimizes $c^2(x^2)$ over its day-to-day variables x^2, and marketing minimizes $c^3(x^3)$ over its day-to-day variables x^3. Furthermore, headquarters attempts to maximize firm profits, which we specify as $p^1(x) = p^1(x^1, x^2, x^3)$. Note that profits depend upon all variables of the firm x^1, x^2, and x^3, although headquarters controls only x^1.

The overall firm is then solving three subproblems at once. Headquarters, given (x^2, x^3), solves

$$\max_{x^1} p^1(x^1, x^2, x^3)$$

$$u_0^2 + u_0^3 \leq d$$

$$v_0^2 + v_0^3 \leq e$$

$$w_0^2 + w_0^3 \leq f,$$

all variables nonnegative.

In this manner headquarters decides the monthly allotments to components $i = 2, 3$.

Each component i, $i = 2, 3$, given its monthly allotment, that is, given x^1, must solve

$$\min_{x^i} c^i(x^i)$$

$$\sum_{k=1}^{30} u_k^i \leq u_0^i$$

$$\sum_{k=1}^{30} v_k^i \leq v_0^i$$

$$\sum_{k=1}^{30} w_k^i \leq w_0^i$$

all variables nonnegative.

We obtain an EP again. Notice how this formulation differs from letting headquarters optimize all variables. That would be a nonlinear programming approach. As previously discussed in nonlinear programming, a single objective function consisting of all variables would be optimized. The optimal solution would then be imposed on production and marketing.

The EP model is more realistic, as it lets each component optimize the variable it controls. Headquarters handles the big picture, the monthly considerations. Then, given the monthly allocations, each component handles its own day-to-day decisions. An EP is obtained and an equilibrium point would be its solution.

These examples have attempted to introduce the variety and richness of EP as a means to model various systems; later chapters present other examples of EP. Before that, however, we must establish a proof of existence of equilibrium.

5.4 PROOF OF EQUILIBRIUM-POINT EXISTENCE

In this section we verify that under certain assumptions an equilibrium point to EP exists and can be calculated. Consider a given system of M components, each of which is solving its own subproblem. We now prove that an equilibrium point $\bar{x} = (\bar{x}^1, \ldots, \bar{x}^M)$ exists. Specifically, $\bar{x}$ has the property that for each

$i = 1, \ldots, M$, given $\bar{x}^{\bar{i}}$, $\bar{x}^i$ solves subproblem i, and therefore $\bar{x}$ is an equilibrium point.

The proof utilizes the path-following approach and since it is based on the NLP algorithm presented in Chapter 4, the reader is urged to review that proof. Essentially, we start with a trivial EP and gradually deform the EP to the original EP. As the deformation proceeds, the path of equilibrium points is followed until an equilibrium point of the given EP is reached. Later chapters demonstrate that economic equilibrium, games, and networks are special cases of EP. Indeed, the EP algorithm presented in this section encompasses a wide variety of practical equilibrium situations.

DEMONSTRATION

Let us restate subproblem i:

$$\max_{x^i} f^i(x^i, x^{\bar{i}})$$

subject to

$$g^i_j(x^i, x^{\bar{i}}) \geq 0, \qquad j = 1, \ldots, r_i$$
$$h^i_j(x^i, x^{\bar{i}}) = 0, \qquad j = 1, \ldots, s_i \qquad (5.4.1)$$

for $x^{\bar{i}}$ given.

We suppose that all functions are C^3 and that the $f^i(x^i, x^{\bar{i}})$ and $g^i_j(x^i, x^{\bar{i}}) \, j = 1, \ldots, r_i$, are concave in x^i for $x^{\bar{i}}$ fixed, and the $h^i_j \, j = 1, \ldots, s_i$, are linear in x^i for $x^{\bar{i}}$ fixed.

Subproblem i is actually an NLP where the optimization is over x^i for $x^{\bar{i}}$ fixed. Several results follow immediately from Chapter 4. Specifically, the K–T equations for subproblem i are, where $x^{\bar{i}}$ is fixed,

$$\nabla_{x^i} f^i(x)^\tau + \sum_{j=1}^{r_i} (\alpha^i_j)^+ \nabla_{x^i} g^i_j(x)^\tau + \sum_{j=1}^{s_i} \mu^i_j \nabla_{x^i} h^i_j(x)^\tau = 0$$
$$(\alpha^i_j)^- \qquad \quad - g^i_j(x) = 0, \qquad j = 1, \ldots, r_i$$
$$h^i_j(x) = 0, \qquad j = 1, \ldots, s_i. \qquad (5.4.2)$$

Here $\nabla_{x^i} = (\partial/\partial x^i_1, \partial/\partial x^i_2, \ldots, \partial/\partial x^i_{n_i})$ is the gradient with respect to the x^i variables only. Also,

$$(\alpha^i_j)^+ = [\max\{0, \alpha^i_j\}]^3 \qquad \text{and} \qquad (\alpha^i_j)^- = [\max\{0, -\alpha^i_j\}]^3. \qquad (5.4.3)$$

Note in (5.4.2) that there are $n_i + r_i + s_i$ equations and $n_i + r_i + s_i$ variables, namely x^i, α^i, and μ^i.

Recalling the concavity and linearity assumptions, suppose also that an appropriate constraint qualification (CQ) holds (a specific one will be given shortly). Then from Chapter 4 the K–T equations are necessary and sufficient to solve subproblem i. In detail, for x^i fixed, x^i optimizes subproblem i if and only if (x^i, α^i, μ^i) satisfies (5.4.2) for some α^i, μ^i. Note especially that since there are M subproblems, there are M systems of K–T equations like (5.4.2), (one for each subproblem i, $i = 1, \ldots, M$).

THE BIG SYSTEM

At equilibrium point $\bar{x}$ all M subproblems are optimized at once, which means for all i, $\bar{x}^i$ solves (5.4.1) given $\bar{x}^i$. In fact, and this is important, at $\bar{x}$ all M systems of K–T equations hold simultaneously. Explicitly, $\bar{x}^i$ and some $\bar{\alpha}^i$, $\bar{\mu}^i$ solve (5.4.2) given $\bar{x}^i$, and this holds for all i. Thus $\bar{x}$ is an equilibrium point if and only if $(\bar{x}^i, \bar{\alpha}^i, \bar{\mu}^i)$ satisfies (5.4.2) given $\bar{x}^i$ for all i. We reiterate, at $\bar{x}$, $(\bar{x}^i, \bar{\alpha}^i, \bar{\mu}^i)$ satisfies (5.4.2) for all i.

Understand that at $\bar{x}$ (5.4.2) must hold simultaneously for all i, $i = 1, \ldots,$ M. But look at the result. The EP has been transformed into one giant system (5.4.2), $i = 1, \ldots, M$. Define

$$k = \sum_{i=1}^{M} (n_i + r_i + s_i). \tag{5.4.4}$$

Then this large system has k equations and k unknowns, and solving it for a solution, $(\bar{x}, \bar{\alpha}, \bar{\mu})$ will yield an equilibrium point $\bar{x}$, where $\bar{\alpha} = (\bar{\alpha}^1, \ldots, \bar{\alpha}^M)$ and $\bar{\mu} = (\bar{\mu}^1, \ldots, \bar{\mu}^M)$.

Intuitively, with NLP we had to solve one K–T system. However, EP has M simultaneous optimization problems, so we must solve M systems of K–T equations simultaneously.

AN EXAMPLE

Consider the EP problem given earlier, (5.3.1) and (5.3.2), with objective functions (5.3.3). Restated, subproblem 1 is: given x^2 to

$$\max_{x^1} x_1^1 x_1^2 - x_2^1 x_2^2$$

subject to

$$x_1^1 \qquad \geq 0$$
$$x_2^1 \geq 0$$
$$x_1^1 + x_2^1 = 1 \tag{5.4.5a}$$

and subproblem 2 is: given x^1 to

$$\max_{x^2} 2x_1^1 x_1^2 - x_2^1 x_2^2$$

subject to

$$x_1^2 \quad\;\; \geq 0$$
$$x_2^2 \geq 0$$
$$x_1^2 + x_2^2 = 1. \tag{5.4.5b}$$

To obtain an equilibrium point we would solve the following system of equations for $(x^1, x^2, \alpha^1, \alpha^2, \mu^1, \mu^2)$, where $x^1 = (x_1^1, x_2^1)$, $x^2 = (x_1^2, x_2^2)$, $\alpha^1 = (\alpha_1^1, \alpha_2^1)$, and $\alpha^2 = (\alpha_1^2, \alpha_2^2)$:

$$x_1^2 + [\max\{0, \alpha_1^1\}]^3 \qquad\qquad + \mu^1 \qquad = 0$$
$$-x_2^2 \qquad + [\max\{0, \alpha_2^1\}]^3 \;\; + \mu^1 \qquad = 0$$
$$[\max\{0, -\alpha_1^1\}]^3 \qquad\qquad -x_1^1 \qquad = 0$$
$$[\max\{0, -\alpha_2^1\}]^3 \qquad\quad - x_2^1 = 0$$
$$x_1^1 + x_2^1 = 1$$
$$2x_1^1 + [\max\{0, \alpha_1^2\}]^3 \qquad\qquad + \mu^2 \qquad = 0$$
$$-x_2^1 \qquad + [\max\{0, \alpha_2^2\}]^3 \;\; + \mu^2 \qquad = 0$$
$$[\max\{0, -\alpha_1^2\}]^3 \qquad\qquad - x_1^2 \qquad = 0$$
$$[\max\{0, -\alpha_2^2\}]^3 \qquad\quad - x_2^2 = 0$$
$$x_1^2 + x_2^2 = 1.$$

This system actually consists of two sets of K–T equations, one for each subproblem, and solving both simultaneously yields $(\bar{x}, \bar{\alpha}, \bar{\mu})$.

5.5   THE DYNAMIC EP

In Chapter 4 we introduced a parameter t and obtained a dynamic NLP. Also, as t changed the optimal NLP point $\overset{*}{x}$ changed and a path was thereby generated. Now we use a parameter t to create a dynamic EP. Moreover, as t changes $\bar{x}$ changes, thus creating our path.

For the dynamic EP the subproblem i given t and x^i is

$$\max_{x^i} f^i(x^i, x^{\bar{i}}, t)$$
$$g_j^i(x^i, x^{\bar{i}}, t) \geq 0, \qquad j = 1, \ldots, r_i$$
$$h_j^i(x^i, x^{\bar{i}}, t) = 0, \qquad j = 1, \ldots, s_i. \tag{5.5.1}$$

At any t, solving this EP would yield an equilibrium point at that t. That is, given t, solving (5.5.1) for all $i = 1, \ldots, M$ simultaneously will yield an equilibrium point at that t.

Next, for t fixed and x^i given, examine the K–T equations for subproblem (5.5.1):

$$\nabla_{x^i} f^i(x, t) + \sum_{j=1}^{r_i} (\alpha_j^i)^+ \nabla_{x^i} g_j^i(x, t) + \sum_{j=1}^{s_i} \mu_j^i \nabla_{x^i} h_j^i(x, t) = 0$$

$$(\alpha_j^i)^- - g_j^i(x, t) = 0, \qquad j = 1, \ldots, r_i$$

$$h_j^i(x, t) = 0, \qquad j = 1, \ldots, s_i. \tag{5.5.2}$$

At any t fixed there are M of these K–T subsystems (5.5.2). By solving all M of them simultaneously, we can obtain a solution $(\bar{x}, \bar{\alpha}, \bar{\mu}, t)$, where $\bar{x}$ is an equilibrium point for that t.

Consider now the full system consisting of the M (5.5.2) subsystems $i = 1, \ldots, M$, and denote

$$H(x, \alpha, \mu, t) = 0 \tag{5.5.3}$$

as this full system of equations, which via (5.4.4) has k equations and $k + 1$ unknowns. Thus $H: R^{k+1} \longrightarrow R^k$, each component of H is defined by the appropriate equation of (5.5.2), $i = 1, \ldots, M$, and the $k + 1$ variables are (x, α, μ, t). Further, $(x, \alpha, \mu, t) \in H^{-1}$ means that (x, α, μ, t) solves the full system for t given.

Suppose that the Jacobian H' has rank k at each solution $(\bar{x}, \bar{\alpha}, \bar{\mu}, t)$; then we call H regular. As we know, under regularity the solutions $(\bar{x}, \bar{\alpha}, \bar{\mu}, t) \in H^{-1}$ form a path. Moreover, $\bar{x}$ is an equilibrium point at that t.

In summary, we identify the full system (5.5.2) for $i = 1, \ldots, M$ with H. Then under regularity a solution path $(\bar{x}, \bar{\alpha}, \bar{\mu}, t)$ is generated. Moreover, since $\bar{x}$ is an equilibrium point at t, by following this path we can study the dynamic EP.

5.6 THE ALGORITHM

Chapter 4 employed a special dynamic NLP to solve the original NLP. Now we will present a special dynamic EP to solve the original EP (5.4.1).

To begin, define

$$D = \{x \in R^N \,|\, g_j^i(x) \geq 0, \quad j = 1, \ldots, r_i, \quad h_j^i(x) = 0,$$

$$j = 1, \ldots, s_i, \text{ and } i = 1, \ldots, M\}. \tag{5.6.1}$$

Clearly, D is the set of points that simultaneously satisfies the constraints of all subproblems (5.4.1), $i = 1, \ldots, M$.

Also, pose the following constraint qualification (CQ):

1. For all $x \in D$ and $i = 1, \ldots, M$, the vectors $\nabla_{x^i} h^i_j(x), j = 1, \ldots, s_i,$ are linearly independent.

2. There is a point $\mathring{x} \in D$, $\mathring{x} = (\mathring{x}^1, \ldots, \mathring{x}^M)$ such that for all $x \in D$, $x = (x^i, x^{\bar{i}})$, we have

$$g^i_j(\mathring{x}^i, x^{\bar{i}}) > 0, \qquad j = 1, \ldots, r_i$$

and

$$h^i_j(\mathring{x}^i, x^{\bar{i}}) = 0, \qquad j = 1, \ldots, s_i. \tag{5.6.2}$$

These CQ are direct generalizations of the CQ for the NLP algorithm. To see this, examine subproblem i. CQ (1) requires that the constraints $j = 1, \ldots, s_i$ have their gradients $\nabla_{x^i} h^i_j(x)$ linearly independent for any $x \in D$. This is exactly analogous to NLP. As for CQ (2), $\mathring{x}^i$ reflects the "interior point" that we had for NLP. Essentially, now each subproblem i has its own "interior point" $\mathring{x}^i$. Notice also that (5.6.2) immediately implies [since $\mathring{x} = (\mathring{x}^i, \mathring{x}^{\bar{i}}) \in D$] that

$$g^i_j(\mathring{x}) > 0, \qquad \text{all } i, j. \tag{5.6.3}$$

STATEMENT OF EP

We are now able to state the special dynamic EP used for solving (5.4.1), which is easily seen to be the direct extension of the NLP in Chapter 4. Explicitly, at t its ith subproblem $i = 1, \ldots, M$ is

$$\max_{x^i} t f^i(x^i, x^{\bar{i}}) - \frac{(1-t)}{2} \| x^i - \mathring{x}^i \|^2$$

subject to

$$g^i_j(x^i, x^{\bar{i}}) \geq 0, \qquad j = 1, \ldots, r_i$$
$$h^i_j(x^i, x^{\bar{i}}) = 0, \qquad j = 1, \ldots, s_i \tag{5.6.4}$$

for $x^{\bar{i}}$ given. Here $\mathring{x}$ is from the CQ and assumed known.

At $t = 0$, notice that $\mathring{x} = (\mathring{x}^i, \mathring{x}^{\bar{i}})$ is an equilibrium point for (5.6.4). This is immediate because at $t = 0$ and given $\mathring{x}^{\bar{i}}$, subproblem i is

$$\max -\tfrac{1}{2} \| x^i - \mathring{x}^i \|^2$$
$$g^i_j(x^i, \mathring{x}^{\bar{i}}) \geq 0, \qquad j = 1, \ldots, r_i$$
$$h^i_j(x^i, \mathring{x}^{\bar{i}}) = 0, \qquad j = 1, \ldots, s_i, \tag{5.6.5}$$

which clearly has $\mathring{x}^i$ as its solution. As this holds for all i, $\mathring{x}$ is an equilibrium point. Furthermore, $\mathring{x}$ is the unique equilibrium point to (5.6.4) at $t = 0$, as we now show.

LEMMA 5.6.1   The point $\mathring{x}$ is the unique equilibrium point at $t = 0$.

Proof.   It is easily seen that at $t = 0$, subproblem i has a unique optimizer $\mathring{x}^i$.

We know that $\mathring{x}$ is an equilibrium point at $t = 0$, so suppose there is another equilibrium point, $\bar{x}$, where $\bar{x} \neq \mathring{x}$. Select i so that $\bar{x}^i \neq \mathring{x}^i$. Since $\bar{x} \in D$, by (5.6.2)

$$(\mathring{x}^i, \bar{x}^{\bar{i}}) \in D. \tag{5.6.6}$$

Now examine subproblem i at $t = 0$ given $\bar{x}^{\bar{i}}$, which is

$$\max_{x^i} -\tfrac{1}{2}\|x^i - \mathring{x}^i\|^2$$

$$g^i_j(x^i, \bar{x}^{\bar{i}}) \geq 0, \qquad j = 1, \ldots, r_i$$

$$h^i_j(x^i, \bar{x}^{\bar{i}}) = 0, \qquad j = 1, \ldots, s_i. \tag{5.6.7}$$

By (5.6.6), $(\mathring{x}^i, \bar{x}^{\bar{i}})$ satisfies the constraints for this subproblem and clearly $\mathring{x}^i$ optimizes uniquely. Since $\mathring{x}^i \neq \bar{x}^i$, the point $\bar{x}^i$ then cannot optimize (5.6.7). Hence $\bar{x} = (\bar{x}^i, \bar{x}^{\bar{i}})$ cannot be an equilibrium point.

The contradiction is clear and $\mathring{x}$ must be the unique equilibrium point at $t = 0$. $\square$

PATH FOLLOWING

By the lemma, $\mathring{x}$ is the unique equilibrium at $t = 0$ and we use $\mathring{x}$ as the starting point for the path to (5.6.4). The idea is to follow that path until when $t = 1$, we obtain the solution to EP (5.4.1). The path will actually be the solution path (x, α, μ, t) to the system of M K–T equations to (5.6.4). In detail, these equations are

$$t\nabla_{x^i} f^i(x)^\tau - (1 - t)(x^i - \mathring{x}^i) + \sum_{j=1}^{r_i} (\alpha^i_j)^+ \nabla_{x^i} g^i_j(x)^\tau + \sum_{j=1}^{s_i} \mu^i_j \nabla_{x^i} h^i_j(x)^\tau = 0$$

$$(\alpha^i_j)^- - g^i_j(x) = 0, \qquad j = 1, \ldots, r_i$$

$$h^i_j(x) = 0, \qquad j = 1, \ldots, s_i, \tag{5.6.8}$$

$i = 1, \ldots, M$.

Just as before, denote this full system of k equations and $k + 1$ unknowns as

$$H(x, \alpha, \mu, t) = 0. \tag{5.6.9}$$

Then $(x, \alpha, \mu, t) \in H^{-1}$ if and only if (x, α, μ, t) solves the full system (5.6.8). We reiterate, the path following will be done in this full system (5.6.9).

To begin the path, let $t = 0$ and given $\mathring{x}$, let us solve the full system (5.6.8), $i = 1, \ldots, M$, for $\mathring{\alpha}, \mathring{\mu}$. Explicitly, at $t = 0$ system (5.6.8) becomes for $i = 1, \ldots, M$,

$$-(x^i - \mathring{x}^i) + \sum_{j=1}^{r_i} (\alpha_j^i)^+ \nabla_{x^i} g_j^i(x)^\tau + \sum_{j=1}^{s_i} \mu_j^i \nabla_{x^i} h_j^i(x)^\tau = 0$$

$$(\alpha_j^i)^- - g_j^i(x) = 0, \qquad j = 1, \ldots, r_i$$

$$h_j^i(x) = 0, \qquad j = 1, \ldots, s_i. \tag{5.6.10}$$

Since at $t = 0$ we start at $x = \mathring{x}$,

$$(\mathring{\alpha}_j^i)^+ = 0, \quad j = 1, \ldots, r_i, \qquad \mathring{\mu}_j^i = 0, \quad j = 1, \ldots, s_i$$

$$(\mathring{\alpha}_j^i)^- = g_j^i(\mathring{x}), \quad j = 1, \ldots, r_i. \tag{5.6.11}$$

This is the starting point $(\mathring{x}, \mathring{\alpha}, \mathring{\mu}, 0)$.

It remains to confirm as we path–follow from $(\mathring{x}, \mathring{\alpha}, \mathring{\mu}, 0)$ that the path reaches $t = 1$, which we do next.

CONVERGENCE PROOF

The convergence proof that the path from $(\mathring{x}, \mathring{\alpha}, \mathring{\mu}, 0)$ reaches $t = 1$ will differ only slightly from the NLP case. First, assume that H is regular, so that at any solution $(x, \alpha, \mu, t) \in H^{-1}$ the Jacobian H' has full rank. We are then assured that a path

$$(x(p), \alpha(p), \mu(p), t(p)) \in H^{-1} \tag{5.6.12}$$

exists.

Next let us verify that the solution to (5.6.8) is unique at $t = 0$. At $t = 0$ we already know by Lemma 5.6.1 that $\mathring{x}$ is unique. Also, from (5.6.10),

$$(\mathring{\alpha}_j^i)^- = g_j^i(\mathring{x}) > 0, \qquad j = 1, \ldots, r_i, \tag{5.6.13}$$

where the strict positivity follows by (5.6.3). The uniqueness of $\mathring{x}$ and the strict positivity of (5.6.13) forces $\mathring{\alpha}_j^i, j = 1, \ldots, r_i$, to be unique. Examining (5.6.10), the linear independence of the $\nabla_{x^i} h_j^i(x)$ from CQ ensures that the $\mathring{\mu}_j^i, j = 1, \ldots, s_i$, are unique, just as in the NLP case. The starting point $(\mathring{x}, \mathring{\alpha}, \mathring{\mu}, 0)$ is thus the unique point in H^{-1} for $t = 0$.

At the unique start $(\mathring{x}, \mathring{\alpha}, \mathring{\mu}, 0)$ consider the $k \times k$ partial Jacobian matrix H'_{-t}. (This, recall, is the $k \times k$ matrix formed by deleting the last column from

H'.) It is straightforward to show that

$$\det H'_{-t}(\dot{x}, \dot{\alpha}, \dot{\mu}, 0) \neq 0. \tag{5.6.14}$$

The proof of this is solely a manipulation exercise and is given in the chapter appendix. Calling now upon the BDE of Chapter 2, (5.6.14) immediately provides that at $(\dot{x}, \dot{\alpha}, \dot{\mu}, 0)$,

$$\frac{dt}{dp} \neq 0. \tag{5.6.15}$$

Again from Chapter 2, (5.6.15) guarantees that the path must strictly increase from $(\dot{x}, \dot{\alpha}, \dot{\mu}, 0)$.

PATH BOUNDEDNESS

With the path unique at $t = 0$ and strictly increasing from $t = 0$, it can never return to $t = 0$. To prove that it reaches $t = 1$, we need only establish boundedness. That is, if the path cannot run off to infinity, the only choice left is to reach $t = 1$. The reasoning is identical to that for NLP in Chapter 4 so is only outlined; the details are left as an exercise.

By direct analogy to NLP assume that the set D is compact. This immediately ensures that the x are bounded.

Next, let

$$f^i(x, t) = tf^i(x) - (1 - t)\tfrac{1}{2}\|x^i - \dot{x}^i\|^2,$$

so that

$$\nabla_{x^i} f^i(x, t) = t\nabla_{x^i} f^i(x) - (1 - t)(x^i - \dot{x}^i)^\tau,$$

where $\dot{x}^i$ is from the CQ. Then define

$$A^i = \max \{\nabla_{x^i} f^i(x, t)(x^i - \dot{x}^i)\}. \tag{5.6.16}$$

The maximum is over $x \in D$ and $0 \leq t \leq 1$. Here A^i is finite because D is compact.

Just as in Chapter 4, consider the K–T conditions for subproblem i and obtain the following linear programming problem for any $x \in D$ and t fixed:

$$\max_{\lambda^i, \mu^i} \sum_{j=1}^{r_i} \lambda^i_j g^i_j(\dot{x}^i, x^{\bar{i}})$$

$$\sum_{j=1}^{r_i} \lambda^i_j \nabla_{x^i} g^i_j(x^i, x^{\bar{i}})^\tau + \sum_{j=1}^{s_i} \mu^i_j \nabla_{x^i} h^i_j(x^i, x^{\bar{i}})^\tau = -\nabla_{x^i} f^i(x, t)^\tau$$

$$\lambda^i_j g^i_j(x^i, x^{\bar{i}}) = 0, \quad j = 1, \dots, r_i$$

$$\lambda^i_j \geq 0, \quad j = 1, \dots, r_i. \tag{5.6.17}$$

The dual becomes

$$\min - \nabla_{x^i} f^i(x, t) \beta^i$$

$$g_j^i(x^i, x^{\bar{i}}) \gamma_j^i + \nabla_{x^i} g_j^i(x^i, x^{\bar{i}}) \beta^i \geq g_j^i(\dot{x}^i, x^{\bar{i}}), \qquad j = 1, \ldots, r_i$$

$$\nabla_{x^i} h_j^i(x^i, x^{\bar{i}}) \beta^i = 0, \qquad j = 1, \ldots, s_i,$$

where $\beta^i = (\beta_1^i, \ldots, \beta_{n_i}^i)$ and $\gamma^i = (\gamma_1^i, \ldots, \gamma_{r_i}^i)$ are the variables.

By concavity of the g_j^i, we see that $\gamma_j^i = 1$ and $\beta^i = (\dot{x}^i - x^i)$ is dual-feasible. Hence the dual has an objective function value less than or equal to

$$-\nabla_{x^i} f^i(x, t)(\dot{x}^i - x^i),$$

so that by (5.6.16),

$$\sum_{j=1}^{r_i} \lambda_j^i g_j^i(\dot{x}^i, x^{\bar{i}}) \leq -\nabla_{x_i} f^i(x, t)(\dot{x}^i - x^i) \leq A^i. \qquad (5.6.18)$$

Since the x^i are in a compact set, CQ (5.6.2) provides that for some $\epsilon_j^i > 0$,

$$g_j^i(\dot{x}^i, x^{\bar{i}}) \geq \epsilon_j^i > 0$$

for any $x^{\bar{i}}$, where $(x^i, x^{\bar{i}}) \in D$. Using (5.6.18) yields

$$0 \leq \lambda_j^i \leq \frac{A^i}{\epsilon_j^i}, \qquad j = 1, \ldots, r_i.$$

The $\lambda_j^i, j = 1, \ldots, r_i$, are therefore bounded. Since $x \in D$ is bounded, the linear independence of the $\nabla_{x^i} h_j^i$ ensures the $\mu_j^i, j = 1, \ldots, s_i$, are bounded. Then for subproblem i, all λ_j^i, μ_j^i are bounded.

Since this result holds for all subproblems, all λ_j^i, μ_j^i are therefore bounded. We conclude that the path is bounded.

With the path bounded it must reach $t = 1$, and we summarize this conclusion as follows.

THEOREM 5.6.2 For EP (5.4.1) suppose that all functions are C^3, the $f^i(x^i, x^{\bar{i}})$ and $g_j^i(x^i, x^{\bar{i}})$ are concave in x^i for $x^{\bar{i}}$ fixed, the $h_j^i(x^i, x^{\bar{i}})$ are linear in x^i for $x^{\bar{i}}$ fixed, the set D is compact, and CQ (5.6.2) holds. Then, assuming that H is regular, the path leads from the unique start $(\dot{x}, \dot{\alpha}, \dot{\mu}, 0)$ to a solution to EP (5.4.1).

SUMMARY

This chapter introduced EP, a formulation of exceptional power because it can model a system of M interacting components. After stating it, we provided a wide range of examples to illustrate not only its practical importance but the insights it uncovers. A dynamic EP that can change as a parameter changes was also introduced. Finally, a path-following algorithm was given for calculating the equilibrium point itself. Chapter 6 continues our study of equilibrium but focuses on the particular but fascinating field of economics.

APPENDIX

We must verify that (5.6.14) holds:

$$\det H'_{-t}(\mathring{x}, \mathring{\alpha}, \mathring{\mu}, 0) \neq 0.$$

Letting a prime denote the derivative, note that

$$((\alpha_j^i)^+)' = 3[\max\{0, \alpha_j^i\}]^2$$

and

$$((\alpha_j^i)^-)' = -3[\max\{0, -\alpha_j^i\}]^2.$$

Using (5.6.11) and (5.6.13), observe for $j = 1, \ldots, r_i$,

$$((\mathring{\alpha}_j^i)^+)' = 0$$

and

$$((\mathring{\alpha}_j^i)^-)' = -3(\mathring{\alpha}_j^i)^2 < 0. \tag{A.5.1}$$

Also from (5.6.11),

$$\mathring{\mu}_j^i = 0, \quad j = 1, \ldots, s_i.$$

The matrix H'_{-t} is $k \times k$ and its rows are formed by differentiating the left-hand side of (5.6.8) for $i = 1, \ldots, M$. That is, each of the k rows of H is differentiated with respect to the k variables $(x, \alpha, \mu) = (x^1, \ldots, x^M, \alpha^1, \ldots, \alpha^M, \mu^1, \ldots, \mu^M)$.

Notice that H'_{-t} will consist of M sets of rows, each set obtained by differentiating (5.6.8) for i fixed. This means that we may break H'_{-t} up into M sets, each set consisting of $n_i + r_i + s_i$ rows, and each set formed by differentiating (5.6.8) for i fixed. Let us examine one of these sets of rows for i fixed. We will prove at $(\mathring{x}, \mathring{\alpha}, \mathring{\mu}, 0)$ that all of its rows are linearly independent. An examination of all M sets will then easily verify that H'_{-t} will be of full rank and therefore (5.6.14) will hold.

At $(\mathring{x}, \mathring{\alpha}, \mathring{\mu}, 0)$ and using (5.6.11) and (5.6.13), each of these sets of $n_i + r_i + s_i$ rows has the form of Figure A.5.1. In that figure I_i is the $n_i \times n_i$ identity matrix, V_i is the $r_i \times r_i$ diagonal matrix with entries

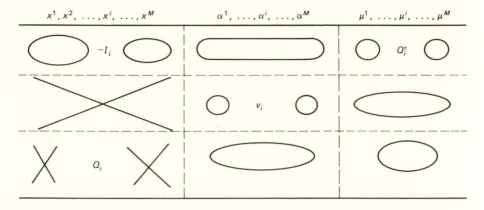

The differentiation with respect to (x, α, μ). The variables with which the differentiation is in respect to are along the top. The X denotes other terms.

Figure A.5.1

$$V_i = -3\begin{pmatrix} (\mathring{\alpha}_1^i)^2 & & \\ & \ddots & \\ & & \ddots \\ & & (\mathring{\alpha}_{r_i}^i)^2 \end{pmatrix},$$

and Q_i is the $s_i \times n_i$ matrix

$$Q_i = \begin{pmatrix} \nabla_{x^i} h_1^i \\ \cdot \\ \cdot \\ \cdot \\ \nabla_{x^i} h_{s_i}^i \end{pmatrix}.$$

Notice that at $t = 0$,

$$\alpha_j^i = \mathring{\alpha}_j^i \neq 0,$$

so V_i is of full rank. Also by CQ (1), matrix Q_i is of full rank.

It is now straightforward to verify that this set of $n_i + r_i + s_i$ equations is linearly independent and, examining all M such sets, that H'_{-t} has full rank (Exercise 5).

EXERCISES/CHAPTER 5

1. Consider the EP

$$\begin{array}{ll} \max\limits_{x_1} x_1 & \max\limits_{x_2} x_2 \\ x_1 + x_2 \leq 2 & 2x_1 + x_2 \leq 3 \\ x_1 \geq 0 & x_2 \geq 0. \end{array} \qquad \text{(E.5.1)}$$

    a. Set up the K–T equations for (E.5.1).

    b. Find the equilibrium points, if any, for (E.5.1).

2. Consider the EP

$$\max_{x_1} x_1 \qquad\qquad\qquad \max_{x_2} x_2$$

$$x_1 + x_2 \leq 2 \qquad\qquad x_1 + x_2 \leq 2$$

$$2x_1 + x_2 \leq 3 \qquad\qquad 2x_1 + x_2 \leq 3$$

$$x_1 \geq 0, \quad x_2 \geq 0 \qquad\qquad x_1 \geq 0, \quad x_2 \geq 0. \qquad\qquad \text{(E.5.2)}$$

    a. Set up the K–T equations for (E.5.2).

    b. Find all equilibrium points for (E.5.2).

3. Consider the EP, where $D^i \subset R^N$:

$$\max f^i(x^i, x^{\bar{i}})$$

$$\text{subject to}$$

$$(x^i, x^{\bar{i}}) \in D^i \qquad\qquad \text{(E.5.3)}$$

given $x^{\bar{i}}$, and the EP

$$\max f^i(x^i, x^{\bar{i}})$$

$$\text{subject to}$$

$$(x^i, x^{\bar{i}}) \in \bigcap_{j=1}^{M} D^j \qquad\qquad \text{(E.5.4)}$$

given $x^{\bar{i}}$. Prove that an equilibrium point for (E.5.3) is an equilibrium point for (E.5.4).

4. For $i = 1, \ldots, M$, consider the EP

$$\max f^i(x^i)$$

$$\text{subject to} \qquad g^i(x^i) \geq 0. \qquad\qquad \text{(E.5.5)}$$

    Prove that (E.5.5) is equivalent to the NLP

$$\max \sum_{i=1}^{M} f^i(x^i)$$

$$g^i(x^i) \geq 0, \qquad i = 1, \ldots, M.$$

5. Prove that the matrix of Figure A.5.1 is full rank at $(\dot{x}, \dot{\alpha}, \dot{\mu}, 0)$. Then prove that $H'_{-t}(\dot{x}, \dot{\alpha}, \dot{\mu}, 0)$ is nonsingular.

6. Consider the EP (5.3.1) and (5.3.2) where

$$A^1 = \begin{pmatrix} 0 & 1 & 2 \\ 1 & 2 & 0 \\ 2 & 0 & 1 \end{pmatrix}, \qquad A^2 = \begin{pmatrix} 1 & 0 & 2 \\ 1 & 0 & 2 \\ 0 & 2 & 1 \end{pmatrix}.$$

To initiate, suppose that person 2 selects action

$$x^2 = (1, 0, 0).$$

    a. Derive the sequence of points $\bar{x}^1, \bar{x}^2, \bar{\bar{x}}^1, \bar{\bar{x}}^2$, and so on, where

$\bar{x}^1$ solves subproblem 1 given x^2,

$\bar{x}^2$ solves subproblem 2 given $\bar{x}^1$,

$\bar{x}^1$ solves subproblem 1 given $\bar{x}^2$,   etc.

b. Show that the equilibrium point for this problem is unique.

7. Consider the EP (5.3.1) and (5.3.2) where A^1 and A^2 are arbitrary $n_1 \times n_2$ matrices. Prove that an equilibrium point for the EP exists.

8. Standard political jargon is a statement that a politician "will maximize social welfare at minimum cost."
   a. Explain why this problem of the politician is not an NLP.
   b. Suppose that we have two senators, one maximizing social welfare and the other minimizing cost. Will the country benefit from their actions? Explain concisely the reason for your answer.

9. The ABC Corp. sells products A, B, and C and the XYZ Inc. sells products X, Y, and Z. There are four manufacturing processes involved. ABC Corp. has the facilities for processes 1 and 2, and XYZ Inc. has the facilities for processes 3 and 4. The table specifies the coefficients of the production process (assumed linear) describing the problem.

| | Product | | | | | | Available |
Processes	A	B	C	X	Y	Z	Time
1	2	4	3	2	3	3	≤ 30
2	20	30	15	35	25	19	≤ 250
3	65	60	71	55	81	70	≤ 800
4	30	35	25	24	32	40	≤ 290
Revenue	$500	$300	$250	$450	$410	$370	

ABC Corp. agrees to process the products of XYZ at $100 for each X, $80 for each Y, and $70 for each Z. Similarly, XYZ Inc. agrees to process the products of ABC at $110 for each A, $60 for each B, and $50 for each C. Formulate the EP for this problem.

10. Three airlines fly to four different cities from Chicago. LJ Airlines own 5 large jets, PD airlines own 10 propeller-driven planes, and SJ airlines own 15 small jets.

Formulate the EP that maximizes profit for each airline to take account of the following:
   a. For LJ Airlines: city D must be served by at least two (not necessarily its own) planes daily.
   b. For PD Airlines: cities A and B must be served by at least three planes daily.
   c. For SJ Airlines: city C must be served at least once daily.
   d. For all airlines: each plane can fly at most 18 hours per day.

Assuming constant flying conditions and passenger usage, the following data is available:

	City	Trip Cost (Round Trip)	Trip Revenue	Average Flying Time (Hours)
Large jet	A	$5,000	$ 6,000	1
	B	6,000	7,000	2
	C	7,000	10,000	5
	D	8,000	12,000	12
Propeller-driven Plane	A	2,000	3,000	2
	B	3,000	5,000	4
	C	4,000	6,000	7
	D	—	—	35
Small jet	A	2,000	3,000	1
	B	3,000	6,000	2
	C	5,000	8,000	7
	D	9,000	12,000	14

11. In the Chicago-Kyoto decentralized firm example, show that the resulting EP may have an infinite number of equilibria.

12. Consider the linear EP

$$\max_{x^i} c^i x^i$$

subject to

$$\sum_{j=1}^{M} A^{ij} x^j \le b^i, \qquad \text{given } x^i,$$

$$x^i \ge 0$$

where c^i is $1 \times n_i$, A^{ij} is $r_i \times n_j$, and b^i is $r_i \times 1$.
    Show that the linear EP is not equivalent to the LP

$$\max_{x} \sum_{i=1}^{M} c^i x^i$$

subject to

$$\sum_{j=1}^{M} A^{ij} x^j \le b^i, \qquad i = 1, 2, \dots, M$$

$$x^i \ge 0.$$

NOTES

This chapter is based on the approach of Zangwill and Garcia [1981].

5.3. The CQ assumption (5.6.2) is quite restrictive. A less restrictive assumption is described in Zangwill and Garcia [1981].

5.6.   The classical way of proving existence of an equilibrium point is by invocation of a fixed-point theroem, and that is demonstrated in Chapter 21. Here, we prove the existence of an equilibrium point directly by path following. The path-following approach is not only easier than the classical approach as applied to EP, but also is more general in that it permits more complicated interactions among the components of the system. In particular, it does not require that the system operate in a cross-product space as does the classical approach; see Zangwill and Garcia [1981].

6

Economic Equilibria

This chapter introduces the *economic equilibrium model*, a powerful means to analyze how people buy and sell in an economy. Essentially, M people get together to buy and sell goods at various prices so that after trading everyone is better off. The economic equilibrium problem truly is basic and a vast number of results in economic theory flow from it. It predates and, in fact, is one of the origins of equilibrium programming given in Chapter 5. Being newer, equilibrium programming is more general. Consequently, economic equilibrium will be presented as a special but ingenious case of it.

6.1 THE ECONOMIC EQUILIBRIUM MODEL

The economic equilibrium problem (EE) applies the concepts of equilibrium programming (EP) to an economic system. Specifically, there are n different types of *goods*, and these n goods are bought and sold among M individuals. Each individual starts with some amount of the n different goods, where we specify w_j^i as the amount of good $j, j = 1, \ldots, n$, that individual i has to begin with. For example, w_1^i might be the amount of wages, w_2^i the number of automobiles, w_3^i the number of houses, and so on, that individual i has initially. The vector $w^i = (w_1^i, \ldots, w_n^i)$ is called the *endowment* of individual i, and we assume that

$$w_j^i > 0 \qquad \text{all } i \text{ and } j. \qquad (6.1.1)$$

THE UTILITY FUNCTION

As would often be the case, individual i may not be satisfied with his or her endowment w^i. For example, he may have a great deal of shelter but little food or fuel. Or he may have many oranges and no milk. Mathematically, individual i has a *utility function* $f^i: R^n \longrightarrow R^1$ and $f^i(w^i)$ may not be the maximum value that f^i can attain.

Quite naturally, each person would then gravitate to a *market* where he or she can exchange w^i for a new vector of goods $x^i = (x_1^i, \ldots, x_n^i)$ that he or she can be happier with. For example, he might sell some shelter in order to buy food and fuel from someone else. Or he may sell some oranges to get milk. By selling the vector of goods w^i and by purchasing the vector of goods x^i, individual i can then increase his or her utility.

BUDGET SET

Let the system for exchanging goods be a pricing mechanism where p_j is the price per unit of good j bought or sold. Also, $p = (p_1, \ldots, p_n)$ is the vector of all prices for all n goods.

When individual i goes to market he brings his initial endowment of goods w^i. If p is the price vector, the money individual i receives for the goods w^i is $pw^i = \sum_{j=1}^n p_j w_j^i$. This amount pw^i is then the *budget* available for purchasing the x^i goods, which themselves cost $px^i = \sum_{j=1}^n p_j x_j^i$. Clearly, to keep within budget, individual i can only choose x^i so that

$$px^i \leq pw^i.$$

Further, let $W_j = \sum_{i=1}^M w_j^i$ be the total amount of good j brought to market by all M people. The vector $W = (W_1, \ldots, W_n)$ indicates the total amount of any good j, $j = 1, \ldots, n$, available in the marketplace. No one can purchase more than that amount, so $x_j^i \leq W_j$, all i and j, or in vector form

$$0 \leq x^i \leq W, \qquad i = 1, \ldots, M. \tag{6.1.2}$$

(Here we are assuming that no borrowing is possible.)

The set

$$\{x^i \,|\, px^i \leq pw^i, \quad 0 \leq x^i \leq W\} \tag{6.1.3}$$

is called the *budget set* of individual i, and clearly x^i must be in that set.

COMPETITIVE EQUILIBRIUM

Let $x = (x^1, \ldots, x^M)$ indicate the amounts purchased by all M individuals, where $x \in R^{Mn}$. The market must determine x; that is, the market must determine what everyone purchases. But the market must also in some manner determine the prices p. The $(M + 1)n$-vector (x, p) therefore comprises the decision variables for the EE.

One of the nice features of our model is that not any quantity and prices (x, p) will do; rather we want everyone to be "satisfied." Call the vector where all people are "satisfied" $(\bar{x}, \bar{p}) = (\bar{x}^1, \ldots, \bar{x}^M, \bar{p})$ a *competitive equilibrium* point. The question now is: How might everyone be satisfied, or in mathematical terms, what characteristics do we want the competitive equilibrium $(\bar{x}, \bar{p})$ to possess?

To be satisfied with purchases $\bar{x}^i$, we mean that individual i's utility should be maximized at $\bar{x}^i$. Mathematically, for all $i = 1, \ldots, M$, and given $p = \bar{p}$, $\bar{x}^i$ must solve

$$\max f^i(x^i)$$

subject to

$$\bar{p}x^i \leq \bar{p}w^i$$
$$0 \leq x^i \leq W. \tag{6.1.4}$$

In this problem x^i is the variable and $p = \bar{p}$ is fixed. Subproblem (6.1.4) ensures that by buying goods $\bar{x}^i$ individual i achieves the best utility possible at the prevailing prices $\bar{p}$.

We also, of course, want our market as a whole to function properly, and certainly the total amount of goods purchased cannot exceed the total amount of goods brought to market. Thus require that

$$\sum_{i=1}^{M} \bar{x}_j^i \leq \sum_{i=1}^{M} w_j^i$$

or, in vector form, that $\bar{x}^i$ satisfies

$$\sum_{i=1}^{M} \bar{x}^i \leq W. \tag{6.1.5}$$

(Here it is assumed that there are no goods brought to market except the endowment goods.)

Our market, we suppose, also permits *free disposal*. Free disposal is convenient and means that if a good j is worthless, it can be thrown away without cost to the owner. (There are goods that cannot be disposed of freely. For example, we may have to pay in order to have a car junked.) Under free disposal all prices are nonnegative. To see this, suppose that $p_j < 0$. Then if individual i sells his or her endowment w_j^i, he receives $p_j w_j^i$ (i.e., he pays $-p_j w_j^i$). But with free disposal he would rather throw away w_j^i, which he can do for free. Therefore, as negative prices need not be considered,

$$p_j \geq 0, \quad j = 1, \ldots, n. \tag{6.1.6}$$

Finally, since there would be no difference if the prices were, say, $(2p_1, 2p_2, \ldots, 2p_n)$ instead of $(p_1, p_2, \ldots, p_n)$, the prices can be normalized, so that

$$\sum_{j=1}^{n} p_j = 1. \tag{6.1.7}$$

STATEMENT OF MODEL

Let us now summarize the way we want our market to operate. Specifically, we require a competitive equilibrium $(\bar{x}, \bar{p})$ to satisfy the following three conditions:

(A) Given $\bar{p}$,
$$f^i(\bar{x}^i) = \max_{x^i} f^i(x^i)$$
$$\bar{p}x^i \leq \bar{p}w^i$$
$$0 \leq x^i \leq W$$

for $i = 1, \ldots, M$.

(B)
$$\sum_{i=1}^{M} \bar{x}^i \leq W.$$

and

(C)
$$\bar{p} \geq 0, \qquad \sum_{j=1}^{n} \bar{p}_j = 1, \tag{6.1.8}$$

where $W = (W_1, \ldots, W_n)$ and $W_j = \sum_{i=1}^{M} w_j^i$.

The condition $\bar{p} \geq 0$ follows from free disposal and ensures that each good always has a nonnegative value. The condition $\sum \bar{p}_j = 1$ is a normalization that bounds the set of prices. Condition (A) provides that each individual achieves the highest utility at the prices $\bar{p}$. Together, conditions (A), (B), and (C) of (6.1.8) comprise the EE model, and any $(\bar{x}, \bar{p})$ that satisfies (A), (B), and (C) is called a competitive equilibrium. Despite the fact that this model is quite simple and bare-bones, it nonetheless captures some of the key features of how nearly all markets operate. The model described by (6.1.8) is usually referred to as the *exchange economy* model.

Before discussing how the EE (6.1.8) relates to EP, let us first present a brief example.

EXAMPLE 6.1.1

Two large international corporations, one in Germany and the other in the United States, are to engage in trade of two items, computer terminals and computer memory diskettes. The German firm has 200 terminals and 100 diskettes, whereas the U.S. firm has 200 units of both. (These are the initial endowments.) Suppose that the two firms engage in trade and let $x^1 = (x_1^1, x_2^1)$, $x^2 = (x_1^2, x_2^2)$ denote the goods of firms 1 and 2 after trading, where

x_1^1 = number of terminals (in hundreds) for the German firm

x_2^1 = number of diskettes (in hundreds) for the German firm

x_1^2 = number of terminals (in hundreds) for the U.S. firm

x_2^2 = number of diskettes (in hundreds) for the U.S. firm.

The utility of the German firm is $100 - (x_1^1 - 2)^2 - (x_2^1 - 3)^2$, and the U.S. firm has utility $50 - (x_1^2 - 4)^2 - (x_2^2 - 2)^2$. At what price $\bar{p} = (\bar{p}_1, \bar{p}_2)$ for the two goods, and for what $\bar{x} = (\bar{x}^1, \bar{x}^2)$, will a competitive equilibrium be obtained?

ANALYSIS The answer is $\bar{p} = (\frac{1}{2}, \frac{1}{2})$, $\bar{x}^1 = (1, 2)$ and $\bar{x}^2 = (3, 1)$. The details of how to calculate this will be discussed in Chapter 7, but let us simply exhibit here that $(\bar{x}, \bar{p})$ is an economic equilibrium.

First notice that the endowment for the German firm is $w^1 = (2, 1)$ and for the U.S. firm it is $w^2 = (2, 2)$. Thus $W = (4, 3)$.

The competitive equilibrium therefore must satisfy the following conditions:

(A.1) $100 - (\bar{x}_1^1 - 2)^2 - (\bar{x}_2^1 - 3)^2 = \max [100 - (x_1^1 - 2)^2 - (x_2^1 - 3)^2]$

subject to

$$\bar{p}_1 x_1^1 + \bar{p}_2 x_2^1 \leq 2\bar{p}_1 + \bar{p}_2$$

$$0 \leq x_1^1 \leq 4$$

$$0 \leq x_2^1 \leq 3.$$

(A.2) $50 - (\bar{x}_1^2 - 4)^2 - (\bar{x}_2^2 - 2)^2 = \max [50 - (x_1^2 - 4)^2 - (x_2^2 - 2)^2]$

subject to

$$\bar{p}_1 x_1^2 + \bar{p}_2 x_2^2 \leq 2\bar{p}_1 + 2\bar{p}_2$$

$$0 \leq x_1^2 \leq 4$$

$$0 \leq x_2^2 \leq 3.$$

(B) $\bar{x}_1^1 + \bar{x}_1^2 \leq 4 \qquad \bar{x}_2^1 + \bar{x}_2^2 \leq 3.$

(C) $\bar{p}_1 \geq 0, \qquad \bar{p}_2 \geq 0, \qquad \bar{p}_1 + \bar{p}_2 = 1.$ (6.1.9)

A quick calculation verifies that if $\bar{p} = (\frac{1}{2}, \frac{1}{2})$, then $\bar{x}^1 = (1, 2)$ optimizes (A.1) and $\bar{x}^2 = (3, 1)$ optimizes (A.2). Also, (B) and (C) are satisfied, so we have a competitive equilibrium.

6.2 TRANSFORMING EE INTO EP

The EE problem is to find a competitive equilibrium point $(\bar{x}, \bar{p})$ where $\bar{x} = (\bar{x}^1, \ldots, \bar{x}^M) \in R^{Mn}$ and $\bar{p} \in R^n$ that satisfies all the conditions of (6.1.8). We now transform EE into an EP. Condition (A) resembles an EP at least for

$i = 1, \ldots, M$, but conditions (B) and (C) do not fit the EP. Also, each x^i, $i = 1, \ldots, M$, has a corresponding subproblem, yet there is no subproblem for p. The idea is to create a "dummy" $(M + 1)$st subproblem for p, which we will see is not naive at all but exceedingly clever. The extra subproblem will replace (B) and (C) and yield the EP.

THE MARKET PLAYER

The $(M + 1)$st subproblem will be optimized by a *market player* whose decision variables are the prices $p = (p_1, \ldots, p_n)$. Think of the market player as an agent hired to determine the prices $\bar{p}$ that will be agreeable to everyone in the sense that (A), (B), and (C) will be simultaneously satisfied.

The first task is to determine the subproblem to be solved by the market player that together with (A) yield both (B) and (C). Since prices are the market player's variables, the constraints of the market player's subproblem are immediate from (6.1.8C). He or she must choose p such that

$$p \geq 0, \qquad \sum_{j=1}^{n} p_j = 1.$$

If $\bar{p}$ satisfies these constraints, then (C) will automatically hold.

Regarding the utility function of the market player, it turns out, and this will be explained in a moment, that given $\bar{x}$, the market player is to select p to maximize

$$p\left(\sum_{i=1}^{M} \bar{x}^i - W\right)$$

or, written out,

$$\sum_{j=1}^{n} p_j\left(\sum_{i=1}^{M} \bar{x}_j^i - \sum_{i=1}^{M} w_j^i\right). \tag{6.2.1}$$

The market player's subproblem therefore, given $\bar{x}$, is to select p to

$$\max p\left(\sum_{i=1}^{M} \bar{x}^i - W\right)$$

$$\sum_{j=1}^{n} p_j = 1$$

$$p \geq 0. \tag{6.2.2}$$

We now add the market player's subproblem to the other subproblems and obtain the EP.

THE EE PROBLEM STATED AS AN EP

The EP that corresponds to EE (6.1.8) is as follows. Determine $(\bar{x}, \bar{p})$ so that:

(A)   For each individual $i = 1, \ldots, M$, given $\bar{p}$, $\bar{x}^i$ solves

$$\max_{x^i} f^i(x^i)$$

subject to

$$\bar{p}x^i \leq \bar{p}w^i$$

$$0 \leq x^i \leq W.$$

(B)   For the market player, given $\bar{x}$, $\bar{p}$ solves

$$\max_p p\left(\sum_{i=1}^M \bar{x}^i - W\right)$$

$$p \geq 0, \qquad \sum_{j=1}^n p_j = 1. \tag{6.2.3}$$

This is clearly an EP with $M + 1$ individuals. We now must verify that an equilibrium point $(\bar{x}, \bar{p})$ for this EP is a competitive equilibrium for EE (6.1.8).

VERIFICATION OF COMPETITIVE EQUILIBRIUM

Let $(\bar{x}, \bar{p})$ be an equilibrium of (6.2.3); then obviously (A) and (C) of (6.1.8) hold. Let us show that (B) holds by contradiction. Suppose (B) does not hold. Then for some j,

$$\sum_{i=1}^M \bar{x}_j^i > W_j. \tag{6.2.4}$$

Yet since $(\bar{x}, \bar{p})$ satisfies (6.2.3B), $\bar{p}$ maximizes, so

$$\bar{p}\left(\sum_{i=1}^M \bar{x}^i - W\right) \geq e^j\left(\sum_{i=1}^M \bar{x}^i - W\right), \tag{6.2.5}$$

where e^j is the vector with a one in the jth position and zeros elsewhere. (Here e^j would be a feasible price for the market player's problem.) Expanding the inner product yields

$$e^j\left(\sum_{i=1}^M \bar{x}^i - W\right) = \left(\sum_{i=1}^M \bar{x}_j^i - W_j\right) > 0 \tag{6.2.6}$$

via (6.2.4). Thus, by (6.2.5) and (6.2.6),

$$\bar{p}\sum_{i=1}^M \bar{x}^i > \bar{p}W = \sum_{j=1}^n \bar{p}_j \sum_{i=1}^M w_j^i. \tag{6.2.7}$$

On the other hand, since $\bar{x}^i$ satisfies (6.2.3A),

$$\bar{p}\bar{x}^i \leq \bar{p}w^i \qquad \text{all } i$$

and summing, we obtain

$$\bar{p}\sum_{i=1}^{M}\bar{x}^i \leq \sum_{i=1}^{M}\bar{p}w^i = \sum_{i=1}^{M}\sum_{j=1}^{n}\bar{p}_j w_j^i. \qquad (6.2.8)$$

Clearly, (6.2.7) and (6.2.8) are contradictory, so for each j

$$\sum_{i=1}^{M}\bar{x}_j^i \leq W_j \qquad (6.2.9)$$

and (6.1.8B) holds. We have demonstrated that the equilibrium point of EP (6.2.3) is the competitive equilibrium of EE (6.1.8).

COMPLEMENTARITY

Examining the competitive equilibrium further, notice that for the market player, $\bar{p} \geq 0$ and also (6.2.9) holds. Suppose $\sum_{i=1}^{M}\bar{x}_j^i = W_j$, some j. If

$$\sum_{i=1}^{M}\bar{x}_j^i < W_j,$$

then $\bar{p}_j = 0$. This is because the market player maximizes

$$\sum_{j=1}^{n}p_j\left(\sum_{i=1}^{M}\bar{x}_j^i - W_j\right).$$

Contrapositively, if $\bar{p}_j > 0$, then $(\sum_{i=1}^{M}\bar{x}_j^i - W_j) = 0$. We arrive at what is termed a *complementarity* relation,

$$\bar{p}_j\left(\sum_{i=1}^{M}\bar{x}_j^i - W_j\right) = 0, \qquad (6.2.10)$$

which must hold at an equilibrium $(\bar{x}, \bar{p})$.

MARKET PLAYER INTERPRETED

Let us now interpret why the market player has as his or her utility

$$p\left(\sum_{i=1}^{M}x^i - W\right) = \sum_{j=1}^{n}p_j\left(\sum_{i=1}^{M}x_j^i - W_j\right). \qquad (6.2.11)$$

The term $W_j = \sum_{i=1}^{M}w_j^i$ is the total amount of good j brought to market by all individuals. Also, $\sum_{i=1}^{M}x_j^i$ is the total demand for good j. The expression

$$\left(\sum_{i=1}^{M} x_j^i - W_j \right)$$

is called the *excess demand* for good j. If the excess demand is positive, people desire more of good j than is available in the market. If it is zero or negative, then a sufficient amount exists to satisfy demand.

Suppose that the excess demand for good j is positive. Via (6.2.11), to increase its utility the market player would then set a price p_j very high. But if the price of good j is high, people buy less of it, reducing the excess demand.

The market player thereby charges a high price for any good with a positive excess demand, thereby driving the demand for any such good down. Adjusting prices in this manner, the market player forces demand down and therefore ensures there is no positive excess demand. That is, he ensures that

$$\sum_{i=1}^{M} x_j^i \leq W_j \qquad \text{for all } j.$$

But then (6.1.8B) must hold.

By maximizing the particular utility function (6.2.1), the market player adjusts prices so that (6.2.3B) must hold. Further, this is accomplished in a very economically reasonable manner—by charging more when demand exceeds supply.

COMPLEMENTARITY INTERPRETED

Because (6.2.3B) holds and the market player maximizes over $p \geq 0$, recall that the complementarity condition (6.2.10) was obtained. Let us also interpret it. If $\sum_{i=1}^{M} \bar{x}_j^i < W_j$, then there is excess supply and with too much in the market, the price of good j must eventually be driven to zero. Consequently, $\bar{p}_j = 0$. Contrapositively, $\bar{p}_j > 0$ only if $\sum_{i=1}^{M} \bar{x}_j^i = W_j$, so supply just equals demand.

Overall, then, by adding a market player the EE can be transformed into an EP. But the EE is more than simply an application of EP, for as we have discussed, it has implications as to how actual markets operate.

6.3 COMPETITIVE EQUILIBRIUM EXTENDED

Since the EE model (6.1.8) is so important, it is valuable to generalize it in several ways.

KEEPING UP WITH THE JONESES

First recall the notation of Chapter 5, $x = (x^i, x^i)$, where in this application x^i means the vector of all goods for all individuals except individual i. One immediate generalization is to let the utility depend not only on x^i but on the other goods $x^{\bar{i}}$. Thus for individual i, let the utility function be

$$f^i(x^i, x^{\bar{i}}), \tag{6.3.1}$$

and here, of course, individual i controls only x^i. A utility function of this type is said to be dependent on *externalities*. Externalities are variables not subject to the control of individual i.

For many people their personal feeling of wealth or utility depends not only on their own possessions x^i but also upon what others have, $x^{\bar{i}}$. To "keep up with the Joneses" you must have more of certain items than your neighbor has. The utility (6.3.1) specifies this more realistic interpretation because it depends upon both x^i and $x^{\bar{i}}$.

PRODUCTION

Another extremely valuable generalization of the model is to permit production. In (6.1.8) there is no production, just an exchange. Individual i starts with an amount w^i_j of good j and after an exchange ends with x^i_j of that good. Now, however, assume there is a production phase prior to bringing the goods to market. Specifically, given w^i_j, suppose that after production by individual i, an amount a^i_j of good j is brought to market. For example, an individual endowed with nails, wood, or other materials may put them together to form furniture. Or given initial amounts of lemons, water, and sugar, some of that may be used to produce lemonade. Thus a^i_j denotes the amount of good j individual i has after production and that amount is brought to the market. Of course, if there is no production, then $a^i_j = w^i_j$ for all i and j.

Next, designate $c^i_j(x)$ as the net change in the amount of good j for individual i after going to the market. Clearly,

$$c^i_j(x) = x^i_j - a^i_j \tag{6.3.2}$$

because a^i_j is taken to market and x^i_j is brought back.

Now define the vectors $a^i = (a^i_1, \ldots, a^i_n)$ and $c^i = (c^i_1, \ldots, c^i_n)$. Since after production a^i is brought to market, the budget restriction on individual i at any price p is

$$px^i \leq pa^i$$

or, by (6.3.2),

$$pc^i(x) \leq 0. \tag{6.3.3}$$

Also, the total amount of goods brought to market is $A = \sum_{i=1}^{M} a^i$, and since no one can purchase more than that amount,

$$0 \leq x^i \leq A. \tag{6.3.4}$$

The set described by both (6.3.3) and (6.3.4) is called the budget set of individual i, and allowing production x^i must be in that set.

OTHER NONBUDGETARY LIMITATIONS

So far the model has largely reflected budgetary considerations. To generalize further, we suppose that there are various limits on the amounts that individual i can produce or purchase and that these are expressed by constraints

$$g_j^i(x^i, x^{\bar{i}}) \leq 0, \qquad j = 1, \ldots, r_i$$
$$h_j^i(x^i, x^{\bar{i}}) = 0, \qquad j = 1, \ldots, s_i. \qquad (6.3.5)$$

For example, government may limit the gasoline purchased by a car owner. A housewife may limit the pounds of potatoes purchased in a week. Or a union contract may restrict what management can do. Also, as we will see later, the constraints (6.3.5) can indicate the technological relations for the production in our model. Constraints (6.3.5) are very useful because they can be used to model a wide variety of restrictions or limitations on an individual.

NEW SUBPROBLEM

At this point we have extended individual i's subproblem to include the utility (6.3.1), production, and the various constraints (6.3.5). Let us now state this more general problem precisely. Specifically, individual i, given prices p and other variables $x^{\bar{i}}$, must select x^i to optimize

$$\max_{x^i} f^i(x^i, x^{\bar{i}})$$

subject to

$$pc^i(x^i, x^{\bar{i}}) \leq 0$$
$$g_j^i(x^i, x^{\bar{i}}) \leq 0, \qquad j = 1, \ldots, r_i$$
$$h_j^i(x^i, x^{\bar{i}}) = 0, \qquad j = 1, \ldots, s_i$$
$$0 \leq x^i \leq A. \qquad (6.3.6)$$

This is the new subproblem individual i, $i = 1, \ldots, M$, would solve.

The various extensions of the individual subproblem do not act in isolation but influence the market as a whole. Now the amount brought to market of good j is

$$A_j = \sum_{i=1}^{M} a_j^i.$$

Also, the total amount of good j purchased is $\sum_{i=1}^{M} x_j^i$, and since that cannot exceed the amount brought to market,

$$\sum_{i=1}^{M} x_j^i \leq A_j$$

or, by (6.3.2),

$$\sum_{i=1}^{M} c_j^i(x) \leq 0. \tag{6.3.7}$$

EXTENDED EE

Bringing all of these considerations together, we now formally state the *extended EE* model. The extended EE is to find a competitive equilibrium $(\bar{x}, \bar{p})$ such that

(A)   For $i = 1, \ldots, M$, given $p = \bar{p}$ and $x^i = \bar{x}^i$, $\bar{x}^i$ solves

$$\max_{x^i} f^i(x^i, \bar{x}^i)$$

subject to

$$\sum_{j=1}^{n} \bar{p}_j c_j^i(x^i, \bar{x}^i) \leq 0$$

$$g_j^i(x^i, \bar{x}^i) \leq 0, \quad j = 1, \ldots, r_i$$

$$h_j^i(x^i, \bar{x}^i) = 0, \quad j = 1, \ldots, s_i$$

$$0 \leq x^i \leq A. \tag{6.3.8}$$

(B)
$$\sum_{i=1}^{M} c_j^i(\bar{x}) \leq 0, \quad j = 1, \ldots, n.$$

and

(C)
$$\sum_{j=1}^{n} \bar{p}_j = 1, \quad \bar{p}_j \geq 0.$$

Clearly, this generalizes the previous exchange economy model (6.1.8) to include the utility function (6.3.1), production, and the constraints (6.3.5).

THE EQUIVALENT EP

It directly follows from our previous reasoning and is left as Exercise 6 to prove that the EP equivalent to the extended EE is as follows. Find an equilibrium point $(x, p) = (\bar{x}, \bar{p})$ such that:

(A)   For $i = 1, \ldots, M$, given $\bar{x}^i$ and $\bar{p}$, $\bar{x}^i$ solves

$$\max_{x^i} f^i(x^i, \bar{x}^i)$$

$$\sum_{j=1}^{n} \bar{p}_j c_j^i(x^i, \bar{x}^i) \leq 0$$

$$g_j^i(x^i, \bar{x}^i) \leq 0, \quad j = 1, \ldots, r_i$$

$$h_j^i(x^i, \bar{x}^i) = 0, \quad j = 1, \ldots, s_i$$

$$0 \leq x^i \leq A.$$

(B)   For $M + 1$, and given $\bar{x}$, the point $\bar{p}$ solves

$$\max_{p} \sum_{j=1}^{n} p_j c_j(\bar{x})$$

$$\sum_{j=1}^{n} p_j = 1, \qquad p_j \geq 0,$$

where

$$c_j(x) = \sum_{i=1}^{M} c_j^i(x), \qquad j = 1, \ldots, n. \tag{6.3.9}$$

Here is an example.

EXAMPLE 6.3.1:   The Story of Robinson–Crusoe

Mr. Robinson and Miss Crusoe live on an island where they are the only inhabitants. Their days are spent growing coconuts, catching crayfish, and producing a delicacy they had discovered called cocofish, which is concocted by mixing two parts crayfish and one part coconut (plus some dry weeds abundantly available on the island).

For the month, Mr. Robinson has 10 coconuts, 5 crayfishes, and 2 coco-fishes, and Miss Crusoe has 4 coconuts, 15 crayfishes, and 3 cocofishes. Mr. Robinson and Miss Crusoe can buy and sell the three items to each other. Additionally, they can produce more cocofishes by mixing 1 coconut to 2 crayfishes. Suppose that $f^1(x_1^1, x_2^1, x_3^1)$ is the utility function for Mr. Robinson, where x_1^1 is the number of coconuts (the first good), x_2^1 the number of crayfishes (the second good), and x_3^1 the number of cocofishes (the third good). Similarly, $f^2(x_1^2, x_2^2, x_3^2)$ is the utility for Miss Crusoe.

MODEL FORMULATION   The endowments for Mr. Robinson and for Miss Crusoe are

$$w^1 = (10, 5, 2) \qquad \text{and} \qquad w^2 = (4, 15, 3) \tag{6.3.10}$$

respectively.

Let $x_j^i, j = 1, 2, 3$, $i = 1, 2$, be the amount of good j brought back by person i from market, and let p_j be the price of good j. Then $\sum_{j=1}^{3} p_j x_j^i$ should not exceed budget.

Suppose that z_1^i is the amount of coconuts and z_2^i is the amount of cray-fishes used by person i in order to produce z_3^i cocofishes. Clearly, person i will bring $w_1^i - z_1^i$ coconuts and $w_2^i - z_2^i$ crayfishes to market and also

$$0 \leq z_1^i \leq w_1^i, \qquad 0 \leq z_2^i \leq w_2^i, \qquad i = 1, 2. \tag{6.3.11}$$

Further, $w_3^i + z_3^i$ will be the number of cocofishes brought by person i to market where

$$z_3^i = z_1^i \qquad \text{and} \qquad 2z_3^i = z_2^i \tag{6.3.12}$$

because each cocofish requires 1 coconut and 2 crayfishes.

As for the budget, person i's budget is

$$p_1(w_1^i - z_1^i) + p_2(w_2^i - z_2^i) + p_3(w_3^i + z_3^i).$$

Thus the goods x^i selected by person i must satisfy

$$\sum_{j=1}^{3} p_j x_j^i \leq p_1(w_1^i - z_1^i) + p_2(w_2^i - z_2^i) + p_3(w_3^i + z_3^i).$$

Since for good j, $j = 1, 2$, there are a maximum of $w_j^1 + w_j^2$ of good j and $z_j^1 + z_j^2$ are consumed in production

$$0 \leq x_j^i \leq w_j^1 + w_j^2 - z_j^1 - z_j^2, \qquad i = 1, 2. \tag{6.3.13}$$

Also, because product 3 is being produced, the total amount of that product brought to market is $w_3^1 + z_3^1 + w_3^2 + z_3^2$, so that

$$0 \leq x_3^i \leq w_3^1 + z_3^1 + w_3^2 + z_3^2, \qquad i = 1, 2. \tag{6.3.14}$$

FORMULATION   Let us now formulate the EP model for this example. The EP is to find $(\bar{x}, \bar{z}, \bar{p})$ that solves:

For $i = 1, 2$, given $\bar{p}$, and $z^i = \bar{z}^i$

$$\max f^i(x_1^i, x_2^i, x_3^i)$$

$$\bar{p}_1 x_1^i + \bar{p}_2 x_2^i + \bar{p}_3 x_3^i \leq \bar{p}_1(w_1^i - z_1^i) + \bar{p}_2(w_2^i - z_2^i) + \bar{p}_3(w_3^i + z_3^i)$$

$$0 \leq z_1^i \leq w_1^i, \qquad 0 \leq z_2^i \leq w_2^i,$$

$$0 \leq x_j^i \leq w_j^1 + w_j^2 - z_j^1 - z_j^2, \qquad j = 1, 2$$

$$0 \leq x_3^i \leq w_3^1 + z_3^1 + w_3^2 + z_3^2$$

$$z_3^i = z_1^i, \qquad 2z_3^i = z_2^i,$$

where the maximum is over the x_j^i and z_j^i.

For the market player given $\bar{x}_j^i$ and $\bar{z}_j^i$,

$$\max_{p} \sum_{i=1}^{2} [p_1(\bar{x}_1^i - w_1^i + \bar{z}_1^i) + p_2(\bar{x}_2^i - w_2^i + \bar{z}_2^i) + p_3(\bar{x}_3^i - w_3^i - \bar{z}_3^i)]$$

$$p_1 + p_2 + p_3 = 1, \qquad p_j \geq 0, \qquad j = 1, 2, 3.$$

Note that the constraints on individual i's subproblem are the technological constraints on production, as previously mentioned.

6.4 THE PRICING-OUT MECHANISM

In EP (6.3.9) the role of the market player is to establish a price p_j for each good j. However, notice that no prices are provided for the constraints g_j^i and h_j^i. Constraints are not traded and hence no price is established for them.

One of the astonishing features of economic equilibrium is that by reformulating the problem, the constraints can also have prices. This does seem incredible since there is no real market for the constraints, yet we are nevertheless able to consider that a hypothetical market exists for the constraints as well.

The prices on the constraints are called *shadow prices* to distinguish them from the other prices. Still, the shadow prices have real meaning economically and they tell us how much a constraint is costing us. For example, suppose the constraint $g_1^i \leq 0$ expresses that a company's warehouse capacity is limited to 1000 square feet. Then the associated shadow price, call it λ_1^i, indicates the marginal value to the company of additional warehouse space. If $\lambda_1^i = \$10$, then one more square foot of warehouse space will benefit the company $10, and one less will cost the company $10. A constraint has an economic effect on the company, and the shadow price tells us that effect. To summarize, the shadow price on a constraint tells us the marginal value of tightening or loosening that constraint.

MODEL REFORMULATION

Associate with each g_j^i a shadow price λ_j^i and with each h_j^i a shadow price μ_j^i, and let $\lambda = (\lambda_j^i)$ and $\mu = (\mu_j^i)$. (For simplicity, we assume that the constraints $0 \leq x^i \leq A$ have been incorporated in the constraints $g_j^i \leq 0$.) The reformulation of EP (6.3.9) that provides the shadow prices is as follows:

(A)   For $i = 1, \ldots, M$, where $\bar{x}^{\bar{i}}$, $\bar{p}$, $\bar{\lambda}$, and $\bar{\mu}$ are given,

$$\max_{x^i} f^i(x^i, \bar{x}^{\bar{i}})$$

subject to

$$\sum_{j=1}^{n} \bar{p}_j c_j^i(x^i, \bar{x}^{\bar{i}}) + \sum_{j=1}^{r_i} \bar{\lambda}_j^i g_j^i(x^i, \bar{x}^{\bar{i}}) + \sum_{j=1}^{s_i} \bar{\mu}_j^i h_j^i(x^i, \bar{x}^{\bar{i}}) \leq 0.$$

(B)   For $M + 1$, where $\bar{x}$ is given,

$$\max \sum_{j=1}^{n} p_j c_j(\bar{x}) + \sum_{i=1}^{M} \left[\sum_{j=1}^{r_i} \lambda_j^i g_j^i(\bar{x}) + \sum_{j=1}^{s_i} \mu_j^i h_j^i(\bar{x}) \right]$$

$$\sum_{j=1}^{n} p_j = 1$$

$$p_j \geq 0, \qquad j = 1, \ldots, n$$

$$\lambda_j^i \geq 0, \qquad \text{all } i \text{ and } j, \tag{6.4.1}$$

where the maximum is over all p_j^i, λ_j^i, and μ_j^i.

Notice in (6.4.1B) that the λ_j^i are nonnegative but the μ_j^i can be positive, negative, or zero. Also, neither the variables λ_j^i nor μ_j^i need sum to 1.

Notationally, let $(\bar{x}, \bar{p}, \bar{\lambda}, \bar{\mu})$ be an equilibrium point for (6.4.1). The next theorem shows that $(\bar{x}, \bar{p})$ is an equilibrium point for (6.3.9).

THEOREM 6.4.1   Let $(\bar{x}, \bar{p}, \bar{\lambda}, \bar{\mu})$ be an equilibrium point for EP (6.4.1). Then $(\bar{x}, \bar{p})$ is an equilibrium point for (6.3.9).

Proof. Let $(\bar{x}, \bar{p}, \bar{\lambda}, \bar{\mu})$ be an equilibrium point for (6.4.1). First observe that

$$h_j^i(\bar{x}) = 0 \qquad \text{all } i \text{ and } j. \tag{6.4.2}$$

Suppose not. Suppose that for some l and k,

$$h_k^l(\bar{x}) \neq 0. \tag{6.4.3}$$

Let

$$\hat{\mu}_j^i = \begin{cases} \bar{\mu}_k^l + \text{sgn } h_k^l(\bar{x}) & \text{if } i = l, j = k \\ \bar{\mu}_j^i & \text{otherwise.} \end{cases} \tag{6.4.4}$$

Thus $\hat{\mu}_j^i$ is the same as the previous $\bar{\mu}_j^i$ except at $i = l, j = k$, where $\hat{\mu}_k^l = \bar{\mu}_k^l + \text{sgn } h_k^l(\bar{x})$. Observe that the new $\hat{\mu}$ is still feasible for (6.4.1B). Denoting f^{M+1} as the objective function for (6.4.1B) yields

$$f^{M+1}(\bar{x}, \bar{p}, \bar{\lambda}, \bar{\mu}) \geq f^{M+1}(\bar{x}, \bar{p}, \bar{\lambda}, \hat{\mu}) = f^{M+1}(\bar{x}, \bar{p}, \bar{\lambda}, \bar{\mu}) + |h_k^l(\bar{x})|$$
$$> f^{M+1}(\bar{x}, \bar{p}, \bar{\lambda}, \bar{\mu})$$

using (6.4.3). This is a contradiction.

Therefore (6.4.2) holds. Using similar reasoning,

$$g_j^i(\bar{x}) \leq 0 \qquad \text{all } i \text{ and } j. \tag{6.4.5}$$

Immediately, from (6.4.2),

$$\sum_{i=1}^{M} \sum_{j=1}^{s_i} \bar{\mu}_j^i h_j^i(\bar{x}) = 0. \tag{6.4.6}$$

Moreover,

$$\sum_{i=1}^{M} \sum_{j=1}^{r_i} \bar{\lambda}_j^i g_j^i(\bar{x}) = 0, \tag{6.4.7}$$

because if $g_j^i(\bar{x}) < 0$, then since (6.4.5) holds and subproblem $M + 1$ is maximizing over $\lambda_j^i \geq 0$, $\bar{\lambda}_j^i = 0$.

Employing (6.4.6) and (6.4.7), we see that the constraints of (6.4.1A) reduce to

$$\sum_{j=1}^{n} \bar{p}_j c_j^i(\bar{x}) \leq 0. \tag{6.4.8}$$

Via (6.4.2), (6.4.5), and (6.4.8), observe that $\bar{x}^i$ is feasible for (6.3.9A) given $\bar{x}^i$ and $\bar{p}$. Also see that any point x^i that is feasible for (6.3.9A) must be feasible for (6.4.1A). Thus $\bar{x}^i$ is optimal for (6.3.9A) given $(\bar{x}^i)$ and $\bar{p}$.

Finally, given $\bar{x}$, let us show that $\bar{p}$ is optimal for (6.3.9B). At $\bar{x}$, if p is feasible for (6.3.9B), then for any $\lambda \geq 0$ and μ, (p, λ, μ) is feasible for (6.4.1B). Conversely, if (p, λ, μ) is feasible for (6.4.1B), p is feasible for (6.3.9B). Hence $\bar{p}$ is optimal for (6.3.9B), given $\bar{x}$.

This completes the proof. $\square$

We have validated that (6.4.1) does indeed obtain shadow prices for the constraints.

A CAVEAT

In both EE models of this chapter, the individual accepts the prices as given. That is, the "market" determines the prices, and given these prices the individual achieves the best bundle of goods within his or her budget. This scenario, that the individual accepts the prices as given, is appropriate when there is a large number of individuals M, each of whom trades a relatively small amount. Then any individual cannot influence the market.

For small M, however, this situation may not be realistic because then by withholding goods or other strategies the individual might be able to alter prices. Situations where individuals are powerful enough to influence prices occur often and are typically studied under the topic of oligopolies or cartels.

In sum, the EE is realistic when there is a fairly "pure" market operating, that is, when there is a large number of small individuals, none of whom can influence the market much. However, oligopolistic situations may require a different form of EP. (See Exercise 7.)

SUMMARY

This chapter introduced the EE problem, which describes the operation of a market in which M individuals buy and sell goods so that they are better off. We demonstrated that the EE was a special case of EP. Also, the chapter appendix explores a convenient property called "efficiency," of the EE solution.

Chapter 7 proves that, under reasonable conditions, a competitive equilibrium $(\bar{x}, \bar{p})$ actually exists. This means that people can buy and sell, and for the going prices $\bar{p}$ each person is actually able to maximize his or her utility.

APPENDIX

PARETO OPTIMUM

A concept central in economic theory and closely related to competitive equilibrium is the notion of "Pareto optimality." An allocation of goods x is *Pareto optimal* if no feasible reallocation of goods could increase the utility of one or more individuals without lowering the utility of another individual. Rephrased, an allocation is Pareto optimal if any different allocation that improves on the utility of one or more persons, also harms someone. However, let us be more precise.

Suppose that the utility functions $f^i(x^i)$ and endowment $w^i > 0$ are given, $i = 1, \ldots, M$. A point $(\overset{*}{x}, \overset{*}{p})$ is *Pareto optimal* if:

1. $(\overset{*}{x}, \overset{*}{p})$ satisfies

$$px^i \leq pw^i, \qquad i = 1, \ldots, M \qquad (A.6.1)$$

$$\sum_{i=1}^{M} x^i \leq W, \qquad x \geq 0 \qquad (A.6.2)$$

$$\sum_{j=1}^{n} p_j = 1, \qquad p \geq 0. \qquad (A.6.3)$$

2. For any x such that $(x, \overset{*}{p})$ satisfies (A.6.1) and (A.6.2), if $f^k(x^k) > f^k(\overset{*}{x}{}^k)$ for some k, then $f^i(x^i) < f^i(\overset{*}{x}{}^i)$ for some i.

Condition 1 requires that a Pareto-optimal point is feasible [i.e., it satisfies (A.6.1) to (A.6.3)]. Rephrasing 2, there cannot be an x such that $(x, \overset{*}{p})$ is feasible and such that $f^i(x^i) \geq f^i(\overset{*}{x}{}^i)$, all i, where $f^k(x^k) > f^k(\overset{*}{x}{}^k)$ for some k. In other words, given $\overset{*}{p}$, no feasible allocation x exists that makes some individual k better off with no individual worse off.

A basic theorem of welfare economics states that a competitive equilibrium to (6.1.8) is a Pareto optimum. This is easily shown by contradiction. Let $(\bar{x}, \bar{p})$ be a competitive equilibrium and suppose that $(\bar{x}, \bar{p})$ is not a Pareto optimum. Then there is an $\overset{*}{x}$ such that $(\overset{*}{x}, \bar{p})$ satisfies (A.6.1) to (A.6.3) and $f^i(\overset{*}{x}{}^i) \geq f^i(\bar{x}^i)$, all $i = 1, \ldots, M$, and $f^k(\overset{*}{x}{}^k) > f^k(\bar{x}^k)$ for some k. Observe that (A.6.2) implies that $0 \leq \overset{*}{x}{}^k \leq W$. Hence for individual k, $\bar{x}^k$ is not the optimum to the problem

$$\max f^k(x^k)$$
$$\bar{p}x^k \leq \bar{p}w^k$$
$$0 \leq x^k \leq W \qquad (A.6.4)$$

given $\bar{p} \geq 0$, $\sum_{j=1}^{n} \bar{p}_j = 1$. This contradicts the fact that $(\bar{x}, \bar{p})$ is a competitive equilibrium.

INTERPRETATION

We have just shown that a competitive equilibrium point for the exchange economy is a Pareto-optimal point. This is highly desirable since surely we would want our concept of a solution to be *efficient*, meaning that no individual can be made better off without someone else being made worse off.

Although Pareto optimality is a very appealing concept, it is not satisfactory by itself. There are many Pareto optima aside from the competitive equilibria. For example, suppose that all goods are desirable and the price $p = (0, 0, \ldots, 0, 1)$. Consider the allocation

$$x_j^1 = W_j, \quad x_j^2 = \ldots = x_j^M = 0, \qquad j = 1, 2, \ldots, n - 1$$
$$x_n^i = w_n^i, \qquad\qquad\qquad\qquad i = 1, \ldots, M. \qquad (A.6.5)$$

In other words, all the first $n - 1$ goods are given to the first individual and no trading takes place on the last good. This allocation is Pareto optimal. The first individual cannot receive more of the last good since for $p = (0, 0, \ldots, 0, 1)$,

$$x_n^i \leq w_n^i, \qquad i = 1, \ldots, M,$$

is the budget constraint (A.6.1). But to take any of the first $n - 1$ goods from the first individual can only harm him or her. Hence (A.6.5) is Pareto optimal, yet clearly it is not a desirable allocation.

In sum, although Pareto optimality is a nice property and a competitive equilibrium possesses it, it is not by itself very satisfactory.

Warning

We also note that for an arbitrary EP there is no need for the equilibrium point $\bar{x}$ to be a Pareto point. That property is true for the exchange economy model (6.1.8), but it is not necessarily true for the extended EE model (6.3.8) or for other models. (See Exercises 3 and 4.)

EXERCISES/CHAPTER 6

1. Suppose that Alice has four loaves of bread and Bob has three bottles of wine. Alice and Bob have identical preferences of

   $$f(x_1, x_2) = x_1 x_2,$$

   where x_1 is the number of loaves of bread and x_2 is the number of bottles of wine. Formulate the EE (6.1.8) for this problem.

2. Two mining companies, Chem I and Chem II, trade in two crude minerals and one product processed from the crude minerals. Chem I has 150 tons of mineral A and 200 tons of mineral B, while Chem II has 100 tons of each mineral. It takes 2 tons of mineral A and 3 tons of mineral B to produce 1 ton of the final product. Because of limited resources, Chem I can process at most 30 tons of the final product, but Chem II can process any amount of it.

   Let $f^i(x_1^i, x_2^i, x_3^i)$, $i = 1, 2$ be the utility function of Chem I and Chem II, respectively, where

   $$x_1^i = \text{tons of crude mineral A}$$
   $$x_2^i = \text{tons of crude mineral B}$$
   $$x_3^i = \text{tons of final product.}$$

   a. Formulate the EE (6.3.8) for this problem.
   b. Formulate the EP corresponding to the EE formulated in part (a).

3. Consider the general EP model of Chapter 5. Define a *Pareto-optimal* point to be a point $\bar{x} \in D$ such that there is no $x \in D$ satisfying

   $$f^i(x) \geq f^i(\bar{x}) \qquad \text{for all } i = 1, \ldots, M$$
   $$f^k(x) > f^k(\bar{x}) \qquad \text{for some } k = 1, \ldots, M. \qquad \text{(E.6.1)}$$

Construct an example of an EP (but not an extended EE) that has an equilibrium point which is not a Pareto-optimal point.

4. Show by example that the competitive equilibrium to the extended EE (6.3.8) need not be a Pareto optimum.

5. Explain the difference between a competitive equilibrium and a Pareto optimum.

6. Prove that an equilibrium point of EP (6.3.9) is a competitive equilibrium of the EE (6.3.8).

7. Formulate an EP to illustrate the oligopolistic situation where, by holding goods off the market, a person can increase his or her utility.

8. a. Give a two-individual example where an individual may alter prices by withholding goods from the market.
   b. Explain why a competitive equilibrium is not an appropriate solution concept for this example.

9. a. Formulate the EP associated with Exercise 1.
   b. Find the competitive equilibrium to Exercise 1 by solving the K–T equations associated with the EP.

NOTES

The economic equilibrium model is central to economic theory. It deals with the oldest problem in economics, that of the exchange of commodities among individuals. The ideas derive from Edgeworth [1881] and Walras [1974]. More recently, valuable contributions have been made by Aumann [1966], Debreu [1970], and Shapley and Scarf [1974]. For references, see Arrow and Hahn [1971], Debreu [1959], Hildenbrand and Kirman [1976], and Scarf and Hansen [1973]. Extensions to abstract economies without ordered preferences have been made by Mas-Colell [1974b], Gale and Mas-Colell [1975], and Shafer and Sonnenschein [1975].

7

Obtaining the Competitive Equilibrium

Two variations of the economic equilibrium (EE) model were introduced in Chapter 6, and for each the competitive equilibrium, $(\bar{x}, \bar{p})$ is the equilibrium point for a corresponding equilibrium programming (EP) problem. This chapter focuses on the existence and calculation of the equilibrium point $(\bar{x}, \bar{p})$ itself. We begin with the Edgeworth box, an ingenious geometrical means of obtaining the equilibrium for two individuals. After that the case of M individuals is probed, and assuming that the utilities are concave, the equilibrium point is again determined.

7.1 THE EDGEWORTH BOX

The *Edgeworth box* is an elegant graphical means to analyze an EE for two individuals. It is a representation in which the two players are superimposed on one another in a special way, and using it we can actually calculate the equilibrium point. Here is this very nice technique.

A TWO-PERSON EXAMPLE

We go back to the very beginning of time when Adam and Eve were the only two individuals in the economy. We suppose that Adam and Eve grew only apples and oranges in their paradise orchard.

Let f^1 and f^2, their respective utilities, be continuous and concave, and let Adam's initial endowment be $w^1 = (w_1^1, w_2^1)$ and Eve's be $w^2 = (w_1^2, w_2^2)$. Here w_1^1 is the number of apples Adam has, and w_2^1 the number of oranges; and similarly for Eve.

To simplify the analysis, assume *greed* or *nonsatiety*, which means that utility increases as the amount of any good increases. In other words, neither Adam nor Eve will refuse more of a good. (Indeed, as any elementary school student knows, it was greed that led to Adam and Eve's downfall.) Greed ensures that an individual will always spend up to his or her budget, since spending less would not maximize utility. Thus given prices $p = (p_1, p_2)$, to maximize utility Adam will always seek $x^1 = (x_1^1, x_2^1)$ such that

$$px^1 = pw^1 \qquad (7.1.1)$$

and Eve will always seek $x^2 = (x_1^2, x_2^2)$ where

$$px^2 = pw^2. \qquad (7.1.2)$$

ADAM AND EVE'S PROBLEMS

First analyze Adam's problem, which for given prices $p = (p_1, p_2)$ is to solve

$$\max f^1(x_1^1, x_2^1)$$
$$p_1 x_1^1 + p_2 x_2^1 \leq p_1 w_1^1 + p_2 w_2^1$$
$$0 \leq x_1^1 \leq W_1, \qquad 0 \leq x_2^1 \leq W_2 \qquad (7.1.3)$$

for

$$W_1 = w_1^1 + w_1^2 \qquad \text{and} \qquad W_2 = w_2^1 + w_2^2.$$

This problem is depicted in Figure 7.1.1. In the figure the *level curves* (or *isoquants*) $f^1(x_1^1, x_2^1) = k$ are given for various k. The level curves denote *indifference curves*. That is, if two points x^1 and y^1 lie on the same curve, Adam is indifferent between x^1 and y^1. In the figure the value of the level curves is increasing to the upper right, since utility is increasing in that direction.

Also the *preference sets*

$$\{(x_1^1, x_2^1) \mid f^1(x_1^1, x_2^1) \geq k\}$$

are all (x_1^1, x_2^1) which give utility at least k. Since f^1 is concave, it is an easy exercise to show that the level curves give rise to convex preference sets as illustrated (Exercise 8).

In the figure, study particularly the *budget line*

$$B^1(p) = \{(x_1^1, x_2^1) \mid p_1 x_1^1 + p_2 x_2^1 = p_1 w_1^1 + p_2 w_2^1\}$$

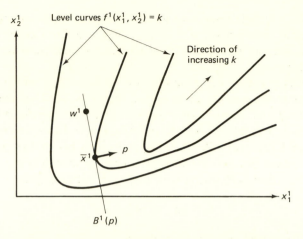

x_2^1 Level curves $f^1(x_1^1, x_2^1) = k$

Direction of
increasing k

w^1

$\overline{x}^1$ p

x_1^1

$B^1(p)$

Adam's problem for a given price p. The budget line $B^1(p)$
passes through w^1. The vector $\overline{x}^1$ optimizes (7.1.3) given
price p.

Figure 7.1.1

for a price p. Examine what happens to the budget line $B^1(p)$ if a different
value of p is given. Figure 7.1.2 illustrates that for two different prices, p_a and
p_b. See that the two budget lines $B^1(p_a)$ and $B^1(p_b)$ both go through w^1. In fact,
as p changes, the corresponding budget line rotates, but it always goes through
w^1.

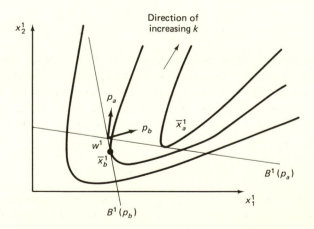

x_2^1 Direction of
increasing k

p_a

p_b $\overline{x}_a^1$

w^1

$\overline{x}_b^1$

$B^1(p_a)$

x_1^1

$B^1(p_b)$

Adam's problem (7.1.3). The budget lines $B^1(p_a)$ and $B^1(p_b)$
go through the point w^1. The point $\overline{x}_a^1$ optimizes (7.1.3) given
p_a, and the point $\overline{x}_b^1$ optimizes (7.1.3) given p_b.

Figure 7.1.2

Finally, notice that for each p there is a corresponding $\bar{x}^1$ which optimizes (7.1.3) at that particular p. Figure 7.1.2, in fact, shows how Adam would obtain his $\bar{x}^1$ for any given value of p.

EVE'S PROBLEM

Eve's problem is to find $\bar{x}^2$ that solves

$$\max f^2(x_1^2, x_2^2)$$
$$p_1 x_1^2 + p_2 x_2^2 \leq p_1 w_1^2 + p_2 w_2^2$$
$$0 \leq x_1^2 \leq W_1, \qquad 0 \leq x_2^2 \leq W_2 \tag{7.1.4}$$

given price p. Figure 7.1.3 illustrates this problem where the budget line for Eve is

$$B^2(p) = \{(x_1^2, x_2^2) \mid p_1 x_1^2 + p_2 x_2^2 = p_1 w_1^2 + p_2 w_2^2\}. \tag{7.1.5}$$

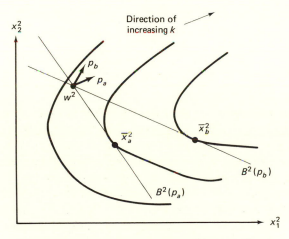

Eve's problem (7.1.4). The budget lines $B^2(p_a)$ and $B^2(p_b)$ go through the point w^2. The point $\bar{x}_a^2$ optimizes (7.1.4) given p_a, and the point $\bar{x}_b^2$ optimizes (7.1.4) given p_b.

Figure 7.1.3

Again as p changes, the budget line rotates, but it always goes through $w^2 = (w_1^2, w_2^2)$. Also, the optimizing point $\bar{x}^2$ to (7.1.4) changes as p changes. Moreover, all preference sets are convex. Eve's problem is thus precisely analogous to Adam's.

To solve the EE we must find a competitive equilibrium $\bar{p}$ and $\bar{x}^i$, $i = 1, 2$, that satisfies (7.1.3) and (7.1.4) simultaneously. The Edgeworth box is a convenient way to represent the situation geometrically and will help us find this equilibrium easily.

Suppose that together, Adam and Eve have 5 apples and 3 oranges (i.e., $W_1 = 5$, $W_2 = 3$). Notice that in Figure 7.1.4, we have drawn a box of width 5 and height 3. The horizontal denotes the total number of apples in the market, and the vertical the total number of oranges.

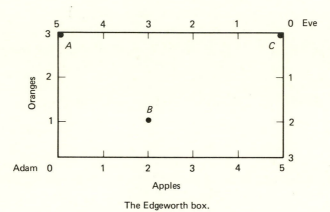

The Edgeworth box.
Figure 7.1.4

A marvelous property of the Edgeworth box is that any point inside the box represents not only what Adam possesses, but also simultaneously, Eve's possessions. To do this, measure Adam's units starting from the lower left corner. However, measure Eve's units from the upper right corner. For example, point B would denote that Adam has 2 apples and 1 orange, whereas Eve has 3 apples and 2 oranges. Notationally, for point B, $x^1 = (2, 1)$ and $x^2 = (3, 2)$. At point A Adam has only oranges ($x^1 = (0, 3)$) and Eve has only apples ($x^2 = (5, 0)$). For point C, $x^1 = (5, 3)$ and $x^2 = (0, 0)$. Note that $x^1 + x^2 = W = (5, 3)$ in all cases.

More generally in an Edgeworth box, the width is W_1 and height is W_2. Given any point x in the box, for Adam $x = x^1$ as measured from the lower left; for Eve, $x = x^2$ as measured from the upper right (and upside down). Further,

$$x^1 + x^2 = W,$$

so that always

$$x_i^1 = W_i - x_i^2, \qquad i = 1, 2.$$

Also, no point outside the box would be a feasible distribution of goods, since a negative amount would then be assigned to Adam or Eve.

Some point w in the box represents the initial endowments of Adam and also of Eve. For Adam (measuring from the lower left),

$$w = (w_1^1, w_2^1)$$

and for Eve (from the upper right),

$$w = (w_1^2, w_2^2).$$

Because the width of the box is W_1 and height W_2, and since $w^1 + w^2 = W$, the single point w represents both endowments simultaneously.

For instance, if point A were the initial endowment, then $w^1 = (0, 3)$ and $w^2 = (5, 0)$, which means Adam would be getting all the oranges and Eve all the apples.

DRAWING THE UTILITY FUNCTIONS

A continuous and concave $f^1(x_1^1, x_2^1)$ is the utility function of Adam, the level curves of which are depicted in Figure 7.1.5.

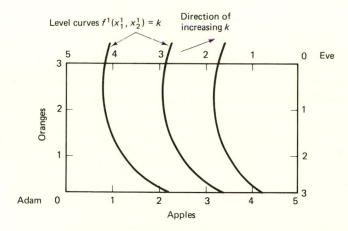

Adam's level curves for $f^1(x_1^1, x_2^1)$ drawn on the Edgeworth box.

Figure 7.1.5

Similarly, Eve has a utility function $f^2(x_1^2, x_2^2)$ continuous and concave. Eve's reference point, recall, is the upper right corner of the box. A typical graph of Eve's level curves $f^2(x_1^2, x_2^2) = k$ is drawn in Figure 7.1.6. Under the assumption of greed, the functions f^i are increasing in the direction shown.

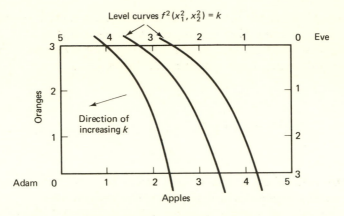

Eve's level curves $f^2(x_1^2, x_2^2) = k$ drawn on the Edgeworth box.

Figure 7.1.6

To emphasize, Adam and Eve produce similar graphs, except that one of the graphs is upside down.

BUDGET LINES

Next, for any $p_1 \geq 0$, $p_2 \geq 0$, $p_1 + p_2 = 1$, consider the budget lines B^1 and B^2 for Adam and Eve, respectively. Notice that in the Edgeworth box these two lines coincide (Figure 7.1.7).

To see this, let p be given and w be a point in the Edgeworth box corresponding to the endowments of Adam and Eve. Select $a = a^1$ a point in B^1.

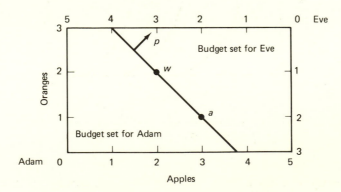

For any $p_1 \geq 0$, $p_2 \geq 0$, $p_1 + p_2 = 1$, the budget lines B^1 and B^2 for Adam and Eve coincide.

Figure 7.1.7

Then relative to Adam,

$$pa^1 = pw^1.\qquad(7.1.6)$$

Let $a = a^2$ be the same point relative to Eve's coordinates. Since

$$a^1 + a^2 = W = w^1 + w^2,$$

we have

$$pa^2 = pw^1 + pw^2 - pa^1,$$

which by (7.1.6) yields

$$pa^2 = pw^2.$$

Thus $a = a^2 \in B^2$.

In short, given an arbitrary point $a = a^1 \in B^1$, we see $a = a^2 \in B^2$. Reversing the reasoning, the converse also holds. Any point on one budget line is on the other. Consequently, in the Edgeworth box the two budget lines coincide and we write

$$B = B^1 = B^2,$$

where B is the budget line in the Edgeworth box.

Let us review the key features of the Edgeworth box:

1. A point x in the box can be expressed both in Adam's coordinates and in Eve's coordinates.

2. The point w in the box is both Adam's endowment w^1 and Eve's w^2.

3. The two budget lines coincide.

In the next section these remarkable properties permit us to easily calculate the competitive equilibrium.

7.2   OBTAINING THE COMPETITIVE EQUILIBRIUM

Suppose that $(\bar{x}^1, \bar{x}^2, \bar{p})$ is the competitive equilibrium for Adam and Eve. Let us underscore several facts about the equilibrium.

1. Since $0 \le \bar{x}^1 \le W$ and $0 \le \bar{x}^2 \le W$, both points $\bar{x}^1$ and $\bar{x}^2$ must be inside the Edgeworth box.

2. From (7.1.1) the point $\bar{x}^1$ must be on the budget line B^1, since $\bar{x}^1$ maxi-

mizes f^1. Similarly, $\bar{x}^2$ is on B^2. But in the box B^1 and B^2 coincide. Thus both $\bar{x}^1$ and $\bar{x}^2$ must be on the line B.

3.   For $i = 1, 2$, the level curves must be tangent to the budget line at $\bar{x}^i$. (This is because $\bar{x}^i$ maximizes f^i.) Thus at equilibrium the level curves of f^1 and f^2 must both be tangent to the line B.

4.   Finally note that (7.1.1), (7.1.2) and $\bar{x}^1 + \bar{x}^2 \leq W$ imply

$$\bar{p}_j(\bar{x}_j^1 + \bar{x}_j^2 - W_j) = 0, \qquad j = 1, 2.$$

Clearly, if $\bar{p}_j > 0$, then

$$\bar{x}_j^1 + \bar{x}_j^2 = W_j. \tag{7.2.1}$$

It turns out, and this will be shown shortly, that at a competitive equilibrium, $\bar{p} > 0$. Thus, by (7.2.1),

$$\bar{x}^1 + \bar{x}^2 = W.$$

Hence in the box, $\bar{x}^1$ and $\bar{x}^2$ must be the *same point* (see Figure 7.2.1).

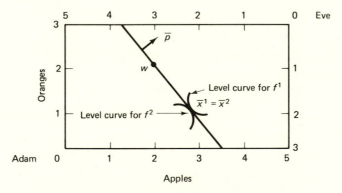

At competitive equilibrium, $\bar{x}^1$ and $\bar{x}^2$ coincide.

Figure 7.2.1

SUMMARY

These observations enable us to find a competitive equilibrium graphically using an Edgeworth box and let us summarize them:

At a competitive equilibrium $(\bar{x}^1, \bar{x}^2, \bar{p})$ the following must hold:

1. $\bar{x}^1$ and $\bar{x}^2$ must be on the budget line B of the Edgeworth box.

2.   The budget line B must be tangent to level curve f^1 at $\bar{x}^1$ and also to level curve f^2 at $\bar{x}^2$.

3. $\bar{p} > 0$ and the points $\bar{x}^1$ and $\bar{x}^2$ coincide.

Return to Example 6.1.1 and let us see how the Edgeworth box obtains the competitive equilibrium. Recall we had two corporations, one in the United States and another in Germany. The endowment for the German firm was $w_1^1 = 2$, $w_2^1 = 1$, and the U.S. firm's endowment was $w_1^2 = 2$, $w_2^2 = 2$. The utility of the German firm was $f^1(x^1) = 100 - (x_1^1 - 2)^2 - (x_2^1 - 3)^2$, and the U.S. firm's utility was $f^2(x^2) = 50 - (x_1^2 - 4)^2 - (x_2^2 - 2)^2$.

To find the competitive equilibrium, first draw the Edgeworth box (Figure 7.2.2). The width of the box is $w_1^1 + w_1^2 = 4$ units and the height is $w_2^1 + w_2^2 = 3$ units. The endowment w is at $(2, 1)$ relative to the German firm [and $(2, 2)$ relative to the U.S. firm].

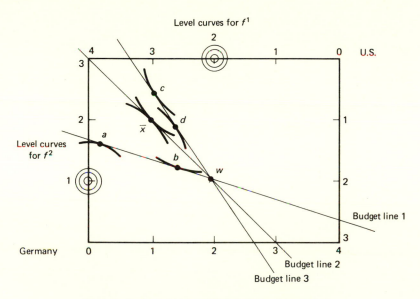

Obtaining the competitive equilibrium via the Edgeworth box.

Figure 7.2.2

Then draw the level curves for the German firm. By the way the utility was chosen, these are concentric circles centered at $(2, 3)$. The value of f^1 decreases as we move away from $(2, 3)$. Next, draw the level curves for the U.S. firm. These are concentric circles centered at $(4, 2)$ in U.S.'s coordinates. Also, f^2 decreases as we move away from $(4, 2)$.

To find the competitive equilibrium, draw budget lines through w. Start with a line through w with zero slope, and gradually rotate the line clockwise until it reaches slope $-\infty$. The line must always pass through w.

At each such line, consider the two points of tangency of the two utility-level curves, and seek for when both tangencies occur at the same point. When the slope is "too little," such as that of budget line 1, the U.S. firm's utility is too high and the German firm's utility is too low. This is because the German firm's maximum is at point b while the U.S. firm's is at point a.

As the budget line is rotated clockwise, the U.S. firm's utility decreases while the German firm's utility increases. Note that the points of tangency are getting closer to each other; that is, the German firm's optimum is moving away from w while the U.S. firm's optimum is moving toward w.

When the budget line is rotated "too much," such as at line 3, the German firm's optimum is at c and is farther from w than the U.S. firm's optimum, which is at point d. Since the points of tangency are gradually changing, they must have crossed somewhere between budget lines 1 and 3. Hence there must be a line 2 such that the two optima coincide. This is the competitive equilibrium point. It occurs at $\bar{p} = (\frac{1}{2}, \frac{1}{2})$, $\bar{x}^1 = (1, 2)$, and $\bar{x}^2 = (3, 1)$.

PROVING $\bar{p} > 0$

We now verify that under the greed assumption $\bar{p}_1 > 0$ and $\bar{p}_2 > 0$ at a competitive equilibrium.

Suppose not; then either $\bar{p} = (1, 0)$ or $\bar{p} = (0, 1)$. Let us suppose that $\bar{p} = (1, 0)$, as the reasoning is similar in the other case. Adam's problem then becomes

$$\max f^1(x_1^1, x_2^1)$$
$$x_1^1 \leq w_1^1$$
$$0 \leq x^1 \leq W. \tag{7.2.2}$$

Eve's problem becomes

$$\max f^2(x_1^2, x_2^2)$$
$$x_1^2 \leq w_1^2$$
$$0 \leq x^2 \leq W. \tag{7.2.3}$$

Under the assumption of greed, the solution to (7.2.2) is $\bar{x}^1 = (w_1^1, W_2)$. Similarly under (7.1.2) the solution to (7.2.3) is $\bar{x}^2 = (w_1^2, W_2)$. But this result cannot be a competitive equilibrium, since the constraint

$$\bar{x}^1 + \bar{x}^2 \leq W$$

is violated. Therefore, $\bar{p} > 0$ at a competitive equilibrium.

The Edgeworth box is extremely useful for analyzing the behavior of a competitive equilibrium, but of course is restricted to two individuals. The more general situation is considered next.

7.3 EQUILIBRIUM FOR *M* INDIVIDUALS

Let us now prove that for *M* individuals the competitive equilibrium exists and can be calculated. [Note that the exchange economy model (6.1.8) is considered here, as the same ideas apply to the extended model, (6.3.8) and is left as Exercises 9 and 10.] The main provisions required are that the utility function f^i of the *M* individuals be concave and that all the initial endowments be strictly positive,

$$w_j^i > 0, \qquad \text{all } i \text{ and } j. \tag{7.3.1}$$

THE INITIAL POINT

The idea is to apply the procedure of Chapter 5 to the EE problem. First find an initial point $(\mathring{x}, \mathring{p})$ satisfying

(A) $$px^i < pw^i$$

(B) $$0 < x^i < W, \qquad i = 1, \ldots, M$$

and

(C) $$\sum_{j=1}^n p_j = 1, \qquad p > 0. \tag{7.3.2}$$

Such a point will be the starting point of the path for the EE. For example, we can take

$$\mathring{p} = \left(\frac{1}{n}, \frac{1}{n}, \ldots, \frac{1}{n}\right)$$

and

$$\mathring{x}^i = \tfrac{1}{2}w^i, \qquad i = 1, \ldots, M. \tag{7.3.3}$$

Clearly, $(\mathring{x}, \mathring{p})$, as defined by (7.3.3), satisfies (7.3.2).

EP FORMULATION

Given $(\mathring{x}, \mathring{p})$ satisfying (7.3.2), next formulate the dynamic EP

(a) For $i = 1, \ldots, M$:

$$\max_{x^i} \; -\frac{(1-t)}{2} \|x^i - \mathring{x}^i\|^2 + tf^i(x^i)$$

$$px^i \leq pw^i$$

$$0 \leq x^i \leq W \tag{7.3.4}$$

given *p*.

(b)
$$\max -\frac{(1-t)}{2}\|p - \mathring{p}\|^2 + tp\left(\sum_{i=1}^{M} x^i - W\right)$$

$$\sum_{j=1}^{n} p_j = 1, \qquad p \geq 0$$

given x.

The process then is exactly as presented in Chapter 5. Starting from $(\mathring{x}, \mathring{p})$ at $t = 0$, follow the corresponding path as t increases from zero to one. At $t = 1$, we then have a competitive equilibrium.

Observe that the CQ (5.6.2) holds for the point $(\mathring{x}, \mathring{p})$. To see this, for subproblems $i = 1, \ldots M$ all constraints can be rewritten in the form $g_j^i(x) \geq 0$ and by (7.3.2A, B) hold with strict inequality at $\mathring{x}^i$. For subproblem $M + 1$, $\mathring{p} = g^{M+1}(\mathring{x}, \mathring{p}) > 0$; and $h^{M+1} = \sum p_j - 1$, so $\nabla_p h^{M+1}$ is a vector of one's which is, of course, linearly independent (Exercise 7). Also the set D of (5.6.1) is compact. Therefore, we can directly apply Theorem 5.6.2 to yield

---

THEOREM 7.3.1   Let $w_j^i > 0$ for all i and j and suppose that the functions f^i are concave and C^3 for all $i = 1, \ldots, M$. Then if the system of Kuhn–Tucker equations (5.6.8) corresponding to (7.3.4) is regular, there exists an equilibrium point $(\bar{x}, \bar{p})$ to the EE.

Proof.   The proof is immediate from Theorem 5.6.2. Simply notice that $w_j^i > 0$ ensures $(\mathring{x}, \mathring{p})$ can be chosen to satisfy (7.3.2), so CQ (5.6.2) holds. All the other hypotheses of Theorem 5.6.2 are also seen to hold.   □

Comment

The proof of Theorem 7.3.1 makes direct use of the path-following approach. As we see in Chapter 21, this is markedly different from the "classical" approach, which transforms the EP into a fixed-point problem. Further note that the regularity condition is easily eliminated in Chapter 22.

SUMMARY

This chapter presented the proofs necessary to validate that the competitive equilibrium exists and can be calculated. Essentially, concavity of the utility functions and positive endowments are all that is required. For the special case of two individuals, the Edgeworth box provides a clever and very helpful technique to calculate the equilibrium. Overall, with the EE model we obtain a powerful means to analyze a wide variety of economic situations.

1. Obtain the competitive equilibrium of Chapter 6, Exercise 1, by using the Edgeworth box.

2. A vector of goods $x = (x^i)$ is Pareto *superior* to another goods vector x' if at least one individual prefers x to x' and all others are indifferent between x and x'. Goods vector x is Pareto *indifferent* to x' if all individuals are indifferent between x and x'. Goods vector x is Pareto *optimal* if there is no possible goods vector that is Pareto superior to it.
   a. Show by the Edgeworth box a diagram with many Pareto-optimal points.
   b. Show a diagram with only one Pareto-optimal point, a diagram with two Pareto-optimal points, and a diagram with no Pareto-optimal point.

3. a. Show by the Edgeworth box that a Pareto-optimal point need not be Pareto superior to a point that is not Pareto optimal.
   b. Show that if there are two Pareto-optimal points, they are either Pareto indifferent or incomparable.

4. Show using the Edgeworth box that a Pareto optimum in which one individual consumes none of one good (the point x lying on the boundary of the Edgeworth box) might not be attainable as competitive equilibrium using nonnegative prices.

5. a. Formulate the K–T equations of the EP corresponding to Example 6.1.1.
   b. Show that the K–T equations are satisfied at competitive equilibrium.

6. Discuss the difficulties that would arise in the Edgeworth box if the assumption of greed were removed.

7. Prove in detail that the CQ (5.6.2) holds for EP (7.3.4) and that the set D of (5.6.1) is compact.

8. Prove that if an individual's utility $f(x)$ is concave, then the preference set
$$\{x \mid f(x) \geq k\}$$
   is convex for any given k.

9. Show how to solve the extended economic equilibrium model of Chapter 6 using:
   a. An Edgeworth box for two individuals.
   b. Equilibrium programming.
   What assumptions are needed?

10. Generalize Theorem 7.3.1 to include the exchange economy model (6.3.8).

NOTES

7.1. The Edgeworth box is due to a development by Bowley of an idea of Edgeworth and should properly be referred to as an Edgeworth–Bowley box. However, we have used the accepted practice of calling it the Edgeworth box.

7.3.   Zangwill and Garcia [1981] proved the existence of competitive equilibrium via the EP approach of Chapter 5. The classical approach, however, makes use of a fixed-point theorem. For example, see Arrow and Hahn [1971], Debreu [1959], and Hildenbrand and Kirman [1976]. Constructive approaches are shown in Scarf and Hansen [1973] and Todd [1976a] by invoking the Kakutani fixed-point theorem. Applications may be found in Shoven [1977] and Whalley [1977].

8

Game Theory

Game theory is the mathematical analysis of conflict and strategy, and probes what occurs when two or more people differ in their goals. It is very wide ranging because a game may simply be a parlor game, such as chess or bridge, or may reflect deep conflicts that arise in social, economic, military, or political situations. In a game, each person selects a strategy or action to take. The game is then played and each player receives a "payoff" dependent upon the success of the strategy he or she has selected.

We shall present game theory as a most interesting application of equilibrium programming. First, the two-person game and the concept of a mixed strategy will be discussed, and then the general M-person game will be investigated. Finally, we consider some difficulties with the equilibrium concept itself and, in particular, two examples where dilemmas can arise.

8.1 TWO-PERSON GAMES

There are a myriad of examples of two-person games, including chess, checkers, GO, backgammon, and tic-tac-toe. In a two-person game, each person i, $i = 1$, 2, chooses from among n_i *strategies*. A strategy is basic to game theory and describes a full sequence of moves that a person can make in any single game. In chess, for example, the sequence of moves made by a player in one game would comprise one strategy. The sequence of moves made in another game would be another strategy. Let strategy j denote a particular sequence of moves in a game by a player. Most games will have a large number of possible strategies j. (Player i has n_i possible strategies.) Nevertheless, we assume that strategy j describes the moves taken in a specific game by a player.

A *play* of the game is a particular selection (j, k) of the strategies where player 1 uses his strategy j and player 2 uses his strategy k. Once player 1 has chosen strategy j and player 2 has chosen k, the moves and outcome of the game are determined. Important to remember is that once a play (j, k) is selected, that determines the *payoffs* to the two players.

Payoffs are the wins or losses incurred due to the particular play. It will be convenient to record payoffs as the elements in the matrices $A^1 = (A^1_{jk})$ and $A^2 = (A^2_{jk})$ of dimension $n_1 \times n_2$. Explicitly, we let

$$A^i_{jk}, \qquad i = 1, 2 \tag{8.1.1}$$

denote the payoff to player i if the play is (j, k). This means that if player 1 selects strategy j and player 2 selects k, the payoff to 1 is A^1_{jk} and to 2 is A^2_{jk}. Should $A^1_{jk} > 0$ and $A^2_{jk} < 0$, player 1 wins amount A^1_{jk} and player 2 loses amount $-A^2_{jk}$.

A game where $A^2_{jk} = -A^1_{jk}$ for all j, k is called *zero-sum* $(A^1 + A^2 = 0)$, which means the winnings of one player will be the losses of the other player. Card games are often zero-sum. However, not all games are zero-sum, as we will see in the examples.

EXAMPLE 8.1.1

Husband and wife are repainting their newly bought apartment. Wife desires that the bedroom and living room both be painted white. Husband, being more flamboyant, agrees that the living room be painted white but wants the bedroom painted fire-engine red, which wife thinks is simply atrocious. To settle this issue, husband is allowed to pick the color (either red or white) for the bedroom while wife decides on the color (red or white) of the living room.

There are four possible outcomes that can arise. On a scale of 1 to 5, husband rates his pleasure as

Bedroom \ Living Room	White	Red
White	1	2
Red	5	4

In other words, he would like best to have a red bedroom and white living room, but he would rather have a red apartment than an all-white apartment, which he feels is too drab-looking. On the other hand, wife's pleasure ratings are

Bedroom \\ Living Room	White	Red
White	5	1
Red	1	2

She wants an all-white apartment, but if this is not possible, she would rather have it all-red than two-toned.

THE PAYOFF MATRICES   Designate husband (H) as the first player, and he selects the bedroom color. He has two strategies: strategy 1—paint the bedroom white; strategy 2—paint the bedroom red. Wife (W) is the second player and selects the living room color. Her two strategies are: strategy 1—paint the living room white; strategy 2—paint the living room red.

The payoff matrices A^1 and A^2 are

$$A^1 = \begin{pmatrix} A^1_{11} & A^1_{12} \\ A^1_{21} & A^1_{22} \end{pmatrix} = \begin{pmatrix} 1 & 2 \\ 5 & 4 \end{pmatrix}$$

and

$$A^2 = \begin{pmatrix} A^2_{11} & A^2_{12} \\ A^2_{21} & A^2_{22} \end{pmatrix} = \begin{pmatrix} 5 & 1 \\ 1 & 2 \end{pmatrix}. \tag{8.1.2}$$

For example, suppose that H selects strategy 1 (paint bedroom white) and W selects strategy 2 (paint living room red). Then he receives A^1_{12} and she receives A^2_{12}. That is, he receives a pleasure of 2 and she receives a pleasure of 1 because the bedroom is white and the living room is red. If H selects red and W white, he receives 5 and she receives 1.

THE EQUILIBRIUM PLAY

A play $(\bar{j}, \bar{k})$ is *equilibrium* if

$$A^1_{\bar{j}k} \geq A^1_{jk} \qquad \text{for } j = 1, 2, \ldots, n_1$$
$$A^2_{\bar{j}\bar{k}} \geq A^2_{\bar{j}k} \qquad \text{for } k = 1, 2, \ldots, n_2. \tag{8.1.3}$$

In more detail, suppose that $(\bar{j}, \bar{k})$ is an equilibrium. Then given that player 2 selects strategy $\bar{k}$, player 1 will maximize his payoff by selecting strategy $\bar{j}$.

Similarly, given that player 1 selects $\bar{j}$, player 2 will maximize his payoff by selecting strategy $\bar{k}$.

Notice that the equilibrium is stable. Suppose that player 2 selects $\bar{k}$. Then no strategy for player 1 is better than $\bar{j}$. In fact, player 1 has no incentive to move from $\bar{j}$. Similarly, if player 1 selects $\bar{j}$, player 2 has no incentive to move from $\bar{k}$. The equilibrium $(\bar{j}, \bar{k})$ is therefore stable. No player can unilaterally improve upon it.

EQUILIBRIUM FOR EXAMPLE 8.1.1

Let us obtain the equilibrium for the game example (8.1.1). First examine play (1, 1), which means that H does white and W does white. Clearly, (1, 1) is not an equilibrium since

$$5 = A_{21}^1 > A_{11}^1 = 1;$$

that is, H can unilaterally select red and improve his position.

The play (2, 2) is an equilibrium, however. Once H selects 2 (paint red), the best strategy for W is 2 (paint red). Similarly, if W selects 2, the best for H is 2. We have obtained an equilibrium and it is stable. Neither party has an incentive to move from (2, 2).

NO EQUILIBRIUM PLAY

Lest we get to the habit of thinking that games have an equilibrium play always, consider the next example.

EXAMPLE 8.1.2   The Bears–Cowboys Big Game

The long-awaited football game between the Bears and the Cowboys is now going on at Soldier Field. It is the last play of the last game of the season and both teams' chances for the playoffs hinge on this final play.

The Bears have the ball on the Cowboys' 1-yard line. Both the Bear quarterback and the Cowboy defensive captain are calling the signals, and clearly, with this play their careers are made or ended. The Bear quarterback is undecided whether to run or pass, whereas the Cowboy defensive captain is undecided whether to use the flex defense or the 3–4 defense.

Both teams are clients of a well-known statistical firm based in the Chicago Loop. The firm supplies the two teams weekly with volumes of statistical data. Page 1099 of this week's data reads: If the Bears run the ball and the Cowboys use the flex defense, or if the Bears pass and the Cowboys use the 3–4 defense, the Bears are expected to gain 1 yard. Otherwise, they are expected to lose 1 yard.

Both coaches are total believers in the Chicago statistical firm and have in fact ordered their respective teams to memorize the data, *especially* page 1099.

What play should the Bear quarterback call? What play should the Cowboy defense captain call?

FORMULATION Let us formulate the situation as a game and determine the equilibrium play, if one exists. Specify the Bear quarterback as player 1, where for him strategy 1 is run and strategy 2 is pass. The Cowboys' defensive captain is player 2, and his strategies are the two possible defenses, strategy 1 being the flex, strategy 2 being the 3–4.

The payoff matrices then are

$$A^1 = \begin{pmatrix} 1 & -1 \\ -1 & 1 \end{pmatrix} \quad \text{and} \quad A^2 = \begin{pmatrix} -1 & 1 \\ 1 & -1 \end{pmatrix}. \tag{8.1.4}$$

For example, if the Bears run and the Cowboys are in a 3–4, $A^1_{12} = -1$ and $A^2_{12} = +1$, as the Bears lose 1 yard. However, $A^2_{22} = -1$, since the Cowboys will lose 1 yard if the Bears pass and the Cowboys are in a 3–4. Note also that $A^1 + A^2 = 0$, which means that this game is zero-sum. Any yard the Bears gain is a yard the Cowboys lose.

Examine the payoff matrices and see that there is no equilibrium strategy. Suppose that the Bears use strategy 1 (run). The Cowboys should counter with strategy 2 (3–4 defense), as then the Cowboys gain 1 yard. But should the Cowboys select strategy 2, the Bears should implement 2. However, if the Bears do 2, Cowboys should do 1. Yet if the Cowboys do 1, Bears should do 1. But notice that the entire cycle then repeats. If the Bears do 1, Cowboys should do 2, and so on. No pair of strategies is stable or, in mathematical terms, no strategy pair $(\bar{j}, \bar{k})$ exists that satisfies (8.1.3). Thus there is no equilibrium play.

8.2 MIXED STRATEGIES

The reason why there is no equilibrium play for game 8.1.2 is because the strategies we have used until now are *pure strategies*. That is, player 1 selects a single pure strategy $\bar{j}$ and player 2 selects a single pure strategy $\bar{k}$. Pure strategies are too restrictive, and we must expand the possibilities of playing the game.

John von Neumann, one of the greatest mathematicians of this century, introduced the notion of a *mixed strategy*. Under that, the players do not select a single pure strategy, but rather the strategy is determined by a game of chance.

Assume that for player 1, some chance device selects strategy j with *probability* $x^1_j \geq 0$, $\sum_{j=1}^{n_1} x^1_j = 1$. Similarly, let another chance device select strategy k for player 2 with probability $x^2_k \geq 0$, $\sum_{k=1}^{n_2} x^2_k = 1$. The chance device for player 1 could be, say, an *urn* with a^1 white balls, a^2 yellow, ..., a^{n_1} black balls. Associate white balls with strategy 1, yellow balls with strategy 2, ..., black balls with strategy n_1, and pick one ball from the urn. Then strategy j is selected with probability $a^j / \sum a^k$. The urn thus permits a player to select different strategies with different probabilities. More specifically, call $x^i = (x^i_1,$

$\dots, x_{n_i}^i)$ the strategy vector of player i, where

$$\sum_{j=1}^{n_i} x_j^i = 1, x_j^i \geq 0, \qquad j = 1, \dots, n_i, \quad i = 1, 2.$$

Then x_j^i is the probability that player i selects strategy j.

When each player uses mixed strategies, the payoff to the players can no longer be precisely determined. All that can be calculated are *expected* payoffs.

Given strategy vectors x^1 for player 1 and x^2 for player 2, define

$$f^1(x^1, x^2) = (x^1)^\tau A^1 x^2 = \sum_{j=1}^{n_1} \sum_{k=1}^{n_2} x_j^1 A_{jk}^1 x_k^2$$

$$f^2(x^1, x^2) = (x^1)^\tau A^2 x^2 = \sum_{j=1}^{n_1} \sum_{k=1}^{n_2} x_j^1 A_{jk}^2 x_k^2, \qquad (8.2.1)$$

where A^1 and A^2 are the payoff matrices for the two players, respectively. We term f^i the (expected) *payoff function* for player i. To explain the expected payoff functions, consider Example 8.1.2 and several pairs of strategy vectors $x = (x^1, x^2)$.

Suppose that $x^1 = (1, 0)$ and $x^2 = (0, 1)$, meaning that the Bears run and the Cowboys go to a 3–4 defense. In this case both teams chose a pure strategy and the payoff by (8.2.1) is

$$f^1(x^1, x^2) = A_{12}^1 = -1 \qquad \text{to the Bears}$$
$$f^2(x^1, x^2) = A_{12}^2 = +1 \qquad \text{to the Cowboys.}$$

Now consider $x^1 = (1, 0)$, $x^2 = (\frac{1}{3}, \frac{2}{3})$. Again the Bears are in a pure strategy, the run. The Cowboys, however, have a mixed strategy, choosing the flex defense with probability $\frac{1}{3}$, and the 3–4 defense with probability $\frac{2}{3}$. Here

$$f^1(x^1, x^2) = \tfrac{1}{3}A_{11}^1 + \tfrac{2}{3}A_{12}^1 = \tfrac{1}{3}(1) + \tfrac{2}{3}(-1) = -\tfrac{1}{3}$$
$$f^2(x^1, x^2) = \tfrac{1}{3}A_{11}^2 + \tfrac{2}{3}A_{12}^2 = \tfrac{1}{3}(-1) + \tfrac{2}{3}(+1) = +\tfrac{1}{3}.$$

Finally, let $x^1 = (\frac{1}{4}, \frac{3}{4})$, $x^2 = (\frac{1}{3}, \frac{2}{3})$, so that both teams are randomly choosing their strategies. Then

$$f^1(x^1, x^2) = \tfrac{1}{4}\tfrac{1}{3}A_{11}^1 + \tfrac{1}{4}\tfrac{2}{3}A_{12}^1 + \tfrac{3}{4}\tfrac{1}{3}A_{21}^1 + \tfrac{3}{4}\tfrac{2}{3}A_{22}^1$$
$$= \tfrac{1}{12}(1) + \tfrac{2}{12}(-1) + \tfrac{3}{12}(-1) + \tfrac{6}{12}(+1) = +\tfrac{1}{6}$$
$$f^2(x^1, x^2) = \tfrac{1}{4}\tfrac{1}{3}A_{11}^2 + \tfrac{1}{4}\tfrac{2}{3}A_{12}^2 + \tfrac{3}{4}\tfrac{1}{3}A_{21}^2 + \tfrac{3}{4}\tfrac{2}{3}A_{22}^2 = -\tfrac{1}{6}.$$

More generally, denote e^j as the jth unit vector (a 1 in the jth position and zeros elsewhere). Suppose that $x^1 = e^j$ and $x^2 = e^k$, so that player 1 is using pure strategy j and player 2 is using pure strategy k. Then

$$f^i(x^1, x^2) = A^i_{jk},$$

which yields the same result as before under pure strategies, as it must.

Clearly, however, mixed strategies permit greater latitude. Indeed, for $x = (x^1, x^2)$, where x^1 and x^2 are strategy vectors, we see that

$$f^i(x^1, x^2) = \sum_{j=1}^{n_1} \sum_{k=1}^{n_2} x^1_j A^i_{jk} x^2_k$$

is the expected payoff to player i because any particular payoff A^i_{jk} is weighted by the probability that payoff will occur, $x^1_j x^2_k$.

EQUILIBRIUM POINT

Since the definition (8.1.3) pertained only to pure strategies, we must generalize the definition of equilibrium to include mixed strategies. We say that $\bar{x} = (\bar{x}^1, \bar{x}^2)$ is an *equilibrium point* in the two-person game if:

(A) given $\bar{x}^2$,

$$f^1(\bar{x}^1, \bar{x}^2) = \max_{x^1} f^1(x^1, \bar{x}^2) = (x^1)^{\tau} A^1 \bar{x}^2$$

subject to

$$\sum_{j=1}^{n_1} x^1_j = 1$$

$$x^1 \geq 0 \qquad\qquad (8.2.2)$$

and

(B) given $\bar{x}^1$,

$$f^2(\bar{x}^1, \bar{x}^2) = \max_{x^2} f^2(\bar{x}^1, x^2) = (\bar{x}^1)^{\tau} A^2 x^2$$

subject to

$$\sum_{k=1}^{n_2} x^2_k = 1$$

$$x^2 \geq 0.$$

This definition of equilibrium is precisely analogous to that in Chapter 5, and in fact, (8.2.2) is an EP with two subproblems. Given $\bar{x}^2$, player 1 will choose $\bar{x}^1$, and given $\bar{x}^1$, player 2 will choose $\bar{x}^2$. Moreover, $\bar{x} = (\bar{x}^1, \bar{x}^2)$ is stable. The players have no motivation to move from $\bar{x}$ if they are already at $\bar{x}$.

Observe that a play $(\bar{j}, \bar{k})$ which is an equilibrium according to definition (8.1.3) is also an equilibrium relative to (8.2.2). That is, if a pure strategy is an equilibrium, it will also be an equilibrium under mixed strategies (Exercise 1).

EXAMPLE 8.1.2   Revisited

The timeout is over, and let us now return to the Bears–Cowboys football game. Observe that $\bar{x}^1 = (\frac{1}{2}, \frac{1}{2})$, $\bar{x}^2 = (\frac{1}{2}, \frac{1}{2})$ is equilibrium, since then by (8.1.4) $A^1\bar{x}^2 = (\bar{x}^1)^\tau A^2 = 0$. Hence we have

$$(x^1)^\tau A^1 \bar{x}^2 \le (\bar{x}^1)^\tau A^1 \bar{x}^2$$

and

$$(\bar{x}^1)^\tau A^2 x^2 \le (\bar{x}^1)^\tau A^2 \bar{x}^2$$

for any mixed strategies x^1, x^2. It can be easily checked that $\bar{x}$ is the unique equilibrium for this game (Exercise 2).

Let us now argue that the Bear quarterback ought to call $\bar{x}^1$. In other words, he should bring out the coin that he has taped onto his helmet, and flip the coin in the huddle (at the expense of the snickering of his teammates and the roar of laughter from the crowd). If the coin turns out heads, he calls pass; otherwise, he calls run.

To see that $\bar{x}^1 = (\frac{1}{2}, \frac{1}{2})$ is his "best" strategy, suppose that he calls out a strategy $\hat{x}^1$ different from $\bar{x}^1$. Then, from (8.1.4),

$$(\hat{x}^1)^\tau A^2 x^2 = (\hat{x}_1^1 - \hat{x}_2^1)(x_2^2 - x_1^2).$$

Since $\hat{x}^1 \ne \bar{x}^1$, clearly $\hat{x}_1^1 - \hat{x}_2^1 \ne 0$. Therefore, if the Cowboys choose the appropriate x^2, the expected payoff to the Bears can be made negative, and the Bears would be more likely to lose a yard than gain a yard.

Using the same reasoning, it can be argued that the Cowboys, defense ought to play $\bar{x}^2$. If they do not, the Bears could choose an x^1 that provides the Bears with a positive payoff.

EXISTENCE OF AN EQUILIBRIUM POINT

An equilibrium point exists for the two-person game (8.2.2). Although that easily follows from Theorem 5.6.2, rather than verifying it here, in the next section we prove equilibrium existence for M-person games. Equilibrium existence for the two-person game will then follow immediately.

ZERO-SUM GAMES

For the two-person zero-sum game the equilibrium point $\bar{x}$ turns out to be a saddle point. Specifically, $\bar{x} = (\bar{x}^1, \bar{x}^2)$ is a *saddle point for* A^1 if for all strategy vectors x^1 and x^2

$$(x^1)^\tau A^1 \bar{x}^2 \le (\bar{x}^1)^\tau A^1 \bar{x}^2 \le (\bar{x}^1)^\tau A^1 x^2. \tag{8.2.3}$$

We know by (8.2.2a) that if $\bar{x}$ is an equilibrium point

$$(\bar{x}^1)^\tau A^1 \bar{x}^2 \ge (x^1)^\tau A^1 \bar{x}^2, \tag{8.2.4}$$

and, by (8.2.2b),

$$(\bar{x}^1)^\tau A^2 \bar{x}^2 \geq (\bar{x}^1)^\tau A^2 x^2. \tag{8.2.5}$$

But $A^2 = -A^1$ in a zero-sum game, so (8.2.3) holds. Thus for the zero-sum game an equilibrium point is also a saddle point. Moreover, because the equilibrium point exists, so does a saddle point. This fact that a saddle point exists was originally discovered by von Neumann and is often called the *fundamental theorem of two-person zero-sum games.*

Be clear that any two-person game (8.2.2) possesses an equilibrium point. Moreover, if the two-person game is zero-sum, the equilibrium point is also a saddle point.

8.3 *M*-PERSON GAMES

In two-person games each player has n_i possible strategies, $i = 1, 2$. Also, a play of the game (j, k) resulted when both players selected a specific strategy for actual use. *M*-person games are precisely analogous except that there are *M* players. Each person i has n_i possible strategies, $i = 1, \ldots, M$. A play of the game occurs when each player selects a specific strategy $j_i, i = 1, \ldots, M$. The vector $(j_1, \ldots, j_M)$ then mathematically describes all the strategies and is called a *play* of the game because it enumerates the strategies of all *M* players.

As for the payoffs to the *M* players, these are given by *M* matrices

$$A^1, A^2, \ldots, A^M, \tag{8.3.1}$$

each of dimension $n_1 \times n_2 \times \ldots \times n_M$. For matrix A^i a particular entry is denoted $A^i_{j_1, \ldots, j_M}$. Given the play $(j_1, \ldots, j_M)$, the payoff to player i is

$$A^i_{j_1, \ldots, j_M}, \qquad i = 1, \ldots, M. \tag{8.3.2}$$

Thus the payoff to player i is simply the $(j_1, \ldots, j_M)$th element of A^i.

To consider mixed strategies, let $x^i = (x^i_1, \ldots, x^i_{n_i})$, where

$$\sum_{j=1}^{n_i} x^i_j = 1 \qquad x^i_j \geq 0, \quad \text{all } j$$

be a strategy vector for player i. Explicitly, player i plays strategy j with probability x^i_j. Then given $x = (x^1, \ldots, x^M)$, where x^i is a strategy vector, the expected payoff to player i is

$$f^i(x^1, x^2, \ldots, x^M) = \sum_{j_1=1}^{n_1} \sum_{j_2=1}^{n_2} \sum_{j_M=1}^{n_M} A^i_{j_1, \ldots, j_M} x^1_{j_1} x^2_{j_2}, \ldots, x^M_{j_M}. \tag{8.3.3}$$

The expected payoff f^i is a direct extension of the two-person-game case. More precisely, $x^1_{j_1} x^2_{j_2}, \ldots, x^M_{j_M}$ is the probability that player 1 does strategy j_1, player

2 does strategy j_2, player 3 does j_3, and so on. And if the players do those specific strategies, the payoff is $A^1_{j_1,\ldots,j_M}$. Consequently, f^i weighs each possible outcome $A^i_{j_1,\ldots,j_M}$ by its probability of occurrence, $x^1_{j_1} x^2_{j_2}, \ldots, x^M_{j_M}$.

EXAMPLE 8.3.1

As an example, suppose that there are three players, where player 1 has three strategies, player 2 has two strategies, and player 3 has two strategies. Then $n_1 = 3, n_2 = 2$, and $n_3 = 2$ and the payoff matrices A^i, $i = 1, 2, 3$, are thus $3 \times 2 \times 2$ in dimension.

Let us suppose that A^1 is as follows:

$$A^1_{111} = 2, \quad A^1_{112} = \quad 0, \quad A^1_{121} = -1, \quad A^1_{122} = 0$$
$$A^1_{211} = 1, \quad A^1_{212} = -1, \quad A^1_{221} = \quad 0, \quad A^1_{222} = 0$$
$$A^1_{311} = 0, \quad A^1_{312} = \quad 1, \quad A^1_{321} = -2, \quad A^1_{322} = 0.$$

For example, if player 1 uses strategy 3, player 2 uses strategy 1, and player 3 uses strategy 2, the payoff to player 1 is $A^1_{312} = 1$.

More generally, the players will select mixed strategies, and let us suppose that the strategy vectors are $x^1 = (\frac{1}{3}, \frac{1}{2}, \frac{1}{6})$, $x^2 = (\frac{1}{2}, \frac{1}{2})$, and $x^3 = (\frac{1}{4}, \frac{3}{4})$. Then the expected payoff to player 1 via (8.3.3) becomes

$$f^1(x^1, x^2, x^3) = \tfrac{1}{3}\tfrac{1}{2}\tfrac{1}{4}A^1_{111} + \tfrac{1}{3}\tfrac{1}{2}\tfrac{3}{4}A^1_{112} + \tfrac{1}{3}\tfrac{1}{2}\tfrac{1}{4}A^1_{121} + \tfrac{1}{3}\tfrac{1}{2}\tfrac{3}{4}A^1_{122}$$
$$+ \tfrac{1}{2}\tfrac{1}{2}\tfrac{1}{4}A^1_{211} + \tfrac{1}{2}\tfrac{1}{2}\tfrac{3}{4}A^1_{212} + \tfrac{1}{2}\tfrac{1}{2}\tfrac{1}{4}A^1_{221} + \tfrac{1}{2}\tfrac{1}{2}\tfrac{3}{4}A^1_{222}$$
$$+ \tfrac{1}{6}\tfrac{1}{2}\tfrac{1}{4}A^1_{311} + \tfrac{1}{6}\tfrac{1}{2}\tfrac{3}{4}A^1_{312} + \tfrac{1}{6}\tfrac{1}{2}\tfrac{1}{4}A^1_{321} + \tfrac{1}{6}\tfrac{1}{2}\tfrac{3}{4}A^1_{322}$$
$$= \tfrac{1}{24}(2) + 0 + \tfrac{1}{24}(-1) + 0$$
$$+ \tfrac{1}{16}(1) + \tfrac{3}{16}(-1) + 0 + 0$$
$$+ 0 + \tfrac{3}{48}(1) + \tfrac{1}{48}(-2) + 0$$
$$= -\tfrac{3}{48}.$$

Similarly, A^2 and A^3 will each have 12 entries, and given explicit representations for A^2 and A^3, the expected payoffs to players 2 and 3 can be calculated using (8.3.3).

EQUILIBRIUM POINT

The solution concept of the M-person game is due to John F. Nash. Specifically, $\bar{x}$ is called a (Nash) *equilibrium point* if $\bar{x} = (\bar{x}^i, \bar{x}^i)$ solves the M subproblems for $i = 1, \ldots, M$:

$$f^i(\bar{x}^i, \bar{x}^i) = \max_{x^i} f^i(x^i, \bar{x}^i)$$

subject to

$$\sum_{j=1}^{n_i} x^i_j = 1, \quad x^i_j \geq 0 \qquad (8.3.4)$$

for $\bar{x}^i$ given.

Clearly, the M-person game is an EP. Nash proved equilibrium-point existence to (8.3.4) by a fixed-point argument. We, however, will rely upon the EP Theorem 5.6.2.

Notice that for all i, the constraints of (8.3.4) are linear and do not involve $\bar{x}^i$. Moreover, given $\bar{x}^i$, the function f^i defined by (8.3.3) becomes a *linear* function. Hence, given $\bar{x}^i$, it is a simple matter to solve for x^i in (8.3.4). In fact, it is easily seen that the conditions of Theorem 5.6.2 hold.

THEOREM 8.3.1   There is an equilibrium point for any M-person game (8.3.4).

Proof. Relative to the EP of Chapter 5, we have $g^i(x) = x^i \geq 0$ and $h^i(x) = \sum_{j=1}^{n_i} x_j^i - 1 = 0$. Let

$$\dot{x}^i = \frac{1}{n_i}(1, 1, \ldots, 1) \qquad \text{for } i = 1, \ldots, M. \tag{8.3.5}$$

Then $g^i(\dot{x}) > 0$ and $h^i(\dot{x}) = 0$. Also, for any $x = (x^1, \ldots, x^M), (\dot{x}^i, x^{\bar{i}})$ automatically satisfies $g^i(\dot{x}^i, x^{\bar{i}}) > 0$, $h^i(\dot{x}^i, x^{\bar{i}}) = 0$ since the constraints are dependent only on x^i. Notice that

$$\nabla_{x^i} h^i(x) = (1, 1, \ldots, 1),$$

being a single vector, is clearly linearly independent. Moreover, given $\bar{x}^i$, f^i defined by (8.3.3) is linear in x^i, hence f^i is concave in x^i. Thus assuming regularity the conditions of Theorem 5.6.2 hold. However, regularity, being a minor assumption, will be eliminated in Chapter 22. Thus the M-person game has an equilibrium point.   □

THE PATH-FOLLOWING APPROACH FOR THE M-PERSON GAME

The proof of Theorem 5.6.2 was by path following, and that approach can be utilized to calculate an equilibrium point for the M-person game. Start at the equilibrium point $\dot{x} = (\dot{x}^1, \ldots, \dot{x}^M)$, where $\dot{x}^i$ is the mixed strategy (8.3.5). This point is the unique equilibrium point for the "game"

$$\max - \tfrac{1}{2}\|x^i - \dot{x}^i\|^2$$

$$\text{subject to}$$

$$\sum_{j=1}^{n_i} x_j^i = 1$$

$$x^i \geq 0. \tag{8.3.6}$$

Set up the "dynamic game," where f^i is the expected payoff (8.3.3) to player i:

$$\max_{x^i} \frac{-(1-t)}{2} \| x^i - \mathring{x}^i \|^2 + tf^i(x^i, x^{\hat{\imath}})$$

subject to

$$\sum_{j=1}^{n_i} x_j^i = 1$$

$$x^i \geq 0 \qquad\qquad (8.3.7)$$

given $x^{\hat{\imath}}$. Then write down the M sets of K–T equations corresponding to the M subproblems (8.3.7). It is easy to obtain $(\mathring{x}, \mathring{\alpha}, \mathring{\mu}, 0)$ which solves (5.5.3) for $t = 0$. Then starting from $(\mathring{x}, \mathring{\alpha}, \mathring{\mu}, 0)$, path-follow in this full system until $t = 1$. In this manner we can calculate an equilibrium point.

EXAMPLE 8.3.2

Suppose that there are three candidates for alderman in a local election, Mr. One, Mr. Two, and Ms. Three. A debate was arranged by a civic organization where the political issues can be discussed and made known to the voters. Naturally, each candidate desires to hold the debate in a place and at a time where the number of his constituents present would be largest.

After some inquiries, the three candidates were able to concur on several possible locations, dates, and times. They all agreed that the debate could be held either in Hyde Park or Kenwood, on October 4 or October 8, and at either 6: 00 PM or 9: 00 PM. For each location, date, and time, each candidate made a study of the number of his or her constituents that would likely be present. The result of their studies is as follows.

Let $A^i_{j_1 j_2 j_3}$ as listed in Table 8.3.1 be the number of constituents of candi-

Table 8.3.1

$j_1 j_2 j_3$	A^1	A^2	A^3
111	1000	2000	3000
112	1500	1000	2000
121	2000	1500	500
122	4000	2000	1000
211	2100	1800	900
212	900	1500	3000
221	3000	4000	3000
222	1800	1700	1900

date i if the debate were held in location j_1, on date j_2, and at hour j_3. For example, $A^1_{111} = 1000$ indicates the number of Mr. One's constituents if the debate were held in Hyde Park on October 4 at 6: 00 PM. Also, $A^2_{221} = 4000$ would be the number of candidate 2's constituents at Kenwood on October 8 at 6: 00 PM. Similarly, $A^3_{121} = 500$ would be the number of candidate 3's constituents at Hyde Park on October 8 at 6: 00 PM.

It ought to be clear from Table 8.3.1 that it would be very hard indeed for the three candidates to agree on a suitable place, date, and time. Mr. One would definitely desire Hyde Park on October 8 at 9: 00 PM (122) since 4000 of his constituents are expected to be present then. On the other hand, he would like Kenwood, October 4, 9: 00 PM (212) least of all. Ms. Three would be indifferent between Hyde Park, October 4, 6: 00 PM (111); Kenwood, October 4, 9: 00 PM (212); and Kenwood, October 8, 6: 00 PM (221). But she would strongly object to Hyde Park, October 8, 6: 00 PM (121).

The problem of settling where and when the debate should be held was brought to the local judge, who was very good at settling disagreements. He proposed the following solution: Mr. One selects the place, Mr. Two selects the date, and Ms. Three chooses the time of the debate. All three candidates were agreeable to this arrangement, and the problem appeared to be solved.

A few days later, the three candidates slowly realized that the choices were not so easy. For instance, if Mr. One selected Hyde Park as the location, he might be fortuitous in having 4000 constituents present, but also he could very well have only 1000 attending. Unable to solve the dilemma, one by one, the three candidates covertly ran to Mr. Games (who was the best consultant in town) and asked for counsel. Mr. Games told each candidate to come back the next day at three different scheduled times, when he would have the answer ready. All three candidates promised to follow his advice.

That night, Mr. Games fretfully worried about the problem, for tomorrow, his reputation would definitely be at stake. Would Mr. Games be able to counsel all three candidates so that each candidate would follow his advice?

The answer is yes. Mr. Games merely has to set up problem (8.3.4) and solve it for an equilibrium point $\bar{x} = (\bar{x}^1, \bar{x}^2, \bar{x}^3)$, which must exist by Theorem 8.3.1. The candidates, being "rational individuals," need only be convinced that the strategy for candidate i ought to be $\bar{x}^i$. For example, if $\bar{x}^1 = (\frac{1}{2}, \frac{1}{2})$, then Mr. One flips a coin to decide between Hyde Park or Kenwood. However, if $\bar{x}^1 = (\frac{1}{6}, \frac{5}{6})$, he throws a die and chooses Hyde Park only if the outcome turns up 1. Similarly, Mr. Two selects the date using probabilities given by $\bar{x}^2 = (\bar{x}^2_1, \bar{x}^2_2)$ and Ms. Three selects the time using the probabilities given by $\bar{x}^3 = (\bar{x}^3_1, \bar{x}^3_2)$.

The persuasion is as follows. Mr. Games tells each candidate i that candidate j, $j \neq i$, has sought counsel from him and has promised to follow his advice, namely to choose strategy $\bar{x}^j$. Then, he shows i that i cannot do better than $\bar{x}^i$. If necessary, he writes down the problem faced by i. For example, problem (8.3.4) for $i = 1$ is

$$\max_{x^1} 1000x^1_1\bar{x}^2_1\bar{x}^3_1 + 1500x^1_1\bar{x}^2_1\bar{x}^3_2 + 2000x^1_1\bar{x}^2_2\bar{x}^3_1 + 4000x^1_1\bar{x}^2_2\bar{x}^3_2 + 2100x^1_2\bar{x}^2_1\bar{x}^3_1$$
$$+ 900x^1_2\bar{x}^2_1\bar{x}^3_2 + 3000x^1_2\bar{x}^2_2\bar{x}^3_1 + 1800x^1_2\bar{x}^2_2\bar{x}^3_2$$

subject to $\qquad x^1_1 + x^1_2 = 1, \qquad x^1_1 \geq 0, \qquad x^1_2 \geq 0.$

For candidate 2 the problem (8.3.4) is

$$\max_{x^2} 2000\bar{x}_1^1 x_1^2 \bar{x}_1^3 + 1000\bar{x}_1^1 x_1^2 \bar{x}_2^3 + 1500\bar{x}_1^1 x_2^2 \bar{x}_1^3 + 2000\bar{x}_1^1 x_2^2 \bar{x}_2^3 + 1800\bar{x}_2^1 x_1^2 \bar{x}_1^3$$

$$+ 1500\bar{x}_2^1 x_1^2 \bar{x}_2^3 + 4000\bar{x}_2^1 x_2^2 \bar{x}_1^3 + 1700\bar{x}_2^1 x_2^2 \bar{x}_2^3$$

$$\text{subject to} \qquad x_1^2 + x_2^2 = 1, \qquad x_1^2, x_2^2 \geq 0.$$

And similarly for candidate 3.

Unless the candidates are very foolish, they would agree that the best choice is the equilibrium solution $\bar{x}^i$. Indeed, given that candidates 2 and 3 are going to follow $\bar{x}^2$ and $\bar{x}^3$, respectively, candidate 1 can do no better than $\bar{x}^1$. Similarly, if candidates 1 and 3 are following $\bar{x}^1$ and $\bar{x}^3$, candidate 2 should do $\bar{x}^2$. And similarly for candidate 3.

8.4 SOME DILEMMAS CONCERNING THE EQUILIBRIUM CONCEPT

Although equilibrium is a powerful and very useful solution concept, there are certain examples where equilibrium cannot be called a solution. Here are two such instances.

EXAMPLE 8.4.1 *(The $5 million dilemma)*

After years of testing, the American SST has finally come to fruition, with all the major problems regarding its commercial use solved.

Two major competing airlines, Flyright and Jetstream, are both in the process of making the crucial decision of whether to keep their jumbo jets or replace them with the SSTs.

Analysis of the increase in profits for the next 10 years to the firms under the various decisions were made. (Analyses beyond 10 years proved hard to make and were excluded from consideration.) If Flyright and Jetstream both keep their jumbo jets, there is no expected increase in profits for both firms. However, suppose that Flyright buys the SSTs and Jetstream keeps the jumbo jets. Then Flyright will gain passengers from Jetstream and will increase profits by $10 million per year while Jetstream will decrease profits by $10 million per year. On the other hand, if Jetstream buys the SSTs and Flyright keeps the jumbo jets, Jetstream is expected to increase profits by $10 million per year while Flyright will decrease profits by $10 million per year. Finally, if both airlines buy SSTs, both firms' profits are expected to decrease by $5 million per year, since hardly any additional revenue is expected in this case, although operating costs will increase due to the amortization of the SSTs.

The problem facing the two firms is whether or not they should replace the jumbo jets with the new SSTs.

ANALYSIS   Let the first player be Flyright with strategies:

1.   Keep the jumbo jets.

2.   Replace the jumbo jets with SSTs.

The second player, Jetstream, has the same two strategies. Then the payoff matrices are

$$A^1 = \begin{pmatrix} 0 & -10m \\ 10m & -5m \end{pmatrix}, \qquad A^2 = \begin{pmatrix} 0 & 10m \\ -10m & -5m \end{pmatrix}.$$

Note that this game is *not* zero-sum, since both airlines lose if they both buy SSTs.

Under these conditions, Flyright should clearly purchase SSTs. For example, suppose that Jetstream keeps its jets. Then Flyright can increase profits $10 million/year by buying SSTs. Or suppose that Jetstream buys SSTs. Then since Flyright would lose $10 million/year by keeping the jets and have a profit decrease of only $5 million/year by buying SSTs, Flyright should buy SSTs. Thus no matter which strategy Jetstream selects, Flyright does better by purchasing SSTs.

Identically, Jetstream should buy SSTs because no matter what strategy Flyright chooses, Jetstream does better with SSTs. Hence, both airlines should buy SSTs, and $\bar{x}^1 = (0, 1)$, $\bar{x}^2 = (0, 1)$ is the unique equilibrium point.

We have arrived at a very odd conclusion. At equilibrium, both airlines stand to lose $5 million in profits per year each. Had they maintained the status quo, they would have been better off because there would be no reduction in profits. The two airlines are indeed faced with a dilemma. Since antitrust regulations forbid them to engage in any agreement, the threat of the other buying SSTs forces both airlines to purchase SSTs, at a costly price for both.

Our analysis leads us to the equilibrium point, $\bar{x}^1 = (0, 1)$ and $\bar{x}^2 = (0, 1)$, which is the unique equilibrium and in fact has all the desired properties such as stability. Yet clearly the equilibrium point is unsatisfactory.

PARADOXES

The reasoning above, although formally sound, is a little hard to accept. This is an example of a paradox. A paradox is a statement that seems contrary to common sense and yet is perhaps true. A very famous one in mathematics is *Russell's paradox*, an entertaining interpretation of which is the following.

Call a word "homothetic" if the word describes itself. Otherwise, call a word "heterothetic." For example, "English" is an English word and hence homothetic. But "German" is not German; it is an English word, hence it is heterothetic. Or "short" is a short word, hence it is homothetic, but "long" is not long; it is a short word, hence heterothetic.

Obviously, every word is either homothetic or heterothetic, because it either describes itself or it does not. Now, try "heterothetic." If "heterothetic" is heterothetic, then since it describes itself, it is homothetic, hence not heterothetic. If "heterothetic" is homothetic, then "heterothetic" does not describe itself, so is heterothetic, hence not homothetic.

A paradox of a different sort is the next example.

EXAMPLE 8.4.2   (A Corporate Takeover Dilemma)

A small but highly productive oil company, Bewlitt, with 30,000 acres of oil-rich land 40 miles southwest of Bakersville, California, is the subject of a takeover bid by two giants, Mexon and TexOil.

There are 1 million shares outstanding of Bewlitt owned by two very secretive and publicity-shy families living in Poughkeepsie, New York. Each share was worth only $100 a few years ago, but at current prices is worth $2100 due to the decontrol of domestic oil prices (thus Bewlitt is now worth $2.1 billion on paper).

Since no one in the families seems desirous to take future charge of Bewlitt, the two families, the Bews and the Litts, have both decided to sell all their shares (400,000 owned by the Bews and 600,000 shares owned by the Litts). Not wanting the publicity connected with a public offering, both families decided that they would be agreeable to sell at $2000 per share to either Mexon or TexOil or both.

As we know, all families of such secrecy and wealth have some unexplainable idiosyncracies that make their lives quite interesting reading, and fortunately for us, the Bews and the Litts are blessed with some. They decide to perform a little experiment on greed. They let it be known that if both Mexon and TexOil bid for the Litts' shares, then the deal is off. Also, if both bid on the Bews' shares, then the 1 million shares will be sold equally to the two companies.

Other than these two restrictions, the companies can bid as they see fit. Moreover, Mexon and TexOil estimate that whoever has majority control of the firm has an additional $20 million advantage over the minority stockholder.

Our task is to determine the actions that Mexon and TexOil should take. Of course, antitrust prohibits the firms from making any agreement prior to the bid.

ANALYSIS   Mexon, the first player, has two options:

1. Bid for Bews' shares.

2. Bid for Litts' shares.

And similarly for TexOil.

Since Mexon or TexOil theoretically gains $100 per share (the difference between the market price of $2100 and the bid price of $2000), and since there is an additional $20 million advantage to the majority stockholder, the payoff matrices are

$$A^1 = \begin{pmatrix} \$50 \text{ million} & \$40 \text{ million} \\ \$80 \text{ million} & \$0 \end{pmatrix}, \qquad A^2 = \begin{pmatrix} \$50 \text{ million} & \$80 \text{ million} \\ \$40 \text{ million} & \$0 \end{pmatrix}.$$

Let us explain this:

1. Suppose that one company bids for the Bews' shares and the other bids for the Litts' shares. The bidder for the Litts' shares will then gain $60 million for the 600,000 Litts' shares plus $20 million for control of Bewlitt for a total of $80 million. Also, the bidder of the Bews' shares will gain $40 million for the 400,000 Bews' shares.

2. If both bid for the Bews' shares, then they each get 500,000 shares (and neither gains control), so both have a profit of $50 million.

3. If both bid for the Litts' shares, both end up with nothing.

The strategy for this game is different from that of the previous example. One company could bid for the Litts' shares in the belief that the other would not. But if both end up bidding for the Litts' shares, there would be disaster for both. On the other hand, if both companies are prudent investors and bid for the Bews' shares, they would still end up with a respectable gain of $50 million each.

Analysis shows that there are two equilibria: if Mexon bids for Bews, TexOil should bid for Litts, and conversely; i.e., if one bids for Bews, the other bids for Litts. However, it is not clear here which equilibrium will be the outcome of the game. For example, suppose that Mexon is the "timid" player, such as would be the case if it were nearing Chapter XI (bankruptcy), and TexOil were afloat with cash. Then TexOil might "bully" Mexon into bidding only for the Bews' shares. However, if the "bully" tactic fails, both could bid for the Litts' shares and end up with nothing. In short the fact there there are two equilibria creates a dilemma, and it is not clear what the outcome will be.

DISCUSSION

These examples show some of the complexities in analyzing equilibrium situations. In Example 8.4.1 there was a unique equilibrium and it was stable. Indeed, no other choice of strategies was stable. Yet at equilibrium the payoff was quite undesirable. Because of the instability of any other strategy, the game would likely be driven to the equilibrium, unpleasant as it might be.

Example 8.4.2 had two equilibria and because of the circumstances it was not clear which one to select. Moreover, the confusion over which equilibrium to select could lead to disaster.

The key property of equilibrium is that it creates a stable situation once arrived at. Yet this is often insufficient. One difficulty is arriving at an equilibrium, which, if there is more than one, may be quite hard to accomplish.

Second, even if we can get to an equilibrium, it need not have a desirable payoff. This is true even if the equilibrium is unique. Indeed, we see the major dilemma in equilibrium analysis—although equilibrium is stable, frequently there is conflict between stability and payoff, and the need to sacrifice one to gain the other.

SUMMARY

This chapter introduced game theory by presenting the concept of an equilibrium point for the two-person and for the M-person game. The theory developed in Chapter 5 applied directly, and by following the path from the trivial equilibrium, we obtained an equilibrium to the M-person game. A number of examples were given; moreover, we explored the fact that certain dilemmas and difficulties can arise in applying the equilibrium notion.

EXERCISES/CHAPTER 8

1. Let $(\bar{j}, k)$ be an equilibrium play for the two-person game. Prove that it is an equilibrium point for the game (8.2.2).

2. Prove that Example 8.1.2 has a unique equilibrium point.

3. In a two-person game, let A^1 be a given $n_1 \times n_2$ payoff matrix and let A^2 satisfy
$$A^1 + A^2 = 0.$$
Suppose that A^1_{11} is the maximum along the first column and the minimum along the first row. Prove that A^1_{11} equals the payoff at the equilibrium point of the game.

4. Find the saddle points of the following matrices:

   a. $\begin{pmatrix} 10 & -2 \\ 3 & 1 \end{pmatrix}$ b. $\begin{pmatrix} 6 & 5 \\ 2 & 3 \\ 1 & 4 \\ 1 & 2 \end{pmatrix}$ c. $\begin{pmatrix} 3 & 3 & 3 & 3 \\ 2 & 3 & 4 & 5 \end{pmatrix}$

5. "Two-finger morra" is a game played in Italy since classical antiquity. This game is played by two people, each of whom shows one or two fingers, and simultaneously calls his guess as to the number of fingers his opponent will show. If just one player guesses correctly, he wins an amount equal to the sum of the fingers shown by himself and his opponent; otherwise, the game is a draw.
   a. Formulate the two-person game for "two-finger morra."
   b. Describe the optimal strategy for playing the game.

6. (*Dominance*) Given a two-person game, suppose that the first row of A^1 is strictly greater than its second row. Show that at the equilibrium point, player 1 will never choose his second strategy.

7. In an M-person game, suppose that the same real number is added to every

entry of the payoff matrix A^1 for player 1. Prove that the equilibrium point for the game remains unchanged after the addition.

8. Prove that for any matrix A,

$$\max_i \min_j A_{ij} \leq \min_j \max_i A_{ij}.$$

9. Show that if $(\overset{*}{j}, \overset{*}{k})$, $(\bar{j}, \bar{k})$ are pure equilibria of a two-person zero-sum game, then so are $(\overset{*}{j}, \bar{k})$ and $(\bar{j}, \overset{*}{k})$.

10. Calculate the equilibrium point of Example 8.1.2 by path following.

11. What is the equilibrium point for Example 8.3.2?

12. Analyze Examples 8.4.1 and 8.4.2 and discuss what is happening. What other solution concepts can you develop that might be applicable here?

NOTES

8.2. Many different proofs of the fundamental theorem of two-person zero-sum games have been given. The earliest was given by von Neumann [1928]. This proof depended on the use of the Brouwer fixed-point theorem. Ville [1938] gave the first elementary proof. Other proofs have been presented by Gale, Kuhn and Tucker [1951], Dantzig [1951], von Neumann and Morgenstern [1947], and Weyl [1950].

8.3. The concept of an equilibrium point for M-person games was introduced by Nash [1950a, b]. The method of proof of Theorem 8.3.1 is taken from Zangwill and Garcia [1981]. An algorithm for solving the game may be found in Garcia, Lemke, and Lüthi [1973]. Also, Harsanyi [1973, 1975] describes a "tracing procedure" which closely relates to the path-following approach.

9

Network Equilibrium and Elasticity

This chapter presents two different but both quite valuable applications of our previous concepts. The first, network equilibrium, concerns flows in networks and their equilibria. Since a flow can represent a wide variety of data: items, money, information people, and so on—the network model is extremely general. Traffic equilibrium, for example, is an interesting and widely used application that we discuss in detail. After formulating the underlying model, we demonstrate that the defining equations are precisely the Kuhn–Tucker equations of an equilibrium programming problem.

The second application is to elasticity of materials or, as it is often termed, strength of materials. A structure is subjected to a load and the load is slowly increased until the structure buckles. For any house, stadium, bridge, or building, how much load can it withstand? Elasticity analysis can evaluate this, thereby informing us if the structures we use are safe.

9.1 NETWORK EQUILIBRIUM INTRODUCED

The network equilibrium model helps us understand a wide variety of situations that can be interpreted as flows in a network. The flow, for instance, can be people, cars, boxes, government memoranda, information, fluid, machines, or cash. The system being analyzed could be the routes and expressways of a city, a computer network of time-sharing facilities, airline routes, the assembly line at an automated automobile plant, and so on.

Solving the network equilibrium problem highlights bottlenecks that could arise, scheduling problems, plus the remedies that can be taken to alleviate delays and difficulties. It is important for the air traffic controller at O'Hare to know the effect of closing a runway for example. Would it cause a major delay on east-west flights? Would there be more planes to supervise than a controller can handle? Would an air collison become more likely? These and numerous questions on similar-type systems can be analyzed via network equilibrium.

Let us begin with an example.

EXAMPLE 9.1.1

A major airline has just been authorized to service Tokyo, Acapulco, Houston, and Chicago and wants to commence operations early next year. The authorization requires that all passengers must originate from Tokyo or Acapulco. They then can fly nonstop to Chicago, or they can go via Houston and change planes for Chicago. At Houston only a change of planes is permitted. A key portion of the planning is to determine how many flights are required on the following links (using the obvious abbreviations): T to C, A to C, A to H, T to H, and H to C (see Figure 9.1.1). This is to be determined by considering the travelers on the various routes.

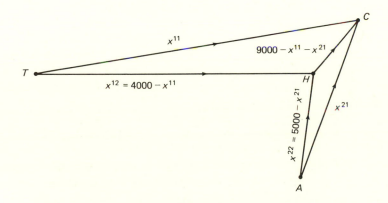

Daily 4000 passengers originate at T with x^{11} flying nonstop to C and $x^{22} = 4000 - x^{11}$ flying to C via H. From A, 5000 passengers originate daily, with x^{21} going to C nonstop and $x^{22} = 5000 - x^{21}$ going to C via H. A total of $9000 - x^{11} - x^{21}$ passengers fly $H \rightarrow C$.

Figure 9.1.1

It was estimated that daily 4000 passengers originate from T and 5000 originate from A. Of course, all end up in C, although any passenger can choose to fly either nonstop or via H.

Let x^{11} be the number of passengers taking the nonstop flight from T to

C, and x^{12} be the number going from T to C via H. Obviously, $x^{12} = 4000 - x^{11}$. Next, specify x^{21} as the number going from A to C nonstop, and x^{22} as the number going $A \longrightarrow H \longrightarrow C$, where $x^{22} = 5000 - x^{21}$. Notice, $9000 - x^{11} - x^{21}$ will then fly from H to C.

The company is concerned about customer service. It believes that a passenger from a foreign city will fly directly to C if it is quicker than by transferring planes at H, and conversely. That is, any T passenger will take the shorter of the $T \longrightarrow C$ flight or the $T \longrightarrow H \longrightarrow C$ flight. Similarly, any A passenger will take the shorter of the $A \longrightarrow C$ flight or the $A \longrightarrow H \longrightarrow C$ flight.

Because of an international treaty, the flight time is identical whether a foreign passenger goes via H or not. That is, the flight time $T \longrightarrow H \longrightarrow C$ is the same as $T \longrightarrow C$. Similarly, the flight time $A \longrightarrow H \longrightarrow C$ is the same as $A \longrightarrow C$. Since the flight time is identical both ways, it can be neglected. The major delay consideration, as any international traveler well knows, is customs. The customs processing delay will therefore determine if a passenger flies via H or nonstop to C.

The customs delay on any route will depend upon the congestion, that is, the number of passengers taking that route. Analysis revealed that a person flying directly $T \longrightarrow C$ will experience a delay of $0.006x^{11}$ minutes due to customs. Thus if he or she were the only person on this route, there would be hardly any delay at all, whereas if all 4000 people were to take this route, the delay per person would be 24 minutes. On the other hand, if a T passenger were to fly via H, the delay would be $0.003(4000 - x^{11})$ on the $T \longrightarrow H$ link, plus an additional $0.0075(9000 - x^{11} - x^{21})$ on the $H \longrightarrow C$ link. For example, suppose that everyone from both T and A were to fly via H. The delay for a T passenger would then be 79.5 minutes per person, since $0.003(4000) + 0.0075(9000) = 79.5$. (Observe that the delay $H \longrightarrow C$ depends also on the number of A passengers traveling that link, as they congest customs, too.)

As for a passenger from A, the customs delay is $0.005x^{21}$ on the nonstop $A \longrightarrow C$ route. If the route is via H, the delay is $0.008(5000 - x^{21})$ on $A \longrightarrow H$ plus $0.006(9000 - x^{11} - x^{21})$ on $H \longrightarrow C$.

Note that the delay on the $H \longrightarrow C$ link is shorter for the A passenger than a T passenger. This is because the T traveler is typically laden with newly purchased cameras and electronic equipment, which slow down customs processing.

The airline wants to determine how many T passengers will take the direct $T \longrightarrow C$ flight versus the $T \longrightarrow H \longrightarrow C$ flight. Also, how many A passengers will go $A \longrightarrow C$ versus $A \longrightarrow H \longrightarrow C$? The traveler will select the route with the minimum customs delay.

EQUILIBRIUM ANALYSIS   To analyze this problem, let $x = (x^{11}, x^{12}, x^{21}, x^{22})$ and specify that

$$V^{11}(x) = 0.006x^{11}$$

as the delay on the $T \longrightarrow C$ route, and

$$V^{12}(x) = 0.003(4000 - x^{11}) + 0.0075(9000 - x^{11} - x^{21})$$

as the delay on the $T \longrightarrow H \longrightarrow C$ route.

Similarly, the delay on the $A \longrightarrow C$ route is

$$V^{21}(x) = 0.005x^{21}$$

and on the $A \longrightarrow H \longrightarrow C$ route is

$$V^{22}(x) = 0.008(5000 - x^{21}) + 0.006(9000 - x^{11} - x^{21}).$$

A passenger from T or A will take the route with the lesser delay. Mathematically, this gives rise to the following network equilibrium conditions.

An equilibrium $\bar{x} = (\bar{x}^{11}, \bar{x}^{12}, \bar{x}^{21}, \bar{x}^{22})$ requires for T passengers that

(A)   if $\bar{x}^{11} > 0$,   then $V^{11}(\bar{x}) \leq V^{12}(\bar{x})$

(B)   if $\bar{x}^{12} > 0$,   then $V^{12}(\bar{x}) \leq V^{11}(\bar{x})$.   (9.1.1)

Condition (9.1.1A) states that a person flies $T \longrightarrow C$ only if the delay on $T \longrightarrow C$ is less than or equal to the delay on $T \longrightarrow H \longrightarrow C$. Condition (9.1.1B) states that the route $T \longrightarrow H \longrightarrow C$ is taken only if the delay on that route is less than or equal to the delay on $T \longrightarrow C$.

Further, equilibrium requires for the A passengers that

(A) $\bar{x}^{21} > 0$ implies $V^{21}(\bar{x}) \leq V^{22}(\bar{x})$

(B) $\bar{x}^{22} > 0$ implies $V^{22}(\bar{x}) \leq V^{21}(\bar{x})$.   (9.1.2)

Obviously, the network equilibrium conditions state that a person selects a route only if the delay on that route is shorter or equal to the delay on any alternative route. (Although these network conditions are intuitively reasonable, later in the chapter we state the equilibrium programming problem which mathematically gives rise to these network equilibrium conditions.)

EQUILIBRIUM SOLUTION   Let $\bar{x}^{11} = 3000$ and $\bar{x}^{21} = 4000$. Then

$$\bar{x}^{12} = 4000 - \bar{x}^{11} = 1000$$

$$\bar{x}^{22} = 5000 - \bar{x}^{21} = 1000.$$

Hence

$$V^{11}(\bar{x}) = 18 \text{ minutes} = V^{12}(\bar{x})$$

$$V^{21}(\bar{x}) = 20 \text{ minutes} = V^{22}(\bar{x}).$$

The point $\bar{x}$ satisfies (9.1.1) and (9.1.2) and is the network equilibrium. Three thousand travelers will go $T \longrightarrow C$, 1000 will go $T \longrightarrow H \longrightarrow C$, 4000 will go $A \longrightarrow C$, and 1000 will go $A \longrightarrow H \longrightarrow C$. At $\bar{x}$, the delay on $T \longrightarrow C$ equals the delay on $T \longrightarrow H \longrightarrow C$, and the delay on $A \longrightarrow C$ equals the delay on $A \longrightarrow H \longrightarrow C$.

That the equilibrium point provides balance is intuitively clear. For instance, suppose that more than 3000 travelers were to go $T \longrightarrow C$. Then delay on $T \longrightarrow C$ will increase while delay on $T \longrightarrow H \longrightarrow C$ will decrease. But that would cause some travelers on $T \longrightarrow C$ to transfer to the $T \longrightarrow H \longrightarrow C$ route. We see, therefore, that no passenger has an incentive to shift from $\bar{x}$, and hence at $\bar{x}$ the system is in equilibrium.

SYSTEM MINIMUM   So far we have concentrated on the delay for each individual. That is, $\bar{x}$ was calculated such that the delay for each individual is minimized. Each individual selected the route to minimize his or her own delay.

A quite different concept is the system delay, which is the sum of the delays for all the individuals. Of course, x^{11} people take route $T \longrightarrow C$ and each experience a delay of V^{11}, and so on. Hence the *system delay* is

$$V(x) = x^{11}V^{11}(x) + x^{12}V^{12}(x) + x^{21}V^{21}(x) + x^{22}V^{22}(x). \qquad (9.1.3)$$

For example, at $\bar{x} = (3000, 1000, 4000, 1000)$,

$$V(\bar{x}) = 3000(18) + 1000(18) + 4000(20) + 1000(20) = 172{,}000 \text{ minutes.}$$

In words, at $\bar{x}$ the system delay is $V(\bar{x}) = 172{,}000$ minutes. This is the sum total delay encountered by all 9000 travelers.

It is essential to note that the system delay is not necessarily minimized at the equilibrium point $\bar{x}$. The system delay is minimized at the point $\overset{*}{x}$ satisfying the NLP

$$\min V(x) = \sum_{i,j} x^{ij}V^{ij}(x)$$

such that

$$x^{11} + x^{12} = 4000$$

$$x^{21} + x^{22} = 5000$$

$$x^{11},\quad x^{12},\quad x^{21},\quad x^{22} \geq 0. \qquad (9.1.4)$$

Understand that at equilibrium, $\bar{x}$, each individual is minimizing his or her own delay, but system delay need not be minimized. At $\overset{*}{x}$, system delay is minimized but individual delay need not be.

The difference between $\bar{x}$ and $\overset{*}{x}$ is important. To explain it, suppose that we are at an equilibrium $\bar{x}$ so each individual minimizes his or her own delay. It is possible that certain individuals, by minimizing their own delay, may greatly inconvenience many other people. A few may gain some time at the expense of the many. The system delay, which is the sum total of the delays for everyone, might then be very large.

On the other hand, suppose that system delay is minimized. Then some passengers might be delayed in order that others go quicker. It is even possible, depending on the situation, that some might be intolerably delayed just so the others can go a little bit faster.

We see that each individual minimizing delay can yield a result quite different from minimizing the sum total delay of everyone. The ideal situation occurs when $\bar{x}$ is fairly close to $\overset{*}{x}$. Then $\bar{x}$ minimizes not only individual delay but also almost minimizes the sum total of all individual delays.

Understand that the airline problem is one of individual delay and that is what the equilibrium conditions (9.1.1) and (9.1.2) express. The airline has no control over which route the passengers select. Each passenger will choose the route that minimizes his or her own delay. The airline must solve the individual delay problem for $\bar{x}$ as given by (9.1.1) and (9.1.2). Solving a system delay problem (9.1.4) for $\overset{*}{x}$ would not give the correct information. We reiterate—the airline must solve the individual delay problem since the passengers themselves select which route they take.

EXAMPLE CALCULATED   In our example, the system delay problem (9.1.4) reduces to

$$\min V(x^{11}, x^{21}) = 0.006(x^{11})^2 + 0.003(4000 - x^{11})^2 + 0.0075(4000 - x^{11})$$
$$\times (9000 - x^{11} - x^{21}) + 0.005(x^{21})^2 + 0.008(5000 - x^{21})^2$$
$$+ 0.006(5000 - x^{21})(9000 - x^{11} - x^{21})$$

$$\text{subject to}$$
$$0 \le x^{11} \le 4000$$
$$0 \le x^{21} \le 5000. \tag{9.1.5}$$

The solution to (9.1.5) is

$$\overset{*}{x}^{11} = 2928, \quad \overset{*}{x}^{12} = 1072, \quad \overset{*}{x}^{21} = 4065, \quad \overset{*}{x}^{22} = 935, \quad \text{and}$$

$$V(\overset{*}{x}) = 171,898 \text{ minutes}.$$

Here $\overset{*}{x}$ is very nearly $\bar{x}$. Thus $\bar{x}$ not only minimizes individual delay but also nearly minimizes system delay. This is a good happenstance because $\bar{x}$ almost solves two problems at once. In Section 9.2 we shall given an example where this does not occur and $\bar{x}$ is quite far from $\overset{*}{x}$. Remember that $\overset{*}{x}$ and $\bar{x}$ are conceptually quite different and there is no reason for them to be the same.

Example 9.1.1 has only one destination, Chicago. A little more complicated example is the next, which has two origins and two destinations.

EXAMPLE 9.1.2

Figure 9.1.2 shows a network of five cities. Every day, 1000 cars leave city B bound for city E, another 1000 cars leave city B for city C, and 1000 more cars leave city E for city D. The figure also gives the delay on each link. For example, if w cars are on the link $A \rightarrow E$, the delay is $0.02w$ per car.

Each of the three trips has two possible alternative routes. The trip from B to E, call it trip T_1, has two routes,

$$B \longrightarrow E \quad \text{or} \quad B \longrightarrow A \longrightarrow E.$$

The trip B to C, call it T_2, has two routes,

$$B \longrightarrow C \quad \text{or} \quad B \longrightarrow E \longrightarrow C$$

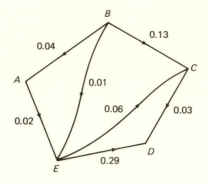

1000 cars leave from B to E,

1000 cars leave from B to C,

1000 cars leave from E to D.

The delays for travel on a link
are also indicated.

Figure 9.1.2

and the E to D trip, designated T_3, also has two routes

$$E \longrightarrow D \qquad \text{or} \qquad E \longrightarrow C \longrightarrow D.$$

ANALYSIS Let

x^{11} = number of cars on trip T_1 taking route $B \longrightarrow E$

x^{12} = number of cars on trip T_1 taking route $B \longrightarrow A \longrightarrow E$

x^{21} = number of cars on trip T_2 taking route $B \longrightarrow C$

x^{22} = number of cars on trip T_2 taking route $B \longrightarrow E \longrightarrow C$

x^{31} = number of cars on trip T_3 taking route $E \longrightarrow D$

x^{32} = number of cars on trip T_3 taking route $E \longrightarrow C \longrightarrow D$.

Notationally, x^{ij} is associated with a unique route j of trip T_i.

We have

$$
\begin{aligned}
x^{11} + x^{12} &= 1000, & x^{11} &\geq 0, & x^{12} &\geq 0 \\
x^{21} + x^{22} &= 1000, & x^{21} &\geq 0, & x^{22} &\geq 0 \\
x^{31} + x^{32} &= 1000, & x^{31} &\geq 0, & x^{32} &\geq 0.
\end{aligned}
\qquad (9.1.6)
$$

This is because the number of cars taking any trip T_i must equal 1000. For example, 1000 cars take the trip T_1 from B to E. Of these, x^{11} go via route $B \longrightarrow E$ and x^{12} go via route $B \longrightarrow A \longrightarrow E$.

Let V^{ij} be the delay time it takes a car to travel on route j of trip T_i. Then we see that

$$V^{11}(x) = 0.01(x^{11} + x^{22})$$
$$V^{12}(x) = 0.04x^{12} + 0.02x^{12}$$
$$V^{21}(x) = 0.13x^{21}$$
$$V^{22}(x) = 0.01(x^{11} + x^{22}) + 0.06(x^{22} + x^{32})$$
$$V^{31}(x) = 0.29x^{31}$$
$$V^{32}(x) = 0.06(x^{22} + x^{32}) + 0.03x^{32}.$$

For example, $V^{22}(x)$, the delay to travel on route $B \longrightarrow E \longrightarrow C$, is defined as above because:

1. There are $x^{11} + x^{22}$ cars using link $B \longrightarrow E$, namely x^{11} from trip T_1 and x^{22} from trip T_2. Hence $0.01(x^{11} + x^{22})$ is the delay for a car to travel link $B \longrightarrow E$.

2. There are $x^{22} + x^{32}$ cars using link $E \longrightarrow C$, x^{22} from trip T_2 and x^{32} from trip T_3. Hence $0.06(x^{22} + x^{32})$ is the delay for a car to travel link $E \longrightarrow C$.

The sum $0.01(x^{11} + x^{22}) + 0.06(x^{22} + x^{32})$ is then the delay for a car to travel on route $B \longrightarrow E \longrightarrow C$. The other delay times are developed identically. Notice that the delay on any link depends on the total number of cars on that link.

EQUILIBRIUM CALCULATION   At an equilibrium $\bar{x}$, of course $\bar{x}$ must satisfy (9.1.6). Moreover, $\bar{x}$ must also satisfy the condition for network equilibrium, which is

$$\bar{x}^{ij} > 0 \quad \text{implies} \quad V^{ij}(\bar{x}) = \min_{l} V^{il}(\bar{x}). \tag{9.1.7}$$

This means for trip T_i that route j is traveled only if the delay on that route is less than or equal to the delay on any alternative route for that trip. Rephrased, people on trip T_i take route j only if it minimizes their delay.

A simple calculation shows that a network equilibrium is

$$\bar{x}^{11} = 800, \quad \bar{x}^{12} = 200, \quad \bar{x}^{21} = 600, \quad \bar{x}^{22} = 400, \quad \bar{x}^{31} = 300, \quad \bar{x}^{32} = 700.$$

Then

$$V^{11}(\bar{x}) = 12 \text{ minutes} = V^{12}(\bar{x})$$
$$V^{21}(\bar{x}) = 78 \text{ minutes} = V^{22}(\bar{x})$$
$$V^{31}(\bar{x}) = 87 \text{ minutes} = V^{32}(\bar{x}).$$

It takes 12 minutes to make the trip T_1, 78 minutes to make the trip T_2, and 87 minutes to make the trip T_3.

EXAMPLE: SYSTEM DELAY   Let us determine the system delay at $\bar{x}$, and then compare it with the minimum possible system delay. System delay is the sum total of all delays, and in this case equals

$$V(x) = \sum_{i=1}^{3} \sum_{j=1}^{2} x^{ij} V^{ij}(x). \tag{9.1.8}$$

Hence

$$V(\bar{x}) = 1000(12) + 1000(78) + 1000(87)$$

$$= 177{,}000 \text{ minutes.}$$

The system delay at $\bar{x}$ is thus 177,000 minutes.

The system delay minimum, $\overset{*}{x}$, is found by solving the NLP

$$\min V(x)$$

subject to

$$x^{i1} + x^{i2} = 1000, \qquad i = 1, 2, 3$$

$$x^{i1} \geq 0, \qquad x^{i2} \geq 0. \tag{9.1.9}$$

The solution to (9.1.9) is $\overset{*}{x} = \bar{x}$.

In this case there is no difference between the system delay at $\bar{x}$ and at $\overset{*}{x}$; however, wait until the next section.

9.2 THE PARADOX OF ADDING OR REMOVING A LINK

It would appear that if we add a new link to a given network, then the system delay would decrease. It is also plausible to surmise that if a link is removed, the system delay would increase since there would be fewer routes to choose. For instance, in Example 9.1.1, suppose that we remove links $T \rightarrow H$ and $A \rightarrow H$. Then all travelers would have to proceed directly to C and the system delay would be

$$V(4000, 5000) = 4000(24) + 5000(25)$$

$$= 221{,}000 \text{ minutes.}$$

This is greater than the previous delay of 177,000 minutes. Similarly, if we add a link $T \rightarrow A$ that experiences very little delay, system delay would decrease (Exercise 5).

A PARADOX

Paradoxically, it is not true that adding a link must decrease the system delay. Indeed, it would be erroneous for the Department of Public Highways to assume that more roads would reduce congestion.

Consider the following example.

EXAMPLE 9.2.1

Suppose that there are 10,000 commuters daily from suburb A to city D and 1000 commuters daily from suburb C to city D (Figure 9.2.1). It takes 35 minutes to go from A to D because of a nice highway, but because of a mountain,

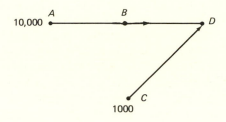

10,000 travelers go from A to D via B, and 1000 go from C to D.

Figure 9.2.1

2 hours are required from C to D. Hence the total system time in this situation is

$$10{,}000(35) + 1000(120) = 470{,}000 \text{ minutes.}$$

Because the travel from C to D was so slow, taking 2 hours over the mountain roads, the State Highway Commission decided to build a new road from C to B. The commuters from C would go to B and then take the highway that links B to D (see Figure 9.2.2).

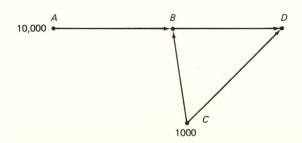

A new road was built from C to B, thereby
shortening the travel time from C to D.

Figure 9.2.2

Soon after the new road from C to B was built, the people at C were delighted, as their new commute time became only 1 hour versus the 2 hours previously. However, the travelers from A to D did not share this joy. The highway was already full with the 10,000 travelers from A to D and with the additional cars it became jammed. As a result, traveling from A to D took 50 minutes instead of 35 minutes. The system time with the new link C to B became

$$10{,}000(50) + 1000(60) = 560{,}000.$$

Building the new link dramatically increased the system time. For most people adding a link made things worse, not better.

ANALYSIS OF THE PARADOX   Let us analyze this paradox mathematically, as it dramatizes the difference between the equilibrium flow $\bar{x}$ (at which each person minimizes his own time) and the total system minimum $\overset{*}{x}$. First consider the new road network with road $C \longrightarrow B$. To minimize their own time, the people at C go via B. The people from A have no choice and must go via B. Thus the new flow pattern is definitely a network equilibrium $\bar{x}$.

The system minimum, $\overset{*}{x}$, however, is quite different. To calculate it exactly requires that we know the travel time delay on each link. Unfortunately, the Highway Department, politically embarrassed over the effect of the new road, refused to divulge that information. Nevertheless, we can still partially estimate $\overset{*}{x}$.

We previously calculated that if everyone from C goes to B, the system time is 560,000 minutes. However, suppose that people from C were to go from C to D directly. In that case, as we know, the system time is 470,000 minutes. Clearly, then, at the optimal $\overset{*}{x}$ at least some and possibly all people from C must go directly to D. Consequently, the system time will be 470,000 minutes or less.

To minimize system delay it is required that some and perhaps all people from C go directly to D. With their new road built, the people from C are unwilling to go directly to D. That would waste an hour going to work, which they absolutely will not countenance. The reality of human nature forces the system to operate at the equilibrium $\bar{x}$ and thereby the system delay is 560,000 minutes.

In sum, with the new road $\bar{x}$ produces a system delay of 560,000 while $\overset{*}{x}$ produces 470,000 minutes or less. Nevertheless, because the people at C want to minimize their time just like everyone else, the system will operate at $\bar{x}$. The paradox is clear, as by adding a link things can actually get worse, and mathematically this paradox is a direct result of $\bar{x} \neq \overset{*}{x}$.

EXAMPLE 9.2.2   Braess Paradox

The previous example is not totally convincing, for it would seem that the people traveling from C really are better off. And it might be possible for some of them to drive from C to D directly, and thereby reduce the overall delay at least somewhat.

We now pose the famous Braess paradox, in which by adding a link everyone is made worse off. Even the situation of the preceding example, that of a few doing better at the price of delaying the many, is eliminated. The added link harms everyone.

Figure 9.2.3 depicts the relevant network, and to keep the numbers simple, suppose that 6 people must travel from A to D. Let x^{11} be the number of people who travel via links 1 and 2, that is, who go via B. Then $x^{12} = (6 - x^{11})$ is the number who travel via links 3 and 4.

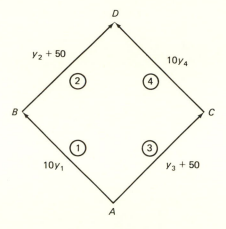

Six people leave A and travel to D. Letting y_i be the flow on link i, the delay on each link is as indicated.

Figure 9.2.3

The equilibrium, as easily seen, is

$$\bar{x}^{11} = \bar{x}^{12} = 3.$$

At equilibrium the delay for the people traveling via B is

$$V^{11}(\bar{x}) = 10\bar{x}^{11} + (\bar{x}^{11} + 50) = 10(3) + (3 + 50) = 83.$$

Similarly, the delay for people traveling via C is 83. The total delay is $6 \cdot 83 = 498$. Also, in this instance the equilibrium $\bar{x}$ and the system optimum $\overset{*}{x}$ are identical.

Let us now add link 5 from B to C as in Figure 9.2.4. Defining flows x^{11} and x^{12} as previously, let x^{13} indicate the flow on the route $A \rightarrow B \rightarrow C \rightarrow D$. Here the delays along each route are

$$V^{11}(x) = 10(x^{11} + x^{13}) + \quad (x^{11} + 50)$$
$$V^{12}(x) = \quad (x^{12} + 50) \ + 10(x^{12} + x^{13})$$
$$V^{13}(x) = 10(x^{11} + x^{13}) + \quad (x^{13} + 10) \ + 10(x^{12} + x^{13}).$$

At equilibrium an easy calculation shows that $\bar{x}^{11} = \bar{x}^{12} = \bar{x}^{13} = 2$ and

$$V^{11}(\bar{x}) = V^{12}(\bar{x}) = V^{13}(\bar{x}) = 92.$$

Also, the total delay at $\bar{x}$ is 552.

Despite adding a link, at equilibrium everyone became worse off. Each person, by pursuing his or her own needs, delayed everyone, including himself or herself.

The system minimum for the network in Figure 9.2.4, $\overset{*}{x}$, does not use the new link and produces the same flows as in Figure 9.2.3.

177

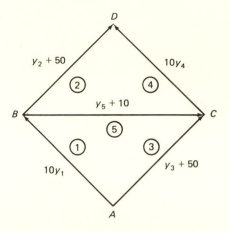

This figure is the same as figure 9.2.3 except that a
link 5 has been added from B to C with delay $y_5 + 10$

Figure 9.2.4

The Braess paradox is justly famous, as it illustrates how, by everyone
trying to do better, everyone can be made worse.

It should be noted that while Example 9.2.2 is hypothetical, the paradox is
not. A very interesting event was reported from Germany a few years ago. A
sizable road-building project in Stuttgart failed to speed traffic flow as expected.
Only after one of the new roads was closed did traffic improve.

COMMENT

We have stressed the difference between $\bar{x}$ and $\overset{*}{x}$, as $\bar{x}$ results when each
individual solves his own subproblem, whereas $\overset{*}{x}$ occurs when the entire system
is optimized at once.

The point $\bar{x}$ is the result of an equilibrium problem and $\overset{*}{x}$ is obtained from
an NLP. The reader will recall the discussion in Chapter 5 on the difference
between $\bar{x}$ and $\overset{*}{x}$ and what can happen if $\bar{x} \neq \overset{*}{x}$. The system actually operates
at an equilibrium $\bar{x}$ because $\bar{x}$ is what the people will do. Yet the fact that $\bar{x} \neq \overset{*}{x}$
can cause a great deal of difficulty. Our network examples provide an excellent
illustration of this.

9.3 FORMULATION OF THE NETWORK EQUILIBRIUM MODEL

The previous examples provided much insight into the network equilibrium
problem. We now formalize the discussion and state the network equilibrium
model in general. For clarity, frequent reference will be given to Example 9.1.2.

Given a network, suppose that M trips are specified, $T_1, T_2, \ldots T_M$. In Example 9.1.2, T_1 represents the trip from B to E, T_2 the trip from B to C, and T_3 from E to D.

Each trip T_i can be taken by r_i different routes. Let $x^{ij} \geq 0$ be the number of travelers on trip T_i who take route j. Also suppose that n_i travelers take trip T_i. Then

$$\sum_{j=1}^{r_i} x^{ij} = n_i \qquad (9.3.1)$$

since there are n_i travelers taking trip T_i.

Define the system flow vector $x = (x^{ij})$ as the number of travelers on trip T_i taking route j for every possible trip and route. The vector x has $\sum_{i=1}^{M} r_i$ components.

THE DELAY FUNCTION

We are primarily interested in the delay suffered by any traveler. Given x, let the delay function $V^{ij}(x)$ denote the delay incurred on route j of trip T_i. The delay function V^{ij} is obtained by summing up the delay on all *links* in the route, and we must develop it carefully.

First, number each link in the network. No special pattern is needed; just make sure that each link has a number, and Figure 9.3.1 provides an illustration for Example 9.1.2. We note that given any link, only certain routes may go through it. In Figure 9.3.1 only route 2 of T_2 and route 2 of T_3 go through link 5. Similarly, only route 1 of T_1 and route 2 of T_2 go through link 3.

Next required is the number of travelers (flow) on link k, and let y_k desig-

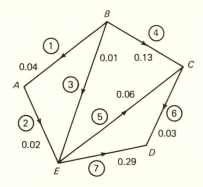

The routes for trip T_1 are $B \to E$ and $B \to A \to E$;
for T_2 they are $B \to C$ and $B \to E \to C$; and for T_3
they are $E \to D$ and $E \to C \to D$. The number of each
link is arbitrary and given simply to identify the link.

Figure 9.3.1

nate that. Specifically,

$$y_k = \sum x^{ij}, \tag{9.3.2}$$

where the sum is over all routes of all trips that use link k. In Figure 9.3.1, for example,

$$y_1 = x^{12}, \quad y_2 = x^{12}, \quad y_3 = x^{11} + x^{22}, \quad y_4 = x^{21},$$
$$y_5 = x^{22} + x^{32}, \quad y_6 = x^{32}, \quad \text{and} \quad y_7 = x^{31}.$$

Here $y_3 = x^{11} + x^{22}$ because travelers on trip T_1 route 1 and trip T_2 route 2 use link 3.

For any link k, let v_k^i be the delay suffered on that link by any traveler taking trip T_i. Explicitly, $v_k^i(y_k)$ is the per person delay on link k for any traveler on trip T_i if y_k travelers are on that link. Observe that the delay on a link depends on the total flow y_k on that link. In Figure 9.3.1,

$$v_1^1(y_1) = 0.04y_1 = 0.04x^{12}$$

since only travelers on $B \longrightarrow A \longrightarrow E$ use link 1. However,

$$v_3^1(y_3) = v_3^2(y_3) = 0.01y_3 = 0.01(x^{11} + x^{22})$$

since a traveler on trip T_1 will take $0.01(x^{11} + x^{22})$ to cross link 3 and similarly for a traveler on trip T_2.

Finally, what is the delay function along a route? The delay function on route j of trip T_i, recall, is the sum of the delays on each link of the route, hence:

$$V^{ij}(x) = \sum_k v_k^i(y_k), \tag{9.3.3}$$

where the sum is over all links in the route. In Figure 9.3.1, for route $B \longrightarrow E \longrightarrow C$:

$$V^{22}(x) = v_3^2(y_3) + v_5^2(y_5)$$
$$= 0.01y_3 + 0.06y_5$$
$$= 0.01(x^{11} + x^{22}) + 0.06(x^{22} + x^{32}).$$

For route $E \longrightarrow C \longrightarrow D$:

$$V^{32}(x) = v_5^3(y_5) + v_6^3(y_6)$$
$$= 0.06y_5 + 0.03y_6$$
$$= 0.06(x^{22} + x^{32}) + 0.03x^{32}.$$

With the delay functions provided, let us now state formally the network equilibrium conditions.

The point $\bar{x} = (\bar{x}^{ij})$ is a *network equilibrium* if

$$\sum_{j=1}^{r_i} \bar{x}^{ij} = n_i, \quad \bar{x}^{ij} \geq 0, \quad \text{all } i \text{ and } j \tag{9.3.4}$$

and

$$\bar{x}^{ij} > 0 \quad \text{implies} \quad V^{ij}(\bar{x}) = \min_l V^{il}(\bar{x}). \tag{9.3.5}$$

Condition (9.3.5) says that for trip T_i, route j is used only if the delay on this route, namely $V^{ij}(\bar{x})$, is no greater than the delay on any other possible route l of trip T_i. Route j of T_i is taken because it is the quickest way to finish trip T_i.

The goal of the network equilibrium problem is to find a network equilibrium point $\bar{x}$. That point, as we have discussed, results when the people on a trip T_i select the route j to minimize their own delay.

9.4 THE EQUILIBRIUM PROGRAMMING FORMULATION

With the network equilibrium conditions presented, this section will demonstrate that existence of a network equilibrium follows from an EP. That the network problem is an EP not only assists in its understanding and use but ensures that it fully fits into our previously developed EP theory.

To describe the EP, first denote $x^i = (x^{i1}, \dots, x^{ir_i})$ as the flow on all routes of trip T_i. Then, for $x = (x^1, x^2, \dots, x^M)$, recall our previous notation $x = (x^i, x^i)$. Here x^i is the flow on all routes of all trips except T_i.

The EP for the network can now be presented. Expressly, subproblem i is, where x^i is given,

$$\min_{x^i} f^i(x^i, x^i) \equiv \sum_k \int_0^{y_k} v_k^i(\theta) \, d\theta$$

subject to

$$\sum_{j=1}^{r_i} x^{ij} = n_i$$

$$x^{ij} \geq 0, \quad j = 1, \dots, r_i. \tag{9.4.1}$$

Here y_k is defined by (9.3.2). Also, the summation in the objective function is over all links k that contain a route from trip T_i. That is, any link k that a route of trip T_i goes through would appear in the summation.

In (9.4.1), x^i is given, and the subproblem is to find the optimal x^i. In other words, the flow on all routes of all trips other than T_i is given. The minimization is over only the flow on the routes of trip T_i. Conceptually, subproblem

i assumes that the flow for all other trips is fixed and worries about the travelers on trip T_i only. For trip T_i, it then adjusts the number of travelers on the routes j of that trip to minimize the objective function.

EXAMPLE

The EP for Example 9.1.2 has three subproblems, as follows:

1. $\min \int_0^{y_1} 0.04\theta \, d\theta + \int_0^{y_2} 0.02\theta \, d\theta + \int_0^{y_3} 0.01\theta \, d\theta$

$$x^{11} + x^{12} = 1000$$

$$x^{11}, x^{12} \geq 0$$

given $x^{\bar{1}} = (x^2, x^3)$.

2. $\min \int_0^{y_3} 0.01\theta \, d\theta + \int_0^{y_4} 0.13\theta \, d\theta + \int_0^{y_5} 0.06\theta \, d\theta$

$$x^{21} + x^{22} = 1000$$

$$x^{21}, x^{22} \geq 0$$

given $x^{\bar{2}} = (x^1, x^3)$.

3. $\min \int_0^{y_5} 0.06\theta \, d\theta + \int_0^{y_6} 0.03\theta \, d\theta + \int_0^{y_7} 0.29\theta \, d\theta$

$$x^{31} + x^{32} = 1000$$

$$x^{31}, x^{32} \geq 0$$

given $x^{\bar{3}} = (x^1, x^2)$.

The first subproblem follows because trip T_1 has routes $B \rightarrow E$ and $B \rightarrow A \rightarrow E$ and so uses links 1, 2, and 3. The second subproblem follows because trip T_2 has routes $B \rightarrow C$ and $B \rightarrow E \rightarrow C$, so uses links 3, 4, and 5. The third subproblem is similar.

EQUIVALENCE

Let us now show the equivalence of the EP (9.4.1) necessary conditions and the network equilibrium conditions (9.3.4) and (9.3.5). A lemma is required that differentiates an integral.

Recall from calculus that for a function $g: R^1 \rightarrow R^1$, where θ is a scalar,

$$\frac{d}{dy} \int_0^y g(\theta) \, d\theta = g(y).$$

Hence

$$\frac{d}{dy_k} \int_0^{y_k} v_k^i(\theta) \, d\theta = v_k^i(y_k). \qquad (9.4.2)$$

Further, using (9.3.2) and the chain rule,

$$\frac{\partial}{\partial x^{ij}} \int_0^{y_k} v_k^i(\theta)\, d\theta = v_k^i(y_k)\frac{\partial y_k}{\partial x^{ij}}. \tag{9.4.3}$$

The next lemma follows quickly.

LEMMA 9.4.1   If f^i is defined by (9.4.1), then

$$\frac{\partial f^i(x^i, x^i)}{\partial x^{ij}} = V^{ij}(x^i, x^i). \tag{9.4.4}$$

Proof.   From (9.4.1) and (9.4.3),

$$\frac{\partial f^i(x^i, x^i)}{\partial x^{ij}} = \sum_k \frac{\partial}{\partial x^{ij}} \int_0^{y_k} v_k^i(\theta)\, d\theta$$

$$= \sum_k v_k^i(y_k)\frac{\partial y_k}{\partial x^{ij}}. \tag{9.4.5}$$

Also by the definition of y_k from (9.3.2),

$$\frac{\partial y_k}{\partial x^{ij}} = \begin{cases} 1 & \text{if route } j \text{ of trip } T_i \text{ goes through link } k \\ 0 & \text{otherwise.} \end{cases} \tag{9.4.6}$$

That is, if link k is part of route j of T_i, then x^{ij} is a term in (9.3.2) and therefore $\partial y_k/\partial x^{ij} = 1$. Otherwise, x^{ij} does not appear in (9.3.2) and $\partial y_k/\partial x^{ij} = 0$.

Combining (9.4.5) and (9.4.6) yields

$$\frac{\partial f^i}{\partial x^{ij}} = \sum_k v_k^i(y_k), \tag{9.4.7}$$

where the sum is over all links k that route j of T_i goes through. But by definition (9.3.3), the right-hand side of (9.4.7) is simply $V^{ij}(x)$. $\square$

We can now prove

THEOREM 9.4.2   The necessary conditions of optimality for the EP (9.4.1) yields a network equilibrium $\bar{x}$, that is,

$$\sum_{j=1}^{r_i} \bar{x}^{ij} = n_i \qquad \bar{x}^{ij} \geq 0 \qquad \text{all } i \text{ and } j$$

and

$$\text{if } \bar{x}^{ij} > 0, \qquad \text{then } V^{ij}(\bar{x}) = \min_l V^{il}(\bar{x}).$$

Proof. Let us write the Kuhn–Tucker conditions for subproblem i of (9.4.1) given $x^{\bar{\imath}}$. (Recall that for subproblem i the variables are x^i only, as $x^{\bar{\imath}}$ is fixed.) These are seen to be, for variables λ^{ij} and μ_i,

$$\sum_{j=1}^{r_i} x^{ij} = n_i, \qquad \lambda^{ij} \geq 0, \quad x^{ij} \geq 0, \quad x^{ij}\lambda^{ij} = 0 \qquad (9.4.8)$$

and

$$\frac{\partial}{\partial x^{ij}} f^i(x^i, x^{\bar{\imath}}) - \mu_i - \lambda^{ij} = 0, \qquad j = 1, \dots, r_i. \qquad (9.4.9)$$

But, by Lemma 9.4.1, (9.4.9) becomes

$$V^{ij}(x^i, x^{\bar{\imath}}) - \mu_i - \lambda^{ij} = 0, \qquad j = 1, \dots, r_i. \qquad (9.4.10)$$

Suppose that $(\bar{x}, \bar{\lambda}, \bar{\mu})$ is a solution to these sets of Kuhn–Tucker conditions, $i = 1, \dots, M$. We now show that $\bar{x}$ satisfies the network equilibrium conditions. Clearly, (9.4.8) corresponds to (9.3.4). Now suppose that $\bar{x}^{ij} > 0$. Then via (9.4.8) $\bar{\lambda}^{ij} = 0$, so from (9.4.10),

$$V^{ij}(\bar{x}^i, \bar{x}^{\bar{\imath}}) = \bar{\mu}_i \leq \bar{\mu}_i + \bar{\lambda}^{il} = V^{il}(\bar{x}^i, \bar{x}^{\bar{\imath}}) \qquad \text{for all } l.$$

Restated, $\bar{x}^{ij} > 0$ implies that

$$V^{ij}(\bar{x}^i, \bar{x}^{\bar{\imath}}) = \min_l V^{il}(\bar{x}^i, \bar{x}^{\bar{\imath}}),$$

which is (9.3.5). $\square$

We have validated that the Kuhn–Tucker conditions for the EP (9.4.1) yield the network equilibrium conditions. EP (9.4.1) is thus equivalent to the network equilibrium problem. Notice, subproblem i has $x^{\bar{\imath}}$ given, which means that travelers taking trip T_i cannot directly influence the travelers on any other trip. They can only choose their own x^{ij}; that is, they can only select route j for trip T_i. Moreover, by optimizing $f^i(x^i, x^{\bar{\imath}})$, that is, by solving their own subproblem, they determine the network equilibrium conditions.

9.5   EXISTENCE OF A NETWORK EQUILIBRIUM

The preceding section proved that if the EP (9.4.1) has an equilibrium point, then this point is also a network equilibrium. We now demonstrate that the EP (9.4.1) actually possesses an equilibrium point. To achieve this, recall Theorem 5.6.2. For the EP (9.4.1) it is easily seen that all the assumptions of the theorem are easily validated, except for the convexity of f^i relative to x^i.

Let us make the following assumption on the mappings $v_k^i(\cdot)$, as that will guarantee the convexity.

$$v_k^i(\cdot) \text{ is a nondecreasing function} \qquad \text{for all } i, k. \qquad (9.5.1)$$

Condition (9.5.1) states that $v_k^i(y_k) \geq v_k^i(y_k')$ whenever $y_k \geq y_k'$. Intuitively, this means that the delay on any link k increases as the number of travelers on it increases. For most applications this is an extremely appropriate assumption.

TWO FACTS ABOUT CONVEXITY

To prove the convexity of f^i assuming condition (9.5.1), recall two facts about convexity.

The first is that if a function f is a differentiable function of one variable y and df/dy is nondecreasing, then f is convex in y. As for the second fact, let f be differentiable and convex in a single variable y. Now suppose that $y = \sum_{i=1}^n c_i z_i$, for given c_i (i.e., y is a linear function of the variables $z_1, \ldots, z_n$). Then f is convex in z.

These facts (Exercise 1) easily yield the following theorem.

THEOREM 9.5.1   Under condition (9.5.1), the function f^i defined by (9.4.1) is convex in x^i.

Proof.   It is sufficient to prove that $\int_0^{y_k} v_k^i(\theta)\, d\theta$ is convex in x^i since the sum of convex functions is convex (Exercise 2).

From (9.4.2) and (9.5.1),

$$\frac{d}{dy_k} \int_0^{y_k} v_k^i(\theta)\, d\theta = v_k^i(y_k)$$

is a nondecreasing function. Hence by the first fact $\int_0^{y_k} v_k^i(\theta)\, d\theta$ is convex in y_k. But y_k is linear in x^i from definition (9.3.2). Hence using the second fact, $\int_0^{y_k} v_k^i(\theta)\, d\theta$ is convex in x^i. $\square$

Theorem 5.6.2 and Theorem 9.5.1 immediately yield:

THEOREM 9.5.2   Assuming that condition (9.5.1) holds, a network equilibrium problem has a network equilibrium point $\bar{x}$.

We are assured that under (9.5.1), a network equilibrium $\bar{x}$ exists. Moreover, $\bar{x}$ can be found by the usual path-following techniques.

9.6 AN IMPORTANT SPECIAL CASE

A very important special case is when

$$v_k^i(\cdot) \equiv v_k(\cdot) \qquad \text{for all } i = 1, \ldots, M. \tag{9.6.1}$$

This means that the delay function on a link is identical for all travelers, no matter which trip T_i they are on. This situation prevails when the travelers are homogeneous and hence indistinguishable on a link. Example 9.1.1 was not of this form because the delay on link $H \to C$ was dependent on whether the passenger started in T or A. Recall that the T passengers took longer in customs. Example 9.1.2 satisfies (9.6.1), however. The delay on a link depended only on the total number of cars on that link. It did not matter where the cars came from or where they were going.

Condition (9.6.1) provides an important situation that arises frequently. Be it cars, people, computer terminals, or production processes, the items in the system can often be assumed identical. The delay on any link then depends only on the number of items on the link and consequently satisfies (9.6.1).

If (9.6.1) holds, the astonishing result is obtained that the network equilibrium conditions can be obtained from an NLP. This is truly impressive, as we no longer need to resort to an EP formulation for the network problem. Rather, a simple and more easily solved NLP can be employed.

Under (9.6.1), the NLP that produces the network equilibrium is as follows:

$$\min \sum_{\text{all } k} \int_0^{y_k} v_k(\theta) \, d\theta$$

subject to

$$\sum_{j=1}^{r_i} x^{ij} = n_i, \qquad \text{all } i$$

$$x^{ij} \geq 0, \qquad \text{all } i \text{ and } j. \tag{9.6.2}$$

The sum is over all links k in the network.

Solving (9.6.2) yields an optimal point that satisfies all the network equilibrium conditions at once. That is, the optimal point for (9.6.2) simultaneously minimizes the delay for all trips. In contrast, for EP formulation (9.4.1), subproblem i determined the routes to minimize delay only for those taking trip T_i.

EXAMPLE

Example 9.1.2 has the same delay function for all trips at any link; hence the EP formulation is unnecessary. The NLP formulation is

$$\min \sum_{k=1}^{7} \int_0^{y_k} v_k(\theta) \, d\theta$$

$$x^{i1} + x^{i2} = 1000, \qquad i = 1, 2, 3$$

$$x^{ij} \geq 0, \quad j = 1, 2, \qquad i = 1, 2, 3. \tag{9.6.3}$$

DEMONSTRATION

We now prove that the Kuhn–Tucker conditions for NLP (9.6.2) are exactly the same as the Kuhn–Tucker conditions for EP (9.4.1), and hence by Theorem 9.4.2 are the network equilibrium conditions. Thus solving (9.6.2) yields the network equilibrium. All that is really needed is a lemma whose proof, being very similar to Lemma 9.4.1, is outlined. The details are left as Exercise 4.

LEMMA 9.6.1

$$\frac{\partial}{\partial x^{ij}} \left(\sum_k \int_0^{y_k} v_k(\theta) \, d\theta \right) = V^{ij}(x). \tag{9.6.4}$$

Proof.

$$\frac{\partial}{\partial x^{ij}} \left(\sum_k \int_0^{y_k} v_k(\theta) \, d\theta \right) = \sum_k v_k(y_k) \frac{\partial y_k}{\partial x^{ij}} = \sum_k v_k(y_k), \tag{9.6.5}$$

where the summation is over all links k that route j of trip T_i goes through. This follows using the same reasoning as in the proof of Lemma 9.4.1. Finally, the right-hand side of (9.6.5) is simply $V^{ij}(x)$. $\square$

The next theorem follows immediately by examining the K–T conditions of (9.6.2) and applying Lemma 9.6.1. The proof is similar to that of Theorem 9.5.2 and is left as Exercise 4.

THEOREM 9.6.2   The necessary conditions of optimality of NLP (9.6.2) yield a point, call it $\bar{x}$, which satisfies the network equilibrium conditions, namely,

$$\sum_{j=1}^{r_i} \bar{x}^{ij} = n_i, \quad \bar{x}^{ij} \geq 0, \qquad \text{all } i \text{ and } j$$

and

$$\text{if } \bar{x}^{ij} > 0, \qquad \text{then } V^{ij}(\bar{x}) = \min_l V^{il}(\bar{x}).$$

Remark

The NLP (9.6.2) objective function is essentially the summation of the objective functions for all M subproblems of EP (9.4.1). Under (9.6.1) the subproblem objective functions f^i have identical terms on any link. Intuitively, (9.6.1) permits the EP to "centralize" to one NLP, and instead of M separate objective functions, we obtain one.

THE SYSTEM MINIMUM IS NOT OBTAINED

Solving NLP (9.6.2) yields a point $\bar{x}$ that satisfies the network equilibrium conditions. That is, $\bar{x}$ yields the individual equilibrium at which each person minimizes his or her own delay. NLP (9.6.2) does not yield the system minimum $\overset{*}{x}$. To obtain the system minimum a totally different NLP must be solved. Explicitly, to obtain $\overset{*}{x}$, which minimizes the sum total of the delays for the system, we must solve the following NLP:

$$\min \sum x^{ij} V^{ij}(x)$$

$$\sum_{j=1}^{r_i} x^{ij} = n_i, \qquad i = 1, \ldots, M$$

$$x^{ij} \geq 0. \tag{9.6.6}$$

The summation is over all routes j in all possible trips T_i.

Although under (9.6.1) the network equilibrium $\bar{x}$ reduces to an NLP, do not be confused. Still, in general, $\bar{x} \neq \overset{*}{x}$ because the two NLPs are different.

PERSPECTIVE AND PHILOSOPHY

The network equilibrium problem is quite powerful because the flows in the links can be interpreted in such a wide variety of ways. The problem is truly an equilibrium problem. The people on a specific trip T_i seek to minimize their own delay given the flows on all other trips. Also, the operation is decentralized, as one would expect in an EP, since each person separately optimizes his or her own subproblem.

The network equilibrium is fascinating, as it highlights the difference between $\bar{x}$, at which each individual minimizes his or her delay, and $\overset{*}{x}$, which minimizes total system delay. As we have seen, the difference between $\bar{x}$ and $\overset{*}{x}$ can lead to paradoxes. Furthermore, it can result in a great deal of difficulty for planning or government. The system will tend to operate at $\bar{x}$, since $\bar{x}$ is what the people will do. Yet if $\bar{x}$ is far from $\overset{*}{x}$, there can be much delay or waste in the system as a whole. On the other hand, forcing the system to operate at $\overset{*}{x}$ when the people naturally operate at $\bar{x}$ might cause even worse problems. For example, unpleasant government regulation or control might result.

Overall, the difference between $\bar{x}$ and $\overset{*}{x}$ is crucial. The people seek $\bar{x}$, even though that may cause waste in the total system or, as in the Braess paradox, harm everyone. The point $\overset{*}{x}$ optimizes the entire system, but often at the price of individual freedom, liberty, and motivation. The difference between $\bar{x}$ and $\overset{*}{x}$ is not merely mathematical but raises fundamental philosophical issues as to how systems should be planned and operated. Perhaps it is this very difficulty that is one of the major reasons why a decision-maker should never be replaced by a computer.

9.7 ELASTICITY AND STRENGTH OF MATERIALS

In this section a quite different type of equilibrium problem is analyzed: How materials respond when a load or stress is placed upon them. Most materials are somewhat elastic and under small loads flex a bit. If the stress placed upon them is sufficient, however, the material breaks.

Millions of dollars are spent annually studying the strength of materials. Continually, headlines blare of roofs collapsing, people being crushed when part of a stadium gives way, bridges becoming unsafe and having to be reinforced, buildings crumbling in earthquakes, and so on. Strength of materials and elasticity analysis are what ensure the safety of every structure we live in or use each and every day.

THE LOAD

To begin, for some constant load b^0 and t a parameter, let tb^0 indicate the load placed on the structure. As t increases, so does the load. The question to be considered is: How does the structure deform as t is increased? Typically, for t small, most materials bend somewhat, but eventually at some $\bar{t}$ that structure will buckle or snap through. The corresponding load $\bar{t}b^0$ is the limiting load the structure can withstand.

In analyzing the problem, usually a system of equations is obtained:

$$H(x, t) = 0, \quad \text{where} \quad x \in R^n \quad \text{and} \quad t \in R^1 \qquad (9.7.1)$$

for $H: R^{n+1} \longrightarrow R^n$. The x will represent the state of the structure at a given load parameter t. In other words, $x \in H^{-1}(t)$ will be the state of the structure under load tb^0. As t changes, so will x, and in general a path is obtained. Note, however, that (9.7.1) is not a homotopy system because our interest is in the entire path H^{-1} and not merely $H^{-1}(1)$. That is, we are interested in the structure's behavior at any t, not just for $t = 1$.

AN EXAMPLE

To illustrate the technique involved, consider the particular structure of Figure 9.7.1. This is a simple two-bar framework in which the two bars are pinned together at A and are pinned to the ground at points B and C, respectively. At point A the system is placed under a load $b = (b_1, b_2)$, where the stress component in the horizontal is b_1 and in the vertical is b_2. Notationally,

$$b = tb^0 = t(b_1^0, b_2^0) \qquad (9.7.2)$$

for some given load b^0.

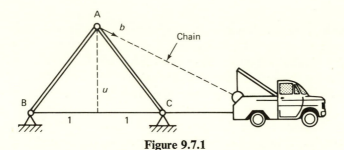

Figure 9.7.1

To simplify, assume that the bars are of uniform density and unit cross-sectional area. Also, let the weights of the bars be negligible relative to the load. In the diagram, the distance from A to the ground is u, the distance from B to C is 2 units, and the distances AB and AC are equal.

The structure will, of course, undergo a deformation when subjected to load changes. Expressly, when b is applied at the point A, the structure will deform as in Figure 9.7.2. The bar AB will elongate to $A'B$, while AC will shorten to $A'C$. (Of course, the drawing is exaggerated, as normally the deformations are hardly noticeable, even though they are crucial.)

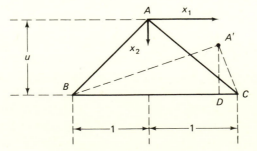

The structure ABC deforms to $A'BC$.

Figure 9.7.2

Let $x = (x_1, x_2)$ denote the displacement from A to A'. As the load parameter t increases, x_1 and x_2 will also increase. When t is sufficiently large, the structure is expected to buckle or snap through. Our goal is to determine how the structure changes under different loads tb^0. More precisely, as t changes, what are the corresponding changes in x?

DISPLACEMENT EQUATION

To conduct this analysis requires that we obtain the equilibrium equations governing the system. Using Figure 9.7.3, first consider $A'BC$, which is the structure at equilibrium after a load b is applied at A. We separate the load b into two vectors, $c = (c_1, c_2)$ and $d = (d_1, d_2)$, where c is pointing in the direction BA' and d is in the direction $A'C$. (Imagine replacing the truck pulling in the direction b with Superman and King Kong. Superman will tug on a chain attached at A' in the direction c, and King Kong will push A' toward C.)

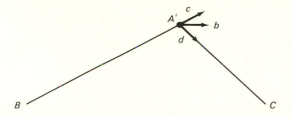

The load vector b is replaced with vectors c and d.

Figure 9.7.3

Vectorially, $c + d = b$; hence

$$c_1 + d_1 = b_1$$
$$c_2 + d_2 = b_2. \qquad (9.7.3)$$

METHOD OF SECTIONS

With c and d obtained, we will now isolate the forces on each bar AB and AC independently. This is called the *method of sections*. Intuitively, the method "disassembles" the load on the structure into loads on single bars, so that each bar can be analyzed independently.

Consider the bar AB as in Figure 9.7.4. A force c is pulling on the bar, thus stretching AB to a longer bar $A'B$. Denote

$$e = |A'B| - |AB| \qquad (9.7.4)$$

as the elongation of the bar AB. (Here $|AB|$ denotes length.)

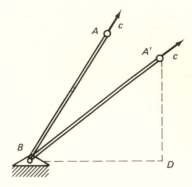

The method of sections isolates the forces on each bar separately.

Figure 9.7.4

Let us consider how to express the elongation e. It is reasonable for the elongation to be proportional to the force c and the length $|AB|$. Double the force c and we expect e to be twice as long. Also, if length $|AB|$ doubles, since there is twice as much material to yield, we would expect e to double. Specified in general, this becomes *Hooke's law*:

stress (force per unit area)

is proportional to strain (deformation per unit length).

In our example, with unit cross-sectional area, Hooke's law states:

$$e = k|c||AB|, \qquad\qquad (9.7.5)$$

where k is termed the modulus of elasticity or *Young's modulus*. The greater scalar k is, the more elastic is the bar, since it would then allow more stretching.

To simplify, let k equal 1. Then uniting (9.7.4) with Hooke's law (9.7.5), we obtain

$$|A'B| - |AB| = |c||AB|. \qquad\qquad (9.7.6a)$$

Similarly,

$$|A'C| - |AC| = -|d||AC|, \qquad\qquad (9.7.6b)$$

where there is a negative sign because $|AC|$ is compressed to $|A'C|$.

THE EQUILIBRIUM EQUATIONS

All that remains is to use our knowledge of trigonometry to consolidate the formulas. So far, we have from (9.7.3),

$$c_1 + d_1 = b_1$$
$$c_2 + d_2 = b_2 \tag{9.7.7}$$

and from (9.7.6),

$$|A'B| - |AB| = |c||AB|$$
$$|A'C| - |AC| = -|d||AC|. \tag{9.7.8}$$

Using the similarity of triangles, as in Figure 9.7.5,

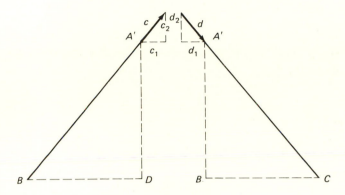

The vector c is pointing in the same direction as BA'. The vector d is pointing in the same direction as $A'C$.

Figure 9.7.5

$$\frac{c_1}{|c|} = \frac{|BD|}{|A'B|}, \qquad \frac{c_2}{|c|} = -\frac{|A'D|}{|A'B|} \tag{9.7.9}$$

$$\frac{d_1}{|d|} = \frac{|DC|}{|A'C|}, \qquad \frac{d_2}{|d|} = \frac{|A'D|}{|A'C|}. \tag{9.7.10}$$

Here, since c_2 is opposite the direction of d_2, there is a negative sign in (9.7.9). Via (9.7.8) and (9.7.9),

$$c_1 = \frac{|c||BD|}{|A'B|} = \frac{|BD|}{|A'B|}\left(\frac{|A'B| - |AB|}{|AB|}\right) = |BD|\left(\frac{1}{|AB|} - \frac{1}{|A'B|}\right)$$

$$c_2 = -\frac{|c||A'D|}{|A'B|} = -\frac{|A'D|}{|A'B|}\left(\frac{|A'B| - |AB|}{|AB|}\right) = -|A'D|\left(\frac{1}{|AB|} - \frac{1}{|A'B|}\right). \tag{9.7.11}$$

Similarly from (9.7.8) and (9.7.10), we obtain

$$d_1 = \frac{|d||DC|}{|A'C|} = |DC|\left(\frac{1}{|A'C|} - \frac{1}{|AC|}\right)$$

$$d_2 = \frac{|d||A'D|}{|A'C|} = |A'D|\left(\frac{1}{|A'C|} - \frac{1}{|AC|}\right). \tag{9.7.12}$$

Further, the geometry of Figure 9.7.2 provides

$$|DC| = 1 - x_1 \qquad\qquad |BD| = 1 + x_1$$

$$|A'D| = u - x_2 \qquad\qquad |A'C| = \sqrt{(1 - x_1)^2 + (u - x_2)^2}$$

$$(9.7.13)$$

$$|A'B| = \sqrt{(1 + x_1)^2 + (u - x_2)^2}, \qquad |AB| = |AC| = \sqrt{1 + u^2}. \qquad (9.7.14)$$

EQUATION STATEMENT

Using (9.7.11), (9.7.13), and (9.7.14), we can now replace (9.7.7) to get our equilibrium equations. Expressly, the equilibrium equations are, recalling that $b = tb^0 = t(b_1^0, b_2^0)$,

$$(1 + x_1)\left[\frac{1}{\sqrt{1 + u^2}} - \frac{1}{\sqrt{(1 + x_1)^2 + (u - x_2)^2}}\right]$$

$$+ (1 - x_1)\left[\frac{1}{\sqrt{(1 - x_1)^2 + (u - x_2)^2}} - \frac{1}{\sqrt{1 + u^2}}\right] = tb_1^0$$

$$-(u - x_2)\left[\frac{1}{\sqrt{1 + u^2}} - \frac{1}{\sqrt{(1 + x_1)^2 + (u - x_2)^2}}\right]$$

$$+ (u - x_2)\left[\frac{1}{\sqrt{(1 - x_1)^2 + (u - x_2)^2}} - \frac{1}{\sqrt{1 + u^2}}\right] = tb_2^0. \qquad (9.7.15)$$

This is our $H(x, t) = 0$ system and it specifies how x changes as t changes. In detail a point $x \in H^{-1}(t)$ describes the displacement x of point A under any load tb^0.

BUCKLING

When t gets large enough, as we know, buckling may result. Buckling occurs when $\dot{t} = 0$ (or, equivalently, H'_x is singular) and the path reverses itself. In Figure 9.7.6 notice that at $\bar{t}$ the path reverses and starts to decrease in t. But t is being increased because the load is being increased. Conceptually, the mathematics is saying: "The value $\bar{t}$ is my top. Watch out! If at $\bar{t}$ you want to maintain the integrity of the system, you must decrease t." Should this warning be unheeded and we persist in increasing t, the system snaps.

An important notion has been introduced here. The path informs us that at $\bar{t}$ we must decrease the parameter t to preserve the system. If, by external means we increase t, then the system must somehow break or collapse. This idea is fundamental and will be examined in a variety of contexts in Chapter 10.

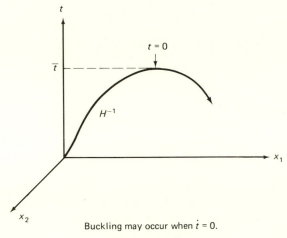

Buckling may occur when $\dot{t} = 0$.

Figure 9.7.6

SUMMARY

This chapter extended the application of our previous ideas by examining network equilibrium and elasticity. Both of these applications are very practical and widely employed. Overall, the ideas of path following and of equilibrium programming are seen to model numerous situations quite well.

EXERCISES/CHAPTER 9

1. Suppose that f is differentiable and convex in a single variable y. Moreover, suppose that $y = \sum_{i=1}^{n} c_i z_i$, where $c_1, \ldots, c_n$ are given numbers. Show that f is convex in z.

2. Prove that the sum of convex functions is convex.

3. Set up the EP formulation for Example 9.1.1.

4. Prove Lemma 9.6.1 and Theorem 9.6.2 completely.

5. In Example 9.1.1, suppose that we add a link T $\longrightarrow$ A, where delay is 0.001 per traveler. (On links $A \longrightarrow H$ and $A \longrightarrow C$, let the delay per person for the $T \longrightarrow C$ route equal the delay per person on the $A \longrightarrow C$ route.)
   a. Compute the new equilibrium.
   b. Calculate the system delay at the new equilibrium.

6. Figure E.9.1 shows a network of four cities. Each day, 1000 cars leave city A bound for city D and a thousand cars leave city C bound for city D. The figure also gives the delay on each link. For example, if x cars use link $A \longrightarrow C$ then

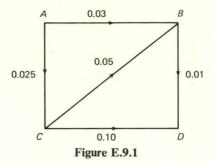

Figure E.9.1

the delay on the link is $0.025x$. Observe that there are three possible routes on the trip from A to D and two possible routes on the trip from C to D.

a. Calculate the equilibrium point for the system.

b. Calculate the system minimum.

7. Formulate the equilibrium programming problem for Figure E.9.1. Can the problem be formulated as a nonlinear programming problem? If yes, give the formulation. If no, explain why not.

8. In Figure E.9.1, suppose that a link from B to C were added such that the delay factor on the link is 0.02. Prove that at equilibrium, no car would travel from B to C and directly back to B.

9. Figure E.9.2 shows a two-bar structure subjected to a load $b = tb^0$. Formulate the equilibrium equations for this structure.

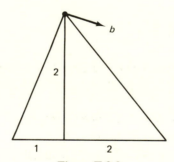

Figure E.9.2

NOTES

9.1. The research in network equilibrium theory is large. See, for example, Florian [1976], Magnanti and Golden [1978]. The concept of network equilibrium was first recognized by Wardrop [1952]. A set-theoretic formulation is given by Asmuth [1978].

9.2. The Braess paradox has generated much literature; see for example, Murchland [1970]. Its extension to general networks and a proof that the paradox can hold half of the time (and is not just a rare occurence) is in Steinberg and Zangwill [1981].

9.5. The formulation of a network equilibrium as an equilibrium programming problem is new.

9.6. For the special case (9.6.1), Beckman, McGuire, and Winston [1956] noted that the problem is simply the Kuhn–Tucker conditions of an NLP.

9.7. The two-bar structure we have presented here was studied by Rheinboldt [1977]. Other examples are given in Oden [1972].

10

Catastrophe Theory

Catastrophe theory is a new and exciting technique for modeling sudden, severe, or disruptive changes in a system. Most of the time a system operates in a fairly stable manner, but occasionally it undergoes a large or severe change, namely a *catastrophe*.

Most systems exhibit catastrophe points. Balls roll off cliffs, war occurs, stock markets plunge, divorces occur. The power of catastrophe theory is that it presents a model for the understanding and analysis of such sudden changes.

10.1 UNCONSTRAINED CATASTROPHES

For $f: R^{n+1} \longrightarrow R^1$ consider the problem

$$\min_x f(x, t). \tag{10.1.1}$$

We want to determine the behavior of the optimizing $\overset{*}{x}$ to (10.1.1), as the parameter t varies. Clearly (10.1.1) is a special form of the dynamic NLP formulated in Chapter 4. [The reader is strongly advised to review Chapter 4. Note also that, instead of maximizing, problem (10.1.1) minimizes, as that is more natural in the study of catastrophes.]

Chapter 4 stressed solving an NLP via homotopy techniques. This chapter, however, develops (10.1.1) in a deeper way, so that it can exhibit catastrophes. Now $f(x, t)$ is not a homotopy at all. Rather it represents something that has political, economic, social or physical meaning, for example the GNP of a country, the potential energy of an atomic particle, the evolutionary change of a living organism, or the likelihood function that it will rain tomorrow. Also, the parameter t reflects the dynamic aspect of the situation, and could be time, temperature, tension, torque, or any thing that can be controlled parametrically.

Usually, as the parameter t changes, the optimal point $\overset{*}{x}$ changes smoothly. That is, small changes in t produce small changes in $\overset{*}{x}$. Sometimes, however, small changes in t produce large changes or jumps in $\overset{*}{x}$. That means the system has had a large change because $\overset{*}{x}$ has jumped. The system then is said to undergo a catastrophe.

Let us now develop these ideas more precisely, and we begin with some definitions.

STABLE POINTS

Given $t \in R^1$ fixed, a point $\bar{x} \in R^n$ is termed a *local minimum* of f if there is a neighborhood of $\bar{x}$ such that for all x in the neighborhood, $f(x, t) \geq f(\bar{x}, t)$. The point $\bar{x}$ is a *strict* local minimum if for all $x \neq \bar{x}$, x in the neighborhood, $f(x, t) > f(\bar{x}, t)$.

If x is a local minimum, as is well known from calculus,

$$\nabla_x f(x, t)^\tau = 0. \tag{10.1.2}$$

Here the derivative is taken only with respect to x, because t is assumed fixed.

We are interested in all points satisfying (10.1.2) and for that define

$$H(x, t) \equiv \nabla_x f(x, t)^\tau = 0, \tag{10.1.3}$$

which is a system of n equations in $n + 1$ unknowns. Any solution (x, t) to (10.1.3) [or, equivalently, (10.1.2)] is said to be a *stationary point*. Expressly,

$$H^{-1} = \{(x, t) \mid H(x, t) = 0\}$$

is the set of all stationary points to (10.1.1).

Equation (10.1.2) is a necessary condition for x to be a local minimum but it is not sufficient. For instance, suppose that $(x, t) \in H^{-1}$. Then although x could be a local minimum, it could also be a local maximum or a saddle point. (In R^1 a saddle point is an inflection point.)

Given $(x, t) \in H^{-1}$, if the point x is a local minimum, we term (x, t) *stable*. Otherwise, (x, t) is called *unstable*. Rephrased, if x is a local minimum, (x, t) is called stable, whereas if x is either a local maximum or a saddle point, (x, t) is unstable (see Figure 10.1.1).

Clearly, H^{-1} can consist of both stable points and unstable points, depending if x is a local minimum or not. In particular, any specific path in H^{-1} may consist of both stable and unstable points. This observation that any path in

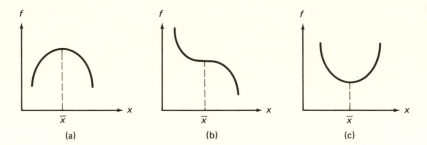

Given a fixed $t \in R^1$, suppose that the graph of f relative to x is as depicted in the figures. In (a) and (b), no (x, t) is stable. In (a), $\bar{x}$ is a local maximum point, so $(\bar{x}, t)$ is not stable. In (b), $\bar{x}$ is an inflection point (saddle point), so $(\bar{x}, t)$ is not stable. In (c), $\bar{x}$ is a local minimum point, so $(\bar{x}, t)$ is stable.

Figure 10.1.1

H^{-1} may consist of both stable points and unstable points is what will produce the catastrophic behavior. If the system moves from a stable point to an unstable point, a catastrophe can occur. Let us explain this with the next two examples.

EXAMPLE 10.1.1

Picture a ball high on a mountaintop (Figure 10.1.2). This is an unnatural situation because a slight wind or tremor will cause the ball to roll downhill. Instead, the ball seeks local minima, such as the valleys and canyons, as there the ball cannot be shaken loose and is stable. In this situation the function f of (10.1.1) represents the potential energy of the ball, and the ball attempts to

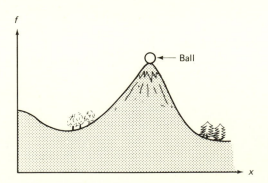

A ball on a mountaintop is an unnatural occurrence.

Figure 10.1.2

minimize potential energy. Stability is therefore quite a natural concept for the ball, and only local minimum points can be stable.

EXAMPLE 10.1.2

Now let us examine Figure 10.1.3, as it will reveal the catastrophe. In the figure we have constructed a surface such that curves *ab* and *de* are perfectly level. However, *bc* is inclined upward in the direction of *c*. Note that for a given *t*, points along *de* and along *ab* are local minima, hence stable. Also, points along *bc* are local maxima, hence unstable. Point *b* is an inflection point and also unstable.

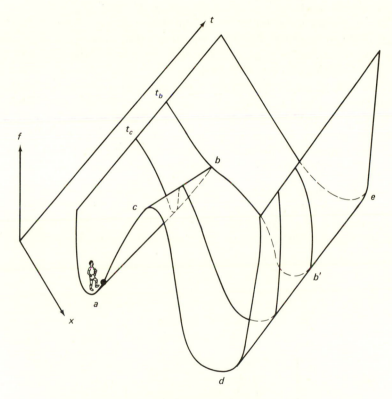

A boy kicking a ball is at point *a*.

Figure 10.1.3

Imagine that a boy placed at *a* is playfully kicking a ball in the direction of increasing *t*. Note that as the boy walks, he will proceed from *a* to *b*. He can be nowhere else because the ball will remain in the canyon *ab*. Mathematically, as *t* gradually increases, the ball gradually moves from *a* to *b*.

Suppose that at $t = t_b$, the boy has walked to point b. Increase t a slight amount and note that the canyon disappears at point b. Suppose that the boy nudges the ball past b. Then the ball will tumble toward point b'. A catastrophe is said to occur at b. The position of the ball, call it x, was changed dramatically for a small change in t. This is a catastrophe. Namely, there is a sudden change in the value of x for a slight change in t.

From b', if the boy continues to move in the direction of increasing t, he and the ball will move to e.

In sum, from a suppose that the ball is pushed in the direction of increasing t. The ball will move from a to b, then rapidly to b', and finally from b' to e.

DIRECTION OF TRAVERSAL IS IMPORTANT   Now suppose that the boy is at e and decides to walk back in the direction of decreasing t, still kicking the ball. Can he reach his starting point a? The answer is no. Notice now that b' is not a catastrophe point. The ball will simply roll along from e toward d. We see that the direction of traversal is very important. What happens going forward need not happen going backward.

To review, at any t the ball seeks the position x at which it minimizes potential energy. For f denoting potential energy, the ball thus moves to a local minimum of (10.1.1). As t changes, and depending upon the contour of the land, the ball may move only slightly or a lot. When it moves a lot, such as at b, a catastrophe is said to occur. Further, depending upon which direction t changes (increase or decrease) we may get different points of catastrophe.

ANALYSIS   The preceding example has some important implications. The ball seeks to minimize potential energy. First, carefully examine Figure 10.1.3 and notice that the set H^{-1} of all stationary points is as depicted in Figure 10.1.4. The set H^{-1} is simply the curves abc and de projected onto the x-t plane. Point

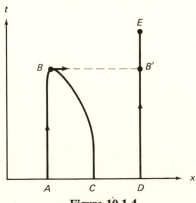

Figure 10.1.4

A is the projection of *a*, *B* the projection of *b*, and similarly for the other points. Recall that the points on the path from *a* to *b* consist of local minima, *b* is a saddle point (inflection point), and *b* to *c* are local maxima. Thus *A* to *B* are local minima, *B* a saddle point, *B* to *C* are local maxima, and so on. Perceive, as the ball moves from *a* to *b* in Figure 10.1.3, that the (x, t) coordinates of the ball move from *A* to *B* in Figure 10.1.4.

Suppose that the (x, t) coordinates of the ball move from *A* to *B*. At point *B* the catastrophe results just as the system loses stability. The (x, t) coordinates then jump to another point in H^{-1}, namely *B'*. The ball seeks a point of stability (a local minimum of the potential energy). But point *B* is not a local minimum; hence with stability lost, the catastrophe occurs.

Overall, the set H^{-1} informs us as to the system behavior. At any *t*, since the ball minimizes potential energy, always $(x, t) \in H^{-1}$. At a catastrophe, however, the system jumps from one point in H^{-1} to a totally different point in H^{-1}.

These observations are fundamental and hold in general for arbitrary systems. Most of the time, a system proceeds stably and smoothly. That means that it moves along stable points of an H^{-1} path. Should a system lose stability, however, a large change (catastrophe) can occur. The system then jumps in H^{-1} to a new stable point. Whenever the system moves out of stability, it will jump to another point in H^{-1} that is stable.

We emphasize that a system always seeks stable points (the local minima in H^{-1}) and in fact will jump to attain stability. Such jumps are the catastrophes.

10.2 SECOND-ORDER CONDITIONS

With the basics of the catastrophe theory introduced, let us now consider the second-order conditions as they specify when a catastrophe can occur.

For *t* fixed, define $\nabla_{xx}f$ as the Hessian matrix of *f*. The matrix $\nabla_{xx}f$ is the $n \times n$ matrix of second partial derivatives

$$\nabla_{xx}f = \begin{pmatrix} \dfrac{\partial^2 f}{\partial x_1\, \partial x_1} & \dfrac{\partial^2 f}{\partial x_1\, \partial x_2} & \cdots & \dfrac{\partial^2 f}{\partial x_1\, \partial x_n} \\ \cdot & \cdot & \cdot & \cdot \\ \cdot & \cdot & \cdot & \cdot \\ \dfrac{\partial^2 f}{\partial x_n\, \partial x_1} & \dfrac{\partial^2 f}{\partial x_n\, \partial x_2} & \cdots & \dfrac{\partial^2 f}{\partial x_n\, \partial x_n} \end{pmatrix}. \tag{10.2.1}$$

Calculus informs us that at *t*, if *x* is a local minimum, the matrix $\nabla_{xx}f(x, t)$ is positive semidefinite. Also, for $(x, t) \in H^{-1}$, if $\nabla_{xx}f(x, t)$ is positive definite, then *x* is a strict local minimum. (Definitions of positive definite and positive semidefinite are provided in Exercise 2.) Further, if $\nabla_{xx}f(x, t)$ is positive definite,

$$\det \nabla_{xx}f(x, t) > 0. \tag{10.2.2}$$

It follows immediately that should $(x, t) \in H^{-1}$ and $\nabla_{xx} f(x, t)$ be positive definite, (x, t) must be a point of stability.

Related conditions also hold for maxima and saddle points. If x is a local maximum at t, $\nabla_{xx} f(x, t)$ is negative semidefinite. If $(x, t) \in H^{-1}$ and $\nabla_{xx} f(x, t)$ is negative definite, x is a strict local maximum. However, if $(x, t) \in H^{-1}$ and $\nabla_{xx} f(x, t)$ is neither negative semidefinite nor positive semidefinite, x must be a saddle point.

EXAMPLE

Considering Example 10.1.2, since $x \in R^1$,

$$\frac{\partial^2 f}{\partial x^2} = \nabla_{xx} f$$

for t fixed. From calculus we know $\partial^2 f / \partial x^2$ is the change in the slope of f. It is then fairly easy to perceive that with t fixed,

$$\frac{\partial^2 f(x, t)}{\partial x^2} > 0$$

for all points on path de, and also from a up to but not including b. At b,

$$\frac{\partial^2 f(x, t)}{\partial x^2} = 0.$$

Beyond b and on the path to c,

$$\frac{\partial^2 f(x, t)}{\partial x^2} < 0.$$

The second-order conditions validate our previous observations about stability. Specifically, the second-order conditions confirm that all points along de and from a up to but not including b are stable, and that all other points in H^{-1} are unstable.

STABILITY AND HESSIANS

Using the Hessian matrix $\nabla_{xx} f$, we can develop an important theorem about stability which generalizes our observations about Example 10.1.2. Specify

$$\nabla_{xt} f = \begin{pmatrix} \dfrac{\partial^2 f}{\partial x_1 \, \partial t} \\ \cdot \\ \cdot \\ \cdot \\ \dfrac{\partial^2 f}{\partial x_n \, \partial t} \end{pmatrix}$$

as the n vector of second derivatives indicated. Then the $n \times (n + 1)$ Jacobian H' can be written

$$H' = (\nabla_{xx} f, \nabla_{xt} f). \qquad (10.2.3)$$

Now calling on the BDE of Chapter 2,

$$\dot{\imath} = (-1)^{n+1} \det H'_{-t},$$

which by (10.2.3) becomes

$$\dot{\imath} = (-1)^{n+1} \det \nabla_{xx} f(x, t). \qquad (10.2.4)$$

Equation (10.2.4) makes the important declaration that t changes direction if and only if $\det \nabla_{xx} f(x, t)$ changes signs. For example, suppose that at a point (x, t) on the path, $\nabla_{xx} f(x, t)$ is positive definite. Then, as discussed, (x, t) is a point of stability and, by (10.2.2), $\dot{\imath} \neq 0$.

Even stronger, suppose we know that (x, t) is a point of stability and that $\dot{\imath} \neq 0$. Then as the following theorem shows, there cannot be a point of catastrophe in a neighborhood of (x, t). (Recall the notation $(x(p), t(p)) \in H^{-1}$, where p parametrizes the path.)

THEOREM 10.2.1   Let $f : R^{n+1} \longrightarrow R^1$ be C^3 and H as defined by (10.1.3) be regular. Suppose that for some $\bar{p}$, $(x(\bar{p}), t(\bar{p}))$ is a stable point such that $\dot{\imath}(\bar{p}) \neq 0$. Then for all p sufficiently near $\bar{p}$, $(x(p), t(p))$ is also a stable point.

Proof.   Since $(x(\bar{p}), t(\bar{p}))$ is a stable point, $x(\bar{p})$ must be a local minimum at $t(\bar{p})$, and hence $\nabla_{xx} f(x(\bar{p}), t(\bar{p}))$ is positive semidefinite. But $\dot{\imath}(\bar{p}) \neq 0$, so by (10.2.4), $\nabla_{xx} f(x(\bar{p}), t(\bar{p}))$ is positive definite.

Via continuity, $\nabla_{xx} f(x(p), t(p))$ will be positive definite in a neighborhood of $(x(\bar{p}), t(\bar{p}))$. Thus all points on the path near $(x(\bar{p}), t(\bar{p}))$ must be stable.   □

The theorem states that if $\dot{\imath}(\bar{p}) \neq 0$, then the catastrophe cannot occur at $(x(\bar{p}), t(\bar{p}))$. Plus we can move at least a little farther on the path and still encounter only stable points. Indeed, catastrophes can occur only when $\dot{\imath} = 0$. Theorem 10.2.1 thus tells us exactly where to search for catastrophes.

EXAMPLE 10.2.1.

As an example, consider Figure 10.2.1, which depicts the stationary points H^{-1} for a particular function. Notice that $\dot{\imath} = 0$ only at A, B, C and D, so if any catastrophe exists, it must occur at one or more of these points.

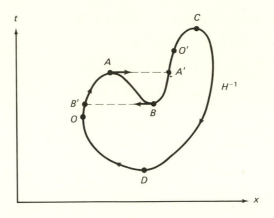

The set H^{-1}

Figure 10.2.1

The point 0, as noted, has $i \neq 0$, and let us presume that it is stable. Then $\nabla_{xx}f$ is positive definite there, and letting n be odd in (10.2.4), we have $i > 0$ at 0. (If n is even, the same reasoning holds with i sign-reversed.) Theorem 10.2.1 implies that the entire open arc DA (excluding points D and A) must be stable points, since $i > 0$ on the entire open arc DA. Similarly, $i > 0$ on the open arc BC, but $i < 0$ on open arcs AB and CD.

For concreteness, let $n = 1$. Then

$$i = (-1)^{1+1} \det \frac{\partial^2 f(x, t)}{\partial x^2}$$

and we have that $\partial^2 f/\partial x^2$ is positive on open arcs DA and BC and negative on open arcs AB and CD. Indeed, it must be that A, B, C, and D are inflection points; DA and BC are local minima; and AB and CD are local maxima.

Let us now examine what happens as t is changed. Suppose that we increase t starting from 0. We remain at stable points until reaching point A. However, at A it is impossible to continue further along the path since AB is unstable. If at A we increase t slightly, a jump must occur. Notice that we cannot land on arc CD since CD is unstable. It can only be that we land at A'. The point A is a catastrophe point and the system leaps from A to A'.

Conversely, suppose that we start at O' and gradually decrease t. Traversing H^{-1} backward, the stable point will move toward A' and then past A' toward B. Notice that A' is not a catastrophe point. However, at B, $i = 0$ and the path near B vanishes when t is decreased. We then jump to B' and, continuing, gradually move back to O. The catastrophe now occurs at B and not at A'.

Again we see how the catastrophe point differs if we increase or decrease t. Also, the example illustrates that analysis of i and of $\nabla_{xx}f$ can help pinpoint where catastrophes occur.

THE BDE SIGNALS THE COMING
OF A CATASTROPHE

As a catastrophe is approached, $\dot{\imath}$ becomes zero. But observe that $dx/dt = \dot{x}/\dot{\imath}$ then grows to infinity. In other words, the change in the value of x relative to a change in t will become faster and faster as the catastrophe is approached. This is the wind before the storm. Just before the coming catastrophe, dx/dt is going infinite. This is a warning that the knowledgeable listener can discern before the actual catastrophe. Note, however, that this warning signal may come so quickly and so near a catastrophe that we may not have enough time to act on it.

EXAMPLE 10.2.2.   Testing a New Drug

A well-known child care facility in the Midwest has developed a new drug, cyanide diptho trichonol (CDT), for treating children with Pushkin's disease. Unfortunately, CDT is a very potent drug and can be administered only with expert supervision. For example, if a child with Pushkin's disease is administered CDT, the probability that he or she lives to age 65 is almost triple that of a sick child not using CDT. However, if a healthy child is given CDT, the probability that he or she lives to age 65 is cut almost in half.

Let t, $0 \leq t \leq 1$, denote the degree of sickness of a child with Pushkin's disease. If $t = 0$, the child is well, but if $t = 1$, he or she is very ill with the disease. Also, let x, $0 \leq x \leq 1$, denote the dosage of CDT administered to the child. If $x = 0$, then no CDT is administered, whereas if $x = 1$, the full dosage of CDT is administered.

Under various (x, t), the probability $f(x, t)$ that the child lives to age 65 was determined after extensive testing to be

$$f(x, t) = \frac{-50x^4 + 100x^3 - 63x^2 + 26tx - 26t + 33}{100}. \qquad (10.2.5)$$

For example:

1. If $x = 0, t = 0$ (healthy child is not given CDT), then $f(0, 0) = 0.33$.
2. If $x = 1, t = 0$ (healthy child is given full dose of CDT), then $f(1, 0) = 0.2$.
3. If $x = 0, t = 1$ (very sick child is not given CDT), then $f(0, 1) = 0.07$.
4. If $x = 1, t = 1$ (very sick child is given full dose of CDT), then $f(1, 1) = 0.2$.

We might suppose that as a sick child's health improves, the dosage of CDT is gradually decreased until no drug is given when the child is completely healthy. This conjecture is patently false, as we will now show.

ANALYSIS   Given t, our objective is to

$$\max_{x} f(x, t). \qquad (10.2.6)$$

In words, given the degree of illness t, we want to select the dosage x that maximizes the child's probability of living to 65. Observe that (10.2.6) is a maximum problem. This means that stable points will now be local maxima. Also recall that $\nabla_{xx}f$ negative definite implies that x is a strict local maximum.

For f defined by (10.2.5),

$$H(x, t) \equiv \nabla_x f(x, t) = \frac{-200x^3 + 300x^2 - 126x + 26t}{100} = 0$$

Hence

$$t = \frac{100x^3 - 150x^2 + 63x}{13} \qquad (10.2.7)$$

defines the set H^{-1}. The graph of (10.2.7) is shown in Figure 10.2.2.

From the BDE,

$$i = \nabla_{xx}f = \frac{-600x^2 + 600x - 126}{100}. \qquad (10.2.8)$$

At $x = 0$, $t = 0$; clearly, $i = \nabla_{xx}f(0, 0) < 0$. Thus the point $(0, 0)$ is stable. (Recall that negative definite now implies stability.) From Theorem 10.2.1, and examining Figure 10.2.2, OB and AC are stable arcs and AB is an unstable arc.

Suppose that we have a very sick child ($t = 1$). The optimal strategy, then, is to administer the full dosage of CDT ($x = 1$). As the child's health

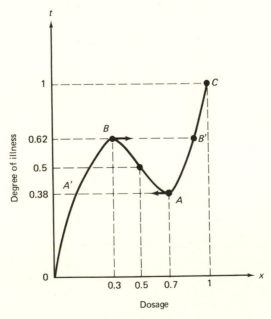

Figure 10.2.2

gets better (t decreases from 1), the dosage is decreased (x decreases from 1) using the strategy described by arc CA. As the child's health gets better and better and, as we near the point A, the rate of decrease of CDT dosage increases. Then suddenly, when the child gets slightly better than at point A ($t < 49/130$ $\doteq 0.38$), the dosage is quickly dropped to A'. In other words, the child has overcome a critical point, and medication must now be drastically curtailed. The point A is a catastrophe, for a slight decrease in t yields a sudden decrease in dosage x. We then continue to decrease the dosage until, finally, no drug is given when the child is completely healthy.

REVERSING THE PATH DIRECTION

Suppose that we traverse the path backward from $(x, t) = (0, 0)$ to (x, t) $= (1, 1)$. Relative to our example, we are seeking the dosage schedule for a child who is getting progressively worse. Let us observe the rather surprising result that the dosage schedule is not the reverse of that for the child who is getting healthier. This is because, starting from O in Figure 10.2.2, $i \neq 0$ until we reach the point B. The catastrophe occurs at B and not at A'. Indeed, the treatment for a child getting progressively worse is quite different from one getting progressively better. (See also Exercise 11 for the analysis of global maxima).

10.3 OTHER FORMS OF CATASTROPHE

In the preceding two sections, we modeled catastrophe as being a sudden change in the optimal point $\overset{*}{x}$ to (10.1.1). There is no reason why the catastrophe model would be limited to optimum-seeking problems. In Chapter 9, for example, we suggested that catastrophic behavior can occur for elasticity problems, where no dynamic NLP was formulated. Let us now investigate these more general situations.

Suppose that we are given a *dynamical system* which is described by the system

$$H(x, t) = 0 \qquad (10.3.1)$$

for $H: R^{n+1} \longrightarrow R^n$, $x \in R^n$ and $t \in R^1$. Here x is the state variable and t is the control parameter. Given any control t, the system H determines the state x of the system. Previously in the chapter, at any t the system sought x to be a local minimum, so $(x, t) \in H^{-1}$. Now x need not be a local minimum. All that is required now is that at t the system seeks x such that $(x, t) \in H^{-1}$.

As t changes, the solution x that satisfies $H(x, t) = 0$ would presumably change. Catastrophes occur at points $(x, t) \in H^{-1}$, where a large change in x results from a small change in t. Necessarily, this occurs at points of H^{-1}, where

$i = 0$. Let us illustrate this with two examples, the first of which deals with elasticity, the second with emotions.

EXAMPLE 10.3.1. Elasticity

Relative to the elasticity discussion of Chapter 9, recall that we placed a structure under stress. Also recall that if $(x, t) \in H^{-1}$, x represented the system displacement at stress level t. Now suppose that H^{-1} is as depicted in Figure 10.3.1.

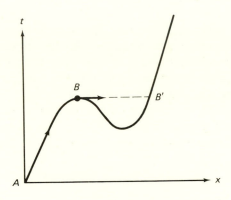

Here x is the system displacement
and t is the stress parameter.

Figure 10.3.1

Starting from point A, assume that the stress as indicated by t is increased. Then the displacement is determined by the curve AB. However at B, $i = 0$. For any further increase in stress, the system must jump to B', and hence B is a point of catastrophe. Notice that the catastrophe at B indicates a very real phenomenon, specifically that some breakage has occurred in the underlying structure. That is, the only way the structure can go from B to B' is to break.

Overall, catastrophe theory represents a quite useful means to study elastic structures, and in particular to analyze such structures under extreme stress. Technically, this case may be considered under (10.1.1) since x minimizes energy, but that is irrelevant. What is important is that the system satisfies $(x, t) \in H^{-1}$ and that we must study the H^{-1} paths to learn its behavior.

The next example is particularly fascinating, as it considers how the underlying emotions of fear or rage can trigger different behaviors, namely fight or flight (escape).

EXAMPLE 10.3.2.   Aggressive Behavior

It has been conjectured that rage and fear are conflicting factors that influence aggression. For example,

(A) an increase in rage causes an increase in aggression.

(B) an increase in fear causes a decrease in aggression.   (10.3.2)

The situation is pictured in Figure 10.3.2. In (a), keeping fear t_2 at a low constant value, as rage t_1 increases, aggression x increases. In (b), keeping rage at a constant low value, as fear t_2 increases, aggression x diminishes.

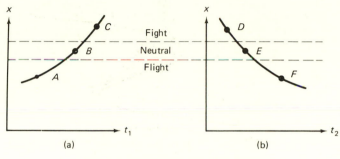

If rage t_1 increases, aggression x increases.

If fear t_2 increases, aggression x decreases.

Figure 10.3.2

To be more precise, consider the situation of a female dog, Spot, whose dinner has just been disturbed by a male dog, Blackie. Point A of Figure 10. 3.2a indicates flight, which means that Spot runs away with her bone if Blackie merely watches from afar. If, however, Blackie approaches Spot, her rage increases and she starts growling at B. Finally, at C, when Blackie lunges for the bone, she fights.

Relative to Figure 10.3.2b, at D, with fear at a minimum she fights for the bone. Should she lose the fight and fear increases, she becomes more reticent at E, until, at F, she runs away.

CATASTROPHE BEHAVIOR   Figures 10.3.2a and 10.3.2b depict how the behavior of fight, neutral, or flight depends upon the amount of rage t_1 or of fear t_2. However, these figures are quite simple, as one control (rage or fear) was varied and the other was held constant. Also note that so far no catastrophe has occurred, since the state x changes smoothly as a control t_1 or t_2 changes.

Let us now study what could happen if both t_1 and t_2 are increased together. To examine this situation, we can design experiments in which the state x is observed for various values of t_1 and t_2. Suppose that the result of such experiments is plotted on a three—dimensional graph, as in Figure 10.3.3.

That is, Figure 10.3.3 indicates the degree of aggression x depending upon the amount of rage t_1 and fear t_2.

Also using a parameter p, the dashed lines indicate particular paths $x(p)$, $t_1(p), t_2(p)$ traced out on the surface. The three different paths indicate three different types of situations. On each path, p is assumed to be increasing in the direction of the arrow.

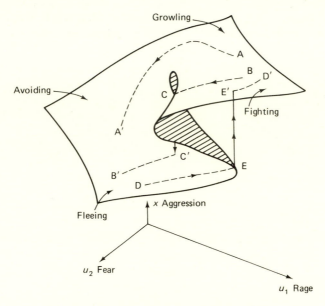

The graph of aggression x as rage u_1 and fear u_2 change.

Figure 10.3.3

ANALYSIS   As long as we remain outside the shaded area, no catastrophe will occur. For example, if the controls t_1 and t_2 are varied such that path AA' is taken, aggression changes smoothly and gradually. No catastrophe occurs because the level of rage or fear is low.

For path BB', rage and fear are both at high levels. Path BB' is drawn so that as p increases, fear increases and rage decreases. At B, Spot may be fighting Blackie ferociously. However, at point C, as she loses and gets hurt, the sense of fear becomes greater than rage, and suddenly Spot runs away. Technically, a catastrophe has occurred because Spot's behavior has jumped from fight at C to flight at C'. As fear increases still more to B', Spot remains in the flight mode of behavior.

Path DD' represents high rage and fear, with rage getting even higher while fear is decreasing. At D, Spot is fleeing. But as her rage increases, she moves toward E. Suddenly, the catastrophe occurs and Spot's behavior jumps to E', where she fights.

One-dimensional projections of the three paths are depicted in Figure 10.3.4.

Conceptually, Figure 10.3.3 represents an H^{-1} surface in which behavior x depends upon the emotions of rage t_1 and fear t_2. The position on that surface indicates whether Spot is in the fight mode of behavior, flight mode, or neutral mode. Also, the path taken on the surface illustrates how behavior can change. On path AA', the change in behavior mode is gradual. On paths BB' and DD', the behavior change is sudden.

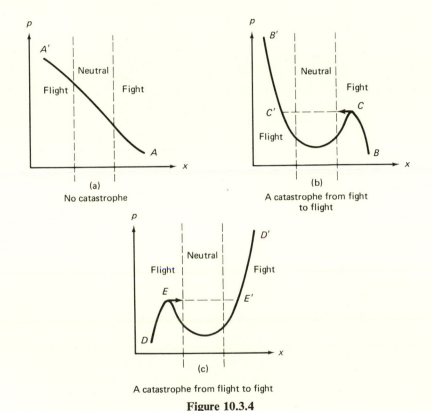

Figure 10.3.4

A catastrophe from flight to fight

OTHER EXAMPLES

Catastrophe theory can also model many other phenomena which exhibit rapid changes. For example, a cycle of the heartbeat is as follows: initially the heart is in a relaxed state (diastole). Then slowly, the heart contracts. This is followed by a rapid contraction, where it remains for a split second (systole). Then it relaxes a little and again follows a rapid relaxation back into diastole. Figure 10.2.1 depicts this behavior and it is left as Exercise 1 to interpret that figure as a heartbeat.

Other examples of sudden changes abound in nature. In the economy, a business cycle of ups and downs has been observed since the invention of money. In politics, we see the change from left to right and back every so many years. In biology, evolutionary biologists are questioning whether a totally gradualistic model of evolution should be enlarged to allow sudden modifications. One's personal life often undergoes many sudden changes. Given any dynamical system, its changes, large or small, can often be modeled by catastrophe theory.

PERSPECTIVE

Catastrophe theory represents a novel and insightful interpretation of systems that, although behaving smoothly much of the time, sometimes undergo sudden changes. Interestingly, as we have observed, the path backward may be quite different from the path forward. This implies that once a sudden change occurs, a mere reversal of what we are doing need not bring us back to the initial state. To return to the initial state may be much more difficult than undoing what has already been done, as is frequently observed in real life. Indeed, catastrophe theory is an extremely useful technique to examine how dynamical systems change and evolve over time.

EXERCISES/CHAPTER 10

1. Construct a model for explaining the heartbeat. (*Hint:* See Figure 10.2.1 and consider the equation

$$x^3 - 3x + 4 \cos \pi t = 0.)$$

2. (*Positive Definite*) Let A be a symmetric $n \times n$ matrix. A is said to be a *positive definite* matrix if $x^\tau A x$ is positive for all nonzero vectors x. (Similarly, we define *positive semidefinite, negative semidefinite,* and *negative definite* if $x^\tau A x \geq 0$ for all x, $x^\tau A x \leq 0$ for all x, and $x^\tau A x < 0$ for all $x \neq 0$, respectively.)
   a. For A positive semidefinite and symmetric, prove that there is a positive semidefinite symmetric (square root) $A^{1/2}$ satisfying $A^{1/2} A^{1/2} = A$.
   b. For A positive definite, prove that $\det A > 0$.

3. Let $f: R^{n+1} \longrightarrow R^1$ be C^2 and suppose that $x \in R^n$ is a local minimum of (10.1.1) given $t \in R^1$. Prove:
   a. $\nabla_x f(x, t) = 0$.
   b. $\nabla_{xx} f(x, t)$ is positive semidefinite.

4. Let $f: R^{n+1} \longrightarrow R^1$ be C^2 and suppose that for a fixed (x, t):
   a. $\nabla_x f(x, t) = 0$.
   b. $\nabla_{xx} f(x, t)$ is positive definite.
   Prove that x is a strict local minimum of (10.1.1) for the given t.

5. Consider the problem

$$\min_x 2x^3 + 3tx^2 - 72t^2x + t + 2, \qquad t \in R^1.$$

   a. Determine the set H^{-1}.
   b. Determine the stable points of H^{-1}.

6. Consider the problem

$$\min_x 2x^4 + 4x^3 - 24x^2 + 20x - tx + 5, \qquad t \in R^1.$$

   a. Determine the set of stable points.
   b. Determine the catastrophe points, if any.
   c. Describe the change in x as t increases from $t = -10$.
   d. Describe the change in x as t decreases from $t = 10$.

7. In Example 10.2.2, show that dx/dt approaches infinity as a catastrophe is approached. Also, suppose that catastrophe is recognized when $dx/dt \geq 100$. At what values of (x, t) will catastrophe be recognized as t is decreased from 1? As t is increased from 0?

8. Let $f: R^{n+1} \longrightarrow R^1$ be C^2 and suppose that for a given $(x, t) \in H^{-1}, \nabla_{xx}f(x, t)$ is neither positive semidefinite nor negative semidefinite. Show that (x, t) is a saddle point by exhibiting a direction where f increases and another direction where f decreases.

9. Suppose that $f: R^{n+1} \longrightarrow R^1$ is C^2 and that as we move along a path exactly one eigenvalue of $\nabla_{xx}f$ changes sign each time we pass through $i = 0$. Prove that to move along a path from a strict local minimum to a strict local maximum, $i = 0$ must occur along the path at least n times.

10. Suppose that $f: R^{n+1} \longrightarrow R^1$ is C^2 and strictly convex in x for any given t. Prove that the path cannot be a loop. Using this f in (10.1.1), can there be a catastrophe?

11. In Example 10.2.2, find the optimal strategy as t decreases from 1 if one is able to find the global maximum at any given t.

NOTES

10.1. Catastrophe theory was invented by Thom [1972]. He proved that for $t \in R^m$ and $m \leq 4$, there are only a few basic forms of catastrophe. This means that behavior near a catastrophe point must be diffeomorphic to one of these forms.

10.2. The path-following approach for catastrophe theory is shown in Garcia and Zangwill [1979e]. This paper also extends the Thom model to handle constraints.

10.3. The examples on aggressive behavior and the heartbeat are discussed at greater length in Zeeman [1977].

III

ALGORITHMS
AND
SOLUTION PROCEDURES

Path-Following Algorithms

OVERVIEW OF PART III

In the book so far we have supposed that a path could be followed without detailing how this could be done. For example, many of the proofs started with a point $x^0 \in H^{-1}(0)$ and assumed that the path could be followed in H^{-1} until reaching a terminal point x^1 in $H^{-1}(1)$.

Now in Part III of the book we present algorithms that numerically follow a path. These algorithms immediately make our previous proofs constructive. Equation solutions, fixed points, optimal points, equilibrium points, and other points can thereby actually be calculated. The existence of such points has been known for years, yet only recently was it discovered how to calculate them. This discovery transformed the field from one that was abstruse, abstract, and theoretical to one of immense practical import. For applications especially, this means that we can determine economic equilibria, optimal points, game solutions, and other forms of solutions. Innumerable problems that only a few years ago we had no way to attack are now easily solved.

There are two fundamental types of algorithms that follow a path in H^{-1}. Simplicial algorithms are the first type and the next four chapters discuss them. The second is by solving differential equations, and that approach is treated in Chapters 15 and 16.

SIMPLICIAL ALGORITHMS

This chapter initiates the discussion of simplicial algorithms. First we review the key ideas ensuring that a path goes to the point desired. Then we introduce the simplex, the simplicial approximation, and finally the basic step in all simplicial-type algorithms.

11.1 GETTING STARTED

The objective of our current discussion is to learn how to follow an underlying path in H^{-1} via an algorithm. The algorithm will typically generate its own path that more or less approximates an underlying path in H^{-1}. Before utilizing any algorithm, however, there is an initial phase in which the problem must be set up appropriately. This means, first, since there are many types of homotopies, we must select a homotopy that is appropriate to solve the specific problem. Second, we must ensure that following a path in H^{-1} does indeed lead to a solution to the problem. These two considerations are essential, irrespective of the algorithm, and must be implemented before utilizing any algorithm. Because of their importance, let us review some concepts developed about them earlier in the book.

1. As for the choice of a homotopy function, suppose, for example, that we want a solution to the system of equations

$$F(x) = 0, \tag{11.1.1}$$

where $F: R^n \longrightarrow R^n$. Several different homotopy functions might be appropriate. For instance, one could employ the Newton homotopy

$$H(x, t) = F(x) - (1 - t)F(x^0), \tag{11.1.2}$$

or the fixed-point homotopy,

$$H(x, t) = (1 - t)(x - x^0) + tF(x), \tag{11.1.3}$$

In both of these the path in H^{-1} would start from x^0.

If it were necessary to determine all solutions to (11.1.1), we would use

$$H(x, t) = (1 - t)G(x) + tF(x), \tag{11.1.4}$$

where G is defined using the ideas in Example 1.2.4 or the concepts developed in Chapter 18. In the all-solutions instance, we would typically have to follow many paths, one from each of the points $x^0 \in H^{-1}(0)$.

Clearly, there are a variety of homotopies to choose from, each with its own properties, and we must select the one that is appropriate for the problem under consideration. Of course, numerous examples of this have been given earlier.

2. Concerning where an underlying path in H^{-1} goes, usually, although not always, we desire it to go to a point $x^1 \in H^{-1}(1)$. Two caveats are quite important here:

   a. The path may not get to an $x^1 \in H^{-1}(1)$ because it may leave from a point $x^a \in H^{-1}(0)$ and return to another point, $x^b \in H^{-1}(0)$. If the fixed-point homotopy is used, only one point $x^0 \in H^{-1}(0)$ exists, so this behavior cannot occur. With other homotopies there may be more than one point in $H^{-1}(0)$, so the possibility of leaving from $x^a \in H^{-1}(0)$ and returning to $x^b \in H^{-1}(0)$ exists. However, Theorem 3.4.1 can help here. For example, suppose that det H'_x is of the same sign at all $x \in H^{-1}(0)$. Then such a return cannot occur.

   b. The second caveat is that the path may never get to $t = 1$ because it might hit a boundary of the set D, or if D is unbounded, the path might run off to infinity. To prevent such occurrences an appropriate boundary-free condition such as presented in Chapter 3 might help. Recall that H is boundary-free at $\bar{t}$ if for any solution $x \in H^{-1}(\bar{t})$ the point x is not in the boundary of D.

A condition related to boundary-free is the norm-coercive condition. We term H *norm-coercive* on $D \times T$ if $\| H(x, t) \| \longrightarrow \infty$ as either x approaches the boundary of D or $\| x \| \longrightarrow \infty$. This means that $\| H(x, t) \|$ gets infinitely large as x gets close to the boundary of D or as x diverges to infinity.

If H is norm-coercive,

$$H(x, t) = 0 \tag{11.1.5}$$

cannot have a solution for any $x \in \partial D$ or for $\| x \|$ sufficiently large. This is because as x gets close to the boundary of D, or as $\| x \|$ gets large, $\| H(x, t) \|$ gets very large. But $\| H(x, t) \|$ large prevents (11.1.5) from holding. In short, if H is norm-coercive, we are assured that the path cannot run off to infinity or hit a boundary.

Overall, then, Parts I and II of this book provide a variety of homotopy functions and properties to ensure that the path behaves appropriately and that by following it we go to the point desired. We reiterate: this aspect must be done before actually initiating the algorithm. Once this is adequately taken cared of, however, we can start following such an underlying path in H^{-1}, and let us now turn to that.

As mentioned, there are two main approaches to path following: solving the differential equation, which is discussed later, and the simplicial algorithm, which we analyze now. We commence study of the simplicial algorithm approach by first closely examining its underlying concept, the simplex.

11.2 SIMPLICES

Although in Latin "simplex" means simple, English provides a more accurate interpretation. A simplex, as its name implies, is neither simple nor complex, but between the two. However one feels about it etymologically, it is the basic building block from which we shall shortly create algorithms, so we must understand it well.

A *j-simplex* is the figure formed by $j + 1$ *independent* points. By independent we mean that no point can lie in the linear surface formed by the j other points. (See Exercise 10.) A 0-simplex is a point, 1-simplices are line segments, 2-simplices are triangles, 3-simplices are tetrahedra, and so on (see Figure 11.2.1). Since the points must be independent, for a 2-simplex the three points must form a triangle and cannot lie on a straight line, for a 3-simplex the four points must form a tetrahedron and cannot lie in a plane, and so on.

Because it is easily forgotten, remember that a *j*-simplex has $j + 1$ points. Also, we often write a simplex in terms of its $j + 1$ points as $s = \{v^0, v^1, \ldots, v^j\}$ (Figure 11.2.1). The v^i are called *vertices*.

There are several helpful ways to understand simplices, but an insightful means is via convex combinations and convex hulls. For readers not already acquainted with these terms, some familiarization exercises are provided at the

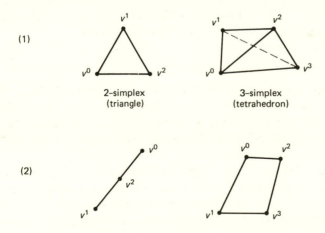

(1)

2–simplex
(triangle)

3–simplex
(tetrahedron)

(2)

In both instances, no simplex is formed because the points are not independent. Instead, each point lies in the linear surface formed by the other points.

(3)

A tetrahedron $s = \{v^0, v^1, v^2, v^3\}$ has 14 faces:
the 4 facets $\{v^0, v^1, v^2\}$, $\{v^0, v^1, v^3\}$, $\{v^0, v^2, v^3\}$, and $\{v^1, v^2, v^3\}$;
the 6 1-simplices $\{v^0, v^1\}$, $\{v^0, v^2\}$, $\{v^0, v^3\}$, $\{v^1, v^2\}$, $\{v^1, v^3\}$, $\{v^2, v^3\}$; and the 4 vertices $\{v^0\}$, $\{v^1\}$, $\{v^2\}$, $\{v^3\}$.

Figure 11.2.1

end of the chapter. A point w is a *convex combination* of the points v^i, $i = 1$, ..., k, if there exist

$$\lambda_i \geq 0, \qquad i = 1, \dots, k,$$

where

$$w = \sum_{i=1}^{k} \lambda_i v^i \tag{11.2.1}$$

and

$$\sum_{i=1}^{k} \lambda_i = 1. \tag{11.2.2}$$

Thus w is a weighted average of the k points v^i. The set of all possible convex combinations of the v^i is termed the *convex hull*. Thus the convex hull of v^i, $i = 0, \dots, j$, is

$$\{w \mid w = \sum_{i=0}^{j} \lambda_i v^i, \ \sum_{i=0}^{j} \lambda_i = 1, \ \lambda_i \geq 0\}.$$

Clearly, a j-simplex is the convex hull of the $j + 1$ independent points v^i, $i = 0, \dots, j$.

Convex hulls are quite useful in describing simplices. For instance, the convex hull of some but not all of vertices v^i is termed a *face* of s. If the face has j vertices, it has a special name and is called a *facet*. Although a droplet is a very small drop, and a roomette is a small room, a facet is not a small face. Rather, facets are the largest faces. Also notice that simplices are in a sense composed of themselves. For instance, any face is itself an i-simplex for some $i < j$, whereas a facet is a $(j - 1)$-simplex.

Simplices can be described in a variety of other ways as well. Examine, for instance, the set

$$\{x = (x_i) \mid \sum_{i=1}^{j} x_i \leq 1, \ x_i \geq 0\} \subset R^j.$$

This forms a j-simplex with $j + 1$ vertices consisting of the origin and the j points $(1, 0, \dots, 0), (0, 1, \dots, 0), \dots, (0, \dots, 0, 1)$. Geometrically, it is the set that results by intersecting the nonnegative orthant with the half-space $\sum_{i=1}^{j} x_i \leq 1$. This particular simplex is encountered frequently and is called a *unit simplex*.

Simplices have many fascinating properties, and we will now see how to construct functions on them. These functions will turn out to be approximations to the homotopy function H and are precisely the functions we will need for the algorithms.

11.3 FUNCTIONS ON SIMPLICES

COORDINATE WEIGHTS

Our initial step in constructing a function on a simplex $s = \{v^i\}$ is to express any point $w \in s$ as a weighted average of the vertices v^i. In particular, let $s = \{v^i\}_0^{n+1}$ be an $(n+1)$-simplex in R^{n+1}. As discussed, any point in it, $w \in s$, can be written as a convex combination of the vertices

$$w = \sum_{i=0}^{n+1} \lambda_i v^i \quad \text{for some} \quad \lambda_i \geq 0, \quad \sum_{i=0}^{n+1} \lambda_i = 1.$$

This way of writing w will be quite useful and to interpret it, consider the case for $n = 1$, so that we are in R^2 as in Figure 11.3.1. Points w on the line segment $[v^1, v^2]$ have $\lambda_0 = 0$, since w is then a convex combination of only v^1 and v^2. As λ_0 increases, the point w moves closer to v^0 until if $\lambda_0 = 1$, $w = v^0$. Clearly, any point $w \in s$ can be expressed by appropriate λ. In fact, the weights λ_i are uniquely determined, since if we change a λ_i, the w changes (Exercise 3). Notice that if w is interior to s, then all λ_i are strictly positive. Conversely, if w is in a facet, then $\lambda_i = 0$ for some i. For instance, if $w \in \{v^i\}_1^{n+1}$, then $\lambda_0 = 0$.

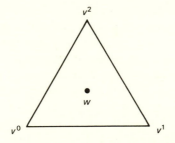

A point w is interior to a simplex $s = \{v^0, v^1, v^2\}$.

Figure 11.3.1

HOMOTOPY FUNCTION APPROXIMATIONS

Suppose that we are given a homotopy function $H: R^{n+1} \longrightarrow R^n$, and let $s = \{v^i\}_0^{n+1}$ be an $(n+1)$-simplex in R^{n+1}. We want to specify another function $G: s \longrightarrow R^n$ that approximates H on the simplex s. To do that, first evaluate H on the vertices v^i so that we have $H(v^i)$, $i = 0, \ldots, n+1$. Then, given a point

$$w = \sum_{i=0}^{n+1} \lambda_i v^i, \tag{11.3.1a}$$

where

$$\sum_{i=0}^{n+1} \lambda_i = 1 \quad \text{and} \quad \lambda_i \geq 0, \quad i = 0, \ldots, n+1,$$

define

$$G(w) = \sum_{i=0}^{n+1} \lambda_i H(v^i). \tag{11.3.1b}$$

Let us examine this definition of G more precisely. Given a simplex $s = \{v^i\}_0^{n+1}$, first calculate the $H(v^i)$, $i = 0, \ldots, n+1$. Next, for any point $w \in s$, obtain the weights λ_i such that

$$w = \sum_{i=0}^{n+1} \lambda_i v^i.$$

Using these $H(v^i)$ and λ_i, define

$$G(w) = \sum_{i=0}^{n+1} \lambda_i H(v^i).$$

Given the simplex $s = \{v^i\}_0^{n+1}$ and a point $w \in s$, we can obtain the weights λ_i and hence the function $G(w)$.

As w changes, the weights λ_i change and thus $G(w)$ changes. For example, suppose that w is a vertex so that $w = v^i$ for some i. Then $G(v^i) = H(v^i)$. This means that G is precisely H at the vertices of s. Also, as the weights λ_i change, observe from equation (11.3.1) that G varies linearly. For example, suppose that $\lambda_i = 0$, $i = 2, \ldots, n+1$. Then $w = \lambda_0 v^0 + \lambda_1 v^1$ and $G(w) = \lambda_0 H(v^0) + \lambda_1 H(v^1)$, so G is clearly linear. Indeed, we have created a function G which is a *linear approximation* to H on the simplex s and, moreover, is equal to H at the vertices (Figure 11.3.2).

This fact—that we can make a linear approximation G to H on any simplex s—is quite important, because as we will show in the next section, it will permit us to follow a path.

EXAMPLE 11.3.1

Let us now provide an example of how to construct G on a simplex s to approximate a given H.

Suppose that $H: R^2 \to R^1$ and let $s \subset R^2$ be the unit simplex. Then $s = \{v^0, v^1, v^2\}$ is the triangle with vertices

$$v^0 = \begin{pmatrix} 0 \\ 0 \end{pmatrix}, \qquad v^1 = \begin{pmatrix} 1 \\ 0 \end{pmatrix}, \qquad v^2 = \begin{pmatrix} 0 \\ 1 \end{pmatrix}.$$

Also suppose that

$$H(v^0) = 2, \qquad H(v^1) = -1, \qquad H(v^2) = 3.$$

Let us now calculate $G(w)$ for several w.

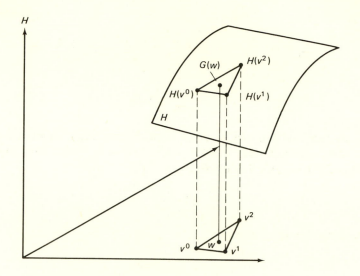

The graph of H: $R^2 \to R^1$ is the curved surface and it passes through $H(v^0)$, $H(v^1)$, and $H(v^2)$. The function G is linear and approximates H on the simplex $s = \{v^0, v^1, v^2\}$.

Figure 11.3.2

If

$$w = \begin{pmatrix} \frac{1}{4} \\ \frac{1}{4} \end{pmatrix},$$

then from (11.3.1a),

$$\begin{pmatrix} \frac{1}{4} \\ \frac{1}{4} \end{pmatrix} = \lambda_0 \begin{pmatrix} 0 \\ 0 \end{pmatrix} + \lambda_1 \begin{pmatrix} 1 \\ 0 \end{pmatrix} + \lambda_2 \begin{pmatrix} 0 \\ 1 \end{pmatrix}$$

$$1 = \lambda_0 + \lambda_1 + \lambda_2$$

$$\lambda_0 \geq 0, \quad \lambda_1 \geq 0, \quad \lambda_2 \geq 0,$$

which yields $\lambda_0 = \frac{1}{2}$, $\lambda_1 = \frac{1}{4}$, and $\lambda_2 = \frac{1}{4}$. Then, using these λ_i, from (11.3.1b),

$$G(w) = \tfrac{1}{2}(2) + \tfrac{1}{4}(-1) + \tfrac{1}{4}(3)$$

$$= \tfrac{3}{2}.$$

If

$$w = \begin{pmatrix} \frac{1}{3} \\ \frac{1}{6} \end{pmatrix},$$

solving (11.3.1a),

$$\begin{pmatrix} \frac{1}{3} \\ \frac{1}{6} \end{pmatrix} = \lambda_0 \begin{pmatrix} 0 \\ 0 \end{pmatrix} + \lambda_1 \begin{pmatrix} 1 \\ 0 \end{pmatrix} + \lambda_2 \begin{pmatrix} 0 \\ 1 \end{pmatrix}$$

$$1 = \lambda_0 + \lambda_1 + \lambda_2$$

$$\lambda_i \geq 0, \qquad i = 0, 1, 2,$$

so $\lambda_0 = \frac{1}{2}$, $\lambda_1 = \frac{1}{3}$, $\lambda_2 = \frac{1}{6}$. Then from (11.3.1b),

$$G(w) = \frac{1}{2}(2) + \frac{1}{3}(-1) + \frac{1}{6}(3)$$
$$= \frac{7}{6}.$$

Similarly, if

$$w = \begin{pmatrix} \frac{1}{4} \\ \frac{2}{3} \end{pmatrix},$$

$\lambda_0 = \frac{1}{12}$, $\lambda_1 = \frac{1}{4}$, $\lambda_2 = \frac{2}{3}$, so

$$G(w) = \frac{1}{12}(2) + \frac{1}{4}(-1) + \frac{2}{3}(3) = \frac{23}{12}.$$

We thus see how we can determine $G(w)$ given any $w \in s$. Further, notice that G is linear and G equals H on the vertices of s. Moreover, only the values of H on the vertices of s were needed in this.

11.4   CREATING A PATH

Our goal is to obtain a path in H^{-1} and hence to solve

$$H(y) = 0. \tag{11.4.1}$$

However, with H generally nonlinear, (11.4.1) is usually very difficult to solve. Consequently, since G approximates H, our approach is to actually solve for $w \in s$ in the system

$$G(w) = 0. \tag{11.4.2}$$

System (11.4.2) being linear in λ will be much easier to solve, and also its solutions will approximate those of (11.4.1). (The notational use of w corresponding to G and y to H should be clear.)

To solve (11.4.2), write it out in detail. Recall that $H(v^i)$, $i = 0, \dots, n+1$, is given. Using (11.3.1), (11.4.2) can then be expressed as

$$\sum_{i=0}^{n+1} \lambda_i H(v^i) = 0$$

$$\sum_{i=0}^{n+1} \lambda_i = 1, \qquad \lambda_i \geq 0, \tag{11.4.3}$$

where

$$w = \sum_{i=0}^{n+1} \lambda_i v^i. \tag{11.4.4}$$

In words, since the $H(v^i)$ are given, we first obtain the λ_i from (11.4.3). Then w from (11.4.4) solves

$$G(w) = 0.$$

Let us probe deeper into this idea. We want all $w \in s$ that satisfy $G(w)$ $= 0$. Written out, the system $G(w) = 0$ becomes system (11.4.3). Notice that (11.4.3) is linear in λ_i but with $n + 2$ variables and $n + 1$ equations, hence under reasonable assumptions will have an entire *line segment* as a solution. We reiterate: an entire line segment in s solves $G(w) = 0$. This line segment is quite important, as it will be part of our path, so we must consider it carefully.

THE EXAMPLE CONTINUED

To clarify what is happening, let us continue our example. Recall that $H: R^2 \to R^1$, where $s \subset R^2$ is the unit simplex. Thus

$$v^0 = \begin{pmatrix} 0 \\ 0 \end{pmatrix}, \qquad v^1 = \begin{pmatrix} 1 \\ 0 \end{pmatrix}, \qquad v^2 = \begin{pmatrix} 0 \\ 1 \end{pmatrix}.$$

Also,

$$H(v^0) = 2, \qquad H(v^1) = -1, \qquad H(v^2) = 3.$$

To solve $G(w) = 0$, we first use equation (11.4.3), which becomes

$$\lambda_0 \begin{pmatrix} 2 \\ 1 \end{pmatrix} + \lambda_1 \begin{pmatrix} -1 \\ 1 \end{pmatrix} + \lambda_2 \begin{pmatrix} 3 \\ 1 \end{pmatrix} = \begin{pmatrix} 0 \\ 1 \end{pmatrix}$$

$$\lambda_0, \lambda_1, \lambda_2 \geq 0.$$

Note that all $(\lambda_0, \lambda_1, \lambda_2)$ between $(\frac{1}{3}, \frac{2}{3}, 0)$ and $(0, \frac{3}{4}, \frac{1}{4})$ solve this system.

Using these λ_i and (11.4.3), we see the corresponding w which solve $G(w) = 0$ form the line segment from the point

$$w^0 = \tfrac{1}{3} v^0 + \tfrac{2}{3} v^1 = \begin{pmatrix} \frac{2}{3} \\ 0 \end{pmatrix}$$

across s to the point

$$w^1 = \tfrac{3}{4} v^1 + \tfrac{1}{4} v^2 = \begin{pmatrix} \frac{3}{4} \\ \frac{1}{4} \end{pmatrix}.$$

Parenthetically, see that $\{v^0, v^1\}$ and $\{v^1, v^2\}$ are the facets in s containing solutions to $G(w) = 0$. This is because $w^0 \in \{v^0, v^1\}$ and $w^1 \in \{v^1, v^2\}$ (Figure 11.4.1).

In the typical case, the line segment will run from one facet, say B^0, of s across s to another facet, say B^1, of s. Note that this result is exactly what we should suspect. Equation (11.4.1), which is nonlinear, yields a path as a solution. Equation (11.4.2) approximates (11.4.1) but is linear; therefore, it should yield a straight-line segment as a solution (Figure 11.4.2).

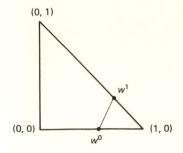

The line segment $[w^0, w^1]$ is the solution
set to $G(w) = 0$ on the simplex s.

Figure 11.4.1

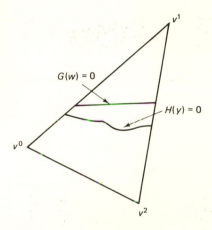

The curve that solves $H(y) = 0$ is approximated on the simplex
by the straight-line solution to $G(w) = 0$. Here, notationally, w
is the approximation to y. Note that the line segment goes from
one facet, $B^0 = \{v^0, v^1\}$, across s to another facet, $B^1 = \{v^1, v^2\}$.

Figure 11.4.2

To sum, the system $G(w) = 0$ is linear and it has $n + 1$ equations but $n + 2$ variables. Therefore, under reasonable assumptions it yields a straight-line segment as a solution.

REGULARITY

Of course, (11.4.2) may not always have a line segment as a solution, although it generally does. Specifically, G is called *regular* if the solution of (11.4.2) is a line segment extending from the interior of a facet B^0 of s to the interior of another facet B^1 of s. (Note the interiority.) Regularity nearly always holds, and if not, perturbing one or more v^i an arbitrarily small amount will make it hold (see Exercise 7 and Exercise 4 of Chapter 12.).

For emphasis we formally state the regularity condition in the following theorem.

THEOREM 11.4.1 Let $H: R^{n+1} \to R^n$ and a simplex $s \subset R^{n+1}$ be given. Define $G: s \to R^n$ as in (11.3.1), where G is regular. Then the solutions $w \in s$ to

$$G(w) = 0,$$

if nonempty, form a straight-line segment from a point w^0 interior to one facet of s across s to a point w^1 interior to another facet.

This theorem expresses the fundamental idea behind all simplicial algorithms. First, approximate H on a simplex s by a linear function G. Under regularity (which almost always holds and if not is easily brought about), the solutions to $G(w) = 0$ form a line segment across s. Furthermore, the solutions form a line segment that crosses s from a point w^0 interior to a facet B^0 of s to a point w^1 interior to another facet B^1 of s.

11.5 SIMPLICIAL ALGORITHMS

Simplicial algorithms operate as follows. First they take a simplex s^0 and generate the line segment across s^0. Then they take an adjacent simplex s^1 and generate a line segment across s^1. (By *adjacent*, we mean that s^0 and s^1 share a common facet and lie on opposite sides of that facet.) The line segment across s^1 will connect with the line segment across s^0. Then they take a third simplex, s^2, where s^2 is adjacent to s^1, and generate the line segment across s^2. Also, the line segment across s^2 will connect with the segment across s^1. In this manner, simplex by simplex and line segment by line segment, a path of line segments is generated.

On each simplex a new approximation G is made of H, and the line segment is generated by solving $G(w) = 0$ on that simplex. This is done simplex by simplex and since G approximates H on each simplex, the resulting path of line segments approximate a path in H^{-1}. Indeed, if the simplices are small, we get a fairly good approximation to a path in H^{-1}. This is the overall concept of how simplicial algorithms operate, and we now must examine the details.

We will start with a point w^0 in a facet B^0 of a simplex s^0, where

$$G(w^0) = 0.$$

Then a path solving $G(w) = 0$ will be generated from w^0 across s^0 to a point w^1 in facet B^1 of simplex s^0.

At this juncture, a new simplex, s^1, also with facet B^1 but on the other side of B^1, will be constructed. The new simplex s^1 will consist of the vertices of B^1 plus a brand new vertex v^{n+2}. Note how in Figure 11.5.1 the facet B^1 is shared by two simplices, one on either side of B^1.

Next, a path will be generated just as before but now across s^1. The path will go from w^1 in B^1 to a new point w^2 in some other facet, call it B^2, of s^1.

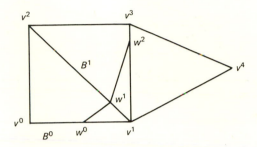

The path starts from w^0 interior to facet $B^0 = \{v^0, v^1\}$.
The path then proceeds across $s^0 = \{v^0, v^1, v^2\}$ to w^1,
a point interior to facet $B^1 = \{v^1, v^2\}$. From w^1 the
path goes across $s^1 = \{v^1, v^2, v^3\}$ to w^2, and so on.

Figure 11.5.1

The process will continue in this manner (see Figure 11.5.2). Using facet B^2 and yet another new vertex v^{n+3}, we will generate a new simplex

$$s^2 = B^2 \cup \{v^{n+3}\}.$$

A path from w^2 to w^3 will go across s^2, and so on.

Let us summarize the central idea. We generate a line segment across a simplex to a facet of that simplex. But that facet (unless we terminate) is shared by another simplex on the other side. We then continue the solution path as a line segment across this new simplex. In this manner a piecewise-linear (PL) path will be generated, simplex by simplex. The path is *piecewise linear* because it is linear on each simplex.

On each simplex we make a new approximation G of H, and we denote the solutions to $G(w) = 0$ across simplices by G^{-1}. If the simplices are fairly small, then, as we discuss in Chapter 12, the path in G^{-1} approximates a path in H^{-1} fairly well. Also note that since we are now more concerned with paths of G, the requirements on H can be relaxed somewhat. Indeed, and as we will later see, simplicial algorithms can operate with H only continuous.

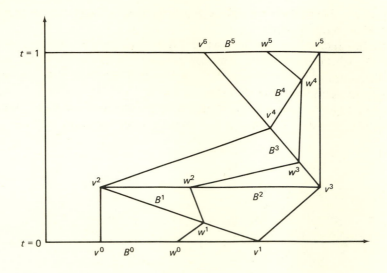

A piecewise-linear path w^0, w^1, ..., w^5 is generated from start w^0 to finish w^5.

Figure 11.5.2

EXTENSION OF EXAMPLE

In the preceding example we generated a path from $w^0 = \begin{pmatrix} \frac{2}{3} \\ 0 \end{pmatrix}$ across the unit simplex s^0 to $w^1 = \begin{pmatrix} \frac{3}{4} \\ \frac{1}{4} \end{pmatrix}$. The point w^1 is in the facet $B^1 = \left\{ \begin{pmatrix} 1 \\ 0 \end{pmatrix}, \begin{pmatrix} 0 \\ 1 \end{pmatrix} \right\}$, Let the new vertex be $v^3 = \begin{pmatrix} 1 \\ 1 \end{pmatrix}$, so that the new simplex on the other side of B^1 is

$$s^1 = B^1 \cup \{v^3\} = \left\{ \begin{pmatrix} 1 \\ 0 \end{pmatrix}, \begin{pmatrix} 0 \\ 1 \end{pmatrix}, \begin{pmatrix} 1 \\ 1 \end{pmatrix} \right\}.$$

Let us also suppose $H(v^3) = -2$. We now must determine G on s^1 and solve

$$G(w) = 0,$$

where $G: s^1 \longrightarrow R^2$. Equation (11.4.3) now becomes, since $s^1 = \{v^1, v^2, v^3\}$ and $H(v^1) = -1$, $H(v^2) = 3$,

$$\lambda_1 \begin{pmatrix} -1 \\ 1 \end{pmatrix} + \lambda_2 \begin{pmatrix} 3 \\ 1 \end{pmatrix} + \lambda_3 \begin{pmatrix} -2 \\ 1 \end{pmatrix} = \begin{pmatrix} 0 \\ 1 \end{pmatrix}.$$

The first two columns yield the solution $w^1 = \frac{3}{4}v^1 + \frac{1}{4}v^2$. The path across s^1 yields $\lambda_2 = \frac{2}{5}$, $\lambda_3 = \frac{3}{5}$ and

$$w^2 = \frac{2}{5}v^2 + \frac{3}{5}v^3 = \frac{2}{5}\begin{pmatrix} 0 \\ 1 \end{pmatrix} + \frac{3}{5}\begin{pmatrix} 1 \\ 1 \end{pmatrix} = \begin{pmatrix} \frac{3}{5} \\ 1 \end{pmatrix}.$$

Thus the path across s^1 goes from w^1 to w^2. Also, w^2 is in the facet B^2
$= \left\{ \begin{pmatrix} 0 \\ 1 \end{pmatrix}, \begin{pmatrix} 1 \\ 1 \end{pmatrix} \right\}$ The path could then be continued in a new simplex s^2,

$$s^2 = B^2 \cup \{v^4\}$$

for a new vertex v^4.

THE BASIC STEP

Given $H: R^{n+1} \longrightarrow R^n$, let us now formally state the basic step of simplicial algorithms, assuming that G is always regular. On the jth iteration, a facet B^j and new vertex $v^{j_{n+1}}$ will be given, forming a simplex $s^j = B^j \cup \{v^{j_{n+1}}\}$. Notationally, let $B^j = \{v^{j_0}, v^{j_1}, \ldots, v^{j_n}\}$. Also, a point $w^j \in B^j$ will be given where w^j solves

$$G(w) = 0. \tag{11.5.1}$$

Then a path that solves (11.5.1) is generated from w^j across s^j to a point w^{j+1} using the equations

$$\sum_{i=0}^{n+1} \lambda_i H(v^{j_i}) = 0$$

$$\sum_{i=0}^{n+1} \lambda_i = 1, \qquad \lambda_i \geq 0, \tag{11.5.2}$$

where

$$w = \sum_{i=0}^{n+1} \lambda_i v^{j_i}.$$

Note that w^j satisfies (11.5.2) with i running from 0 to n (not $n + 1$). Moreover, the point w^{j+1} will be in a facet, call it B^{j+1}, of s^j.

A new vertex $v^{(j+1)_{n+1}}$ will now be generated so that the new simplex $s^{j+1} = B^{j+1} \cup \{v^{(j+1)_{n+1}}\}$ will be on the other side of B^{j+1} from s^j. The iteration then continues in this manner using $j + 1$ instead of j.

To commence the algorithm, an initial point $w^0 \in B^0$ will be given where B^0 is a facet of the initial simplex s^0. In actual operation, the starting point w^0 will be easy to find, as it will be close to the initial point $y^0 \in H^{-1}(0)$. The operation then continues simplex by simplex until a final point w^k is reached that approximates a point $y^1 \in H^{-1}(1)$.

This basic step is employed by all simplicial-type algorithms, and using it they generate a PL path in G^{-1}. The main difference among simplicial algorithms is in how the new vertex $v^{(j+1)_{n+1}}$ is selected on each step, and that will be treated beginning in Chapter 12.

IMPORTANT OBSERVATIONS

1. If we examine the process of solving (11.5.2), notice that it is essentially a pivot operation as in the *simplex method*. For those interested in computational detail, that pivoting process is discussed in this chapter's appendix.

2. The regularity of G ensures that a PL path is generated for G^{-1}. To see this, recall that under regularity the line segment across s^j goes to a point w^{j+1} which must be interior to a facet. This interiority is crucial. Indeed, suppose that w^{j+1} were not interior to a facet. Then the path could conceivably bifurcate into two or more segments (Figure 11.5.3). Regularity of G thus ensures that G^{-1} consists of nice PL paths with no bifurcation or other calamity. (Regularity of G is precisely analogous to H regular, as regularity of H ensures that H^{-1} consists of nice paths as well.) Also, of course, regularity holds virtually always.

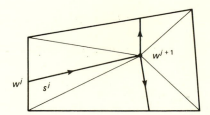

The path crosses s^j from w^j to w^{j+1}, but w^{j+1} is
not interior to a facet. The path then might bifurcate.
If w^{j+1} is interior to a facet, no bifurcation can occur.

Figure 11.5.3

APPENDIX

THE BASIC STEP OF THE SIMPLICIAL
ALGORITHM VIA PIVOTING TECHNIQUES

Let us review the basic step of the algorithm from a viewpoint of pivoting analogous to a linear programming step. Suppose that we have a solution $w^j \in B^j$, where to simplify notation

$$B^j = \{v^0, v^1, \ldots, v^n\}$$

and the current simplex is

$$s^j = B^j \cup \{v^{n+1}\}.$$

Define the $(n+1) \times (n+1)$ matrix A, $(n+1)$-vector b,

$$A = \begin{bmatrix} H(v^0) & H(v^1) & \cdots & H(v^n) \\ 1 & 1 & & 1 \end{bmatrix}$$

$$b = \begin{bmatrix} 0 \\ 1 \end{bmatrix}, \tag{A.11.1}$$

where b is composed of zeros except in the $(n+1)$st position, and also define the $(n+1) \times (n+1)$ matrix V with columns v^i as

$$V = [v^0, v^1, \ldots, v^n].$$

Now consider

$$A\lambda = b, \qquad \lambda \geq 0$$

and let

$$\bar{\lambda} = A^{-1}b$$
$$w^j = V\bar{\lambda}. \tag{A.11.2}$$

We know that the matrix A is invertible because, by regularity of G, w^j is unique, so A must be of full rank. Also by regularity of G, since w^j is interior to B^j,

$$\bar{\lambda}_i > 0, \qquad \text{all } i. \tag{A.11.3}$$

We now must generate the path from w^j across s^j to w^{j+1}. Let the $(n+1)$-vector a be

$$a = \begin{pmatrix} H(v^{n+1}) \\ 1 \end{pmatrix}$$

and note that equation (11.4.3) becomes

$$A\lambda + a\lambda_{n+1} = b$$
$$\lambda_i \geq 0, \qquad i = 0, \ldots, n+1,$$

where, from (A.11.2),

$$\lambda = \bar{\lambda} - A^{-1}a\lambda_{n+1}$$
$$\lambda_i \geq 0, \qquad \text{all } i. \tag{A.11.4}$$

Next we execute the pivot step; that is, we must increase λ_{n+1} until some λ_i, $i = 0, \ldots, n$ hits zero. To determine which column of A drops, utilize the ratio test. Define the vector $\bar{a} = (\bar{a}_i)$ as

$$\bar{a} = A^{-1}a \tag{A.11.5}$$

and

$$\bar{\lambda}_{n+1} = \min_i \left\{ \frac{\bar{\lambda}_i}{\bar{a}_i} \,\middle|\, \bar{a}_i > 0 \right\}.$$

From (A.11.3) we know that $\bar{\lambda}_i > 0$ and also some $\bar{a}_i > 0$ because the simplex s^j is bounded. Moreover, by regularity of G, there will be a unique index h such that

$$\bar{\lambda}_{n+1} = \frac{\bar{\lambda}_h}{\bar{a}_h} > 0. \tag{A.11.6}$$

Then the new facet containing w^{j+1} is

$$B^{j+1} = \{v^0, v^1, \ldots, v^{h-1}, v^{h+1}, \ldots, v^{n+1}\}.$$

In other words, the vertex v^h has dropped but v^{n+1} has come into the basis.

Also, since $a = \begin{pmatrix} H(v^{n+1}) \\ 1 \end{pmatrix}$, define

$$C = \left[\begin{pmatrix} H(v^0) \\ 1 \end{pmatrix}, \begin{pmatrix} H(v^1) \\ 1 \end{pmatrix}, \ldots, \begin{pmatrix} H(v^{h-1}) \\ 1 \end{pmatrix}, a, \begin{pmatrix} H(v^{h+1}) \\ 1 \end{pmatrix}, \ldots, \begin{pmatrix} H(v^n) \\ 1 \end{pmatrix} \right]$$

and

$$W = [v^0, v^1, \ldots, v^{h-1}, v^{n+1}, v^{h+1}, \ldots, v^n].$$

Observe how we have replaced the hth column with the new column.

Letting

$$C\mu = b, \qquad \mu \geq 0$$
$$\mu = C^{-1}b,$$

we see that

$$w^{j+1} = W\mu,$$

which yields the new point w^{j+1} in facet B^{j+1}. The process continues in this manner.

UPDATING THE INVERSE

In the calculation of $\bar{\lambda}$ in (A.11.2) and $\bar{a}$ in (A.11.5), the inverse A^{-1} is necessary. Thus to implement the basic pivot step above, we require a procedure for updating the inverse matrix. In particular, given A^{-1}, the ratio test determines the column h that drops, and we need merely update it using the new column to determine C^{-1}.

From (A.11.5)

$$\bar{a} = A^{-1}a.$$

Also, $\bar{a}_h \neq 0$ by (A.11.6), so simplifying notation and letting A have columns A^i,

$$a = A\bar{a}$$
$$= \bar{a}_0 A^0 + \ldots + \bar{a}_h A^h + \ldots + \bar{a}_n A^n$$

or

$$A^h = -\frac{\bar{a}_0}{\bar{a}_h} A^0 \ldots - \frac{\bar{a}_{h-1}}{\bar{a}_h} A^{h-1} + \frac{a}{\bar{a}_h} - \frac{\bar{a}_{h+1}}{\bar{a}_h} A^{h+1} \ldots - \frac{\bar{a}_n}{\bar{a}_h} A^n.$$

Examining C, observe that the hth column of A becomes

$$A^h = Cg,$$

where

$$g = \left(-\frac{\bar{a}_0}{\bar{a}_h}, \ldots, -\frac{\bar{a}_{h-1}}{\bar{a}_h}, \frac{1}{\bar{a}_h}, -\frac{\bar{a}_{h+1}}{\bar{a}_h}, \ldots, -\frac{\bar{a}_n}{\bar{a}_h} \right)^{\tau}.$$

Now let e^i be a vector of all zeros except a 1 in the ith position. We see that

$$A = CE,$$

where the $(n + 1) \times (n + 1)$ matrix E is

$$E = [e^1, \ldots, e^h, g, e^{h+2}, \ldots, e^{n+1}].$$

Thus

$$C^{-1} = EA^{-1}.$$

Since A^{-1} is known and E is an *elementary* matrix, the computation of C^{-1} is quite easy.

In sum, the basic step of the simplicial algorithm is straightforward when considered as a pivot operation and thus easily implementable on a computer.

1. Let $v^1, v^2, \ldots, v^n$ be nonzero and linearly independent points in R^n. Describe in words each of the following sets:

   a. $\{x \mid x = \sum_{i=1}^{n} \lambda_i v^i, \lambda_i \geq 0 \text{ all } i\}$

   b. $\{x \mid x = \sum_{i=1}^{n} \lambda_i v^i\}$

   c. $\{x \mid x = \sum_{i=1}^{n} \lambda_i v^i, \sum_{i=1}^{n} \lambda_i = 1\}$

   d. $\{x \mid x = \sum_{i=1}^{n} \lambda_i v^i, \sum_{i=1}^{n} \lambda_i = 1, \lambda_i \geq 0 \text{ all } i\}$

2. Prove that if w is in the interior of a simplex s, all the weights on the vertices of s that describe w are positive.

3. Prove that for any point w in a simplex s, w can be expressed uniquely as a convex combination of the vertices of s.

4. Prove that $(j + 1)$-simplices cannot exist in R^j.

5. Consider Example 11.3.1. With $w \in s = \{v^0, v^1, v^2\}$, write $G(w)$ as a function of w.

6. Let $s = \{v^0, v^1, v^2, v^3\}$ where

$$v^0 = \begin{pmatrix} 0 \\ 0 \\ 0 \end{pmatrix}, \quad v^1 = \begin{pmatrix} 2 \\ 0 \\ 0 \end{pmatrix}, \quad v^2 = \begin{pmatrix} 0 \\ 0 \\ 1 \end{pmatrix}, \quad v^3 = \begin{pmatrix} 1 \\ 2 \\ 3 \end{pmatrix}.$$

For $w \in s$, let

$$G(w) = \begin{pmatrix} w_1 & + w_3 - 1 \\ 2w_1 + w_2 + 2w_3 - 1 \end{pmatrix}.$$

   a. Write G as a function of the weights λ_i.
   b. Find the solutions $w \in s$ satisfying $G(w) = 0$.

7. Let the square $0 \leq x \leq 2, 0 \leq t \leq 2$ be subdivided into triangles as shown in Figure E.11.1. Define $H(x, t) = (1 - t)(x - \frac{1}{2}) + t(x - 1)^2$, and let G be defined by (11.3.1).

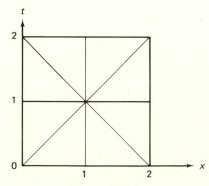

Figure E.11.1

a. Determine the set G^{-1} on the square.

b. Suppose that the vertex $(1, 1)$ is wiggled to $(1, 1 + \epsilon)$, $\epsilon > 0$. Determine the set G^{-1} in this case.

c. Suppose that the vertex $(1, 1)$ is wiggled to $(1 + \epsilon, 1)$, $\epsilon > 0$. Determine the set G^{-1} in this case.

8. Define a function H: $R^2 \longrightarrow R^1$ such that if G is defined by (11.3.1), a path in G^{-1} bifurcates into exactly two segments. Show what would happen to G^{-1} if the vertex where the bifurcation occurred were wiggled.

9. Let G be defined by (11.3.1). Show that the matrix (A.11.1) has rank $n + 1$ if and only if $G'(v^0)$ has rank n. [*Hint:* Subtract the first column from each of the remaining columns. Then, use the identity

$$H(v^i) = G(v^i) = G(v^0) + G'(v^0)(v^i - v^0), \qquad i = 1, 2, \ldots, n.]$$

10. Prove that $v^0, v^1, \ldots, v^n$ are independent points if and only if the points $v^1 - v^0, \ldots, v^n - v^0$ are linearly independent.

NOTES

References on simplicial algorithms may be found in Allgower and Georg [1980], Eaves and Scarf [1976], and Todd [1976a], although many authors have contributed to this field.

Among the earliest simplicial algorithms were those by Eaves [1972], Eaves and Saigal [1972], Jeppson [1972], Kuhn [1968], Hansen and Scarf [1969], and Merrill [1971]. Later, algorithms were developed by Allgower and Jeppson [1973], Fisher and Gould [1974], Fisher, Gould, and Tolle [1976, 1977], Garcia [1975, 1976, 1977a, b], Garcia and Gould [1979], Garcia and Zangwill [1979a, 1980a], Kojima [1978b], Kuhn and MacKinnon [1975], Lüthi [1976b], Peitgen and Prüfer [1979], Saigal [1977], Todd [1976a, c, 1977], Van der Laan and Talman [1979, 1980], and Zangwill [1977].

12

The Flex-Simplicial Algorithm

Chapter 11 presented the basic iteration step of all simplicial algorithms as they proceed to generate a PL path in G^{-1} from simplex to simplex. On each simplex the path in G^{-1} is linear and approximates the underlying path in H^{-1}. The main difference among simplicial algorithms is in how the new vertex forming the new simplex is selected.

This chapter presents the flex-simplicial algorithm. With it the new vertex can be selected almost arbitrarily, and hence the flexibility. Moreover, if the simplices generated are sufficiently small, a PL path in G^{-1} closely follows an underlying path in H^{-1}. There are times, however, when it is desirable to have larger simplices. The PL path can then stray from an underlying path. However, even in that case, the PL path need not be lost but can come home again and still converge. The flex-simplicial algorithm thus operates well under a variety of circumstances. Indeed, it is not only excellent for calculation, but also provides valuable theoretical results that illuminate all simplicial algorithms.

12.1 NOTATION

Let us rephrase the basic step of simplicial algorithms, clarifying notation somewhat by dropping the superscripts. A line segment in G^{-1} is generated across a simplex s from a point w to $\bar{w}$, where $\bar{w} \in \bar{B}$, a facet of s. A new vertex, $\bar{v}$, is generated so that the new simplex

$$\bar{s} = \bar{B} \cup \{\bar{v}\}$$

is adjacent to s. Recall that s and $\bar{s}$ are adjacent if

$$s \cap \bar{s} = \bar{B},$$

so that s and $\bar{s}$ are on opposite sides of the hyperplane through facet $\bar{B}$. Here all simplices are, of course, $(n + 1)$-simplices in R^{n+1}.

This easier notation will help us focus on exactly what is happening in our algorithm because the difference among the algorithms is then in the choice of $\bar{v}$. The flex-simplicial algorithm permits $\bar{v}$ to be selected quite freely. In fact, for the flex-simplicial, $\bar{v}$ can be almost any point on the other side of $\bar{B}$. The only restriction, as we will discuss, is minor, in that the new simplex $\bar{s}$ formed cannot be flat or squashed.

12.2 BEHAVIOR OF THE FLEX SIMPLICIAL

The key feature of the flex-simplicial algorithm is that as long as the simplices generated are sufficiently small, a path in G^{-1} can be made arbitrarily close to an underlying path in H^{-1}. Notationally, let $l(s)$ be the length of the longest edge of s. If $l(s)$ is sufficiently small, say $l(s) \leq l$ for all simplices s in the path, the PL path will hug an underlying path. We will now establish this result.

Recall that for any simplex $s = \{v^0, \ldots, v^{n+1}\}$ on the PL path, $G_s : s \rightarrow R^n$ was defined by

$$G_s(w) = \sum_{i=0}^{n+1} \lambda_i H(v^i), \tag{12.2.1}$$

where

$$w = \sum_{i=0}^{n+1} \lambda_i v^i$$

$$\sum_{i=0}^{n+1} \lambda_i = 1, \quad \lambda_i \geq 0, \quad i = 0, 1, \ldots, n + 1.$$

Here we write G_s for G to focus on the particular simplex s.

From (12.2.1),

$$\| H(w) - G_s(w) \| \leq \sum_{i=0}^{n+1} \lambda_i \| H(w) - H(v^i) \|.$$

If s is sufficiently small, say $l(s) \leq l$, then $\| H(w) - H(v^i) \|$ can be made small by continuity of H. Consequently, for any $\epsilon > 0$, if s is sufficiently small,

$$\| H(w) - G_s(w) \| < \epsilon \tag{12.2.2}$$

for any $w \in s$.

The key point is that if s is sufficiently small, G_s approximates H for all $w \in s$. We now need only one further step to prove this section's main result—that the PL path in G^{-1} is quite close to an H^{-1} path.

THE δ-TUBE

Mathematically, specify the idea of closeness by defining the δ-tube, N_δ, as

$$N_\delta = \{y \in R^{n+1} \mid \|\bar{y} - y\| < \delta, \quad \bar{y} \in H^{-1}\}.$$

The set N_δ is all y within δ of a path in H^{-1} (Figure 12.2.1). Geometrically, the δ-tube N_δ makes a tube around each path in H^{-1}. (If δ is large, more than one path in H^{-1} might be included per tube, but that situation is of little interest to us here.)

The δ-tubes enable us to obtain our result. Exactly stated, if the simplices are sufficiently small, the entire PL path in G^{-1} will be inside the δ-tube of a path in H^{-1}. And that holds for any given $\delta > 0$.

In the proof below, the symbol $\sim$ denotes set subtraction. Thus $y \in D \times T \sim N_\delta$ means all y in $D \times T$ that are not in the δ-tube.

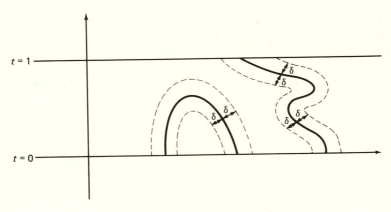

The δ-tubes N_δ around paths in H^{-1}.

Figure 12.2.1

THEOREM 12.2.1 Consider a continuous $H: D \times T \to R^n$ for D the closure of an open bounded set. Given $\delta > 0$, if the simplices along the PL path are sufficiently small and

$$y \in G^{-1},$$

then

$$y \in N_\delta.$$

Proof.   Clearly, if y is not in the δ-tube, $\| H(y) \| > 0$. Notationally,

$$y \in D \times T \sim N_\delta \quad \text{implies} \quad \| H(y) \| > 0.$$

But N_δ is open, so the set $D \times T \sim N_\delta$ is compact. Via compactness the minimum would be achieved and for some $\epsilon > 0$,

$$\min\{\| H(y) \| \mid y \in D \times T \sim N_\delta\} > \epsilon.$$

In words, if y is not in a δ-tube, then $\| H(y) \| > \epsilon$. Stated conversely, this means given any point y such that

$$\| H(y) \| \le \epsilon \tag{12.2.3}$$

that y is inside a δ-tube.

Because H is uniformly continuous on the compact set $D \times T$, using (12.2.2):

$$\| H(y) - G_s(y) \| < \epsilon \tag{12.2.4}$$

for any simplex s sufficiently small.

Finally, any $y \in G^{-1}$ has $G_s(y) = 0$, so from (12.2.4),

$$\| H(y) \| < \epsilon.$$

But by (12.2.3) y must be in a δ-tube of H^{-1}. In words, if s is sufficiently small and $y \in G^{-1}$, then $y \in N_\delta$. $\square$

Theorem 12.2.1 assures us of the following: If the simplices are sufficiently small, say for some $l^* > 0$ all simplices s along the path have $0 < l(s) \le l^*$, then a PL path is very close to an underlying path in H^{-1}.

Note, however, that the PL path need not follow a path in H^{-1}. Theorem 12.2.1 provides only that it is close to an H^{-1} path. A situation such as that shown in Figure 12.2.2b could occur where the PL path remains close to an H^{-1} path but just cycles endlessly. Since the new point $\bar{v}$ can be picked almost arbitrarily, it appears that such cycling might occur and thus destroy convergence. Fortunately, these fears are unfounded.

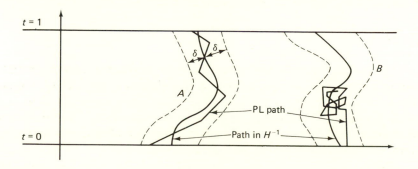

The PL path stays within a δ-tube of H^{-1}

Figure 12.2.2

12.3 PREVENTION OF CYCLING

We will now show that the PL path does not cycle endlessly but instead actually follows along an H^{-1} path. The PL path is not differentiable in general but is differentiable (in fact linear) on any simplex s. Thus the BDE hold for G on any simplex s. Let G' be the $n \times (n + 1)$ Jacobian and G'_{-i} the $n \times n$ matrix formed from G' after deleting the ith column. On any s, we then have, via the BDE,

$$\dot{w}_i = (-1)^i \det G'_{-i}(w), \qquad (12.3.1)$$

where here we employ w to clearly distinguish the PL path and $\dot{w} = dw/dp$.

Solving (12.3.1) piece by piece, that is, simplex by simplex, will in fact generate our PL path in G^{-1}, as we know from Chapter 2.

Recall also from Chapter 2 that $\dot{w}$ from (12.3.1) is the tangent to the PL path at any $w \in G^{-1}$ where w is interior to a simplex along the path. Of course, since the BDE apply to H also,

$$\dot{y}_i = (-1)^i \det H'_{-i}. \qquad (12.3.2)$$

Moreover, $\dot{y}$ is tangent to a path in H^{-1} at any $y \in H^{-1}$.

This fact that both $\dot{w}$ and $\dot{y}$ are the tangents to their respective paths is the clue we need. Given $y \in H^{-1}$ and $w \in G^{-1}$, we must show that if w and y are close, then $\dot{w}$ and $\dot{y}$ are close. But this will provide the proof that the PL path does follow along a path in H^{-1}. If the PL path did not follow along, then as in Figure 12.2.2b, it would have to reverse itself. But in that case $\dot{w}$ and $\dot{y}$ could not always point in approximately the same direction. Rephrased, if the path orientations $\dot{w}$ and $\dot{y}$ are similar, say $\dot{w}\dot{y} > 0$, the PL path cannot reverse itself and must follow along the H^{-1} path.

Examining (12.3.1) and (12.3.2), see that $\dot{y}$ and $\dot{w}$ will be similar if the Jacobians G' and H' are similar. To ensure that the Jacobians G' and H' are similar, however, requires a certain condition on s.

ROBUSTNESS

The problem in proving G' and H' similar is that the simplex s might be rather squashed. For example, consider a triangle that is very flat. Even though it has long sides, it can have a very small area. Choosing $\bar{v}$ too close to the facet $\bar{B}$ will cause this difficulty because the simplex will then be flat and have little volume. Such thin, emaciated simplices must be excluded in order to prove G' and H' similar. Fortunately, that is easy to do.

Let $r(s)$ be the volume of simplex s and recall that $l(s)$ is the largest edge of s. Also note that $(l(s))^{n+1}$ is the volume of an $(n + 1)$-cube with edge $l(s)$. (The power is $n + 1$, since $s \subset R^{n+1}$.) Given $a > 0$, we call s *robust* if

$$\frac{r(s)}{(l(s))^{n+1}} > a.$$

Essentially, if s is robust, its volume is at least the fraction a of the corresponding $(n + 1)$-cube. Furthermore, the simplices along a path are called robust if all the simplices are robust for the same constant $a > 0$.

Robustness has nothing to do with the size of a simplex, only with its shape. Robustness ensures that the simplices along the path do not become excessively flat. In practice, robustness is usually easy to obtain and merely requires that $\bar{v}$ not be selected extremely close to $\bar{B}$.

The robustness condition will ensure that the Jacobians approximate each other well. Specifically, if our simplices are robust and sufficiently small, then, as shown in the appendix to this chapter,

$$G'(y) \quad \text{and} \quad H'(y)$$

can be made arbitrarily close for any $y \in s$.

Consequently, from (12.3.1) and (12.3.2), if two points $y \in H^{-1}$ and $w \in G^{-1}$ are sufficiently close to each other,

$$\dot{w}\dot{y} > 0.$$

This means that $\dot{w}$ will point in an acute angle to $\dot{y}$ and thus that the two paths will point in similar directions. But that, as noted, prevents cycling and provides our convergence proof, which we summarize in the next theorem.

THEOREM 12.3.1   Consider an $H: D \times T \longrightarrow R^n$, $H \in C^2$ and regular, where $D \subset R^n$ is the closure of an open bounded set. Suppose that the flex-simplicial algorithm generates a PL path in G^{-1}, where G is regular. Also, let $\delta > 0$ be given. If the simplices are sufficiently small and robust, the PL path is within δ of a path in H^{-1} and follows along that path.

Proof.   By Theorem 12.2.1, the PL path remains within the δ-tube of an underlying path in H^{-1}. We are also assured that if $y \in H^{-1}$ and $w \in G^{-1}$ are sufficiently close, since the simplices are robust and small,

$$\dot{w}\dot{y} > 0.$$

But then the PL path points in a direction similar to that of a path in H^{-1}, so it cannot cycle. It must then follow along the underlying path. $\square$

12.4 LARGE SIMPLICES

The flex simplicial will thus track an H^{-1} path quite accurately if the simplices are small and robust. In certain applications following the entire path in H^{-1} is essential, and then small simplices are required. For other applications only a point $x^1 \in H^{-1}(1)$ is desired, and it is not necessary to follow the entire path. Then large simplices might be helpful at least for $t < 1$ in an effort to get close to $t = 1$ quickly. Of course, near $t = 1$ small simplices might be employed to obtain greater accuracy.

In certain applications, therefore, we might want to use large simplices, at least some of the time. The difficulty is that Theorem 12.3.1 applies only to small simplices, so if the simplices are large, the PL path can drift away from an underlying path in H^{-1}. And without further analysis we cannot be sure where it might wander.

Fortunately, in most instances the flex simplicial can be adapted to work with large simplices. As before, we require that all simplices be robust; that is, the new point $\bar{v}$ on the other side of the facet $\bar{B}$ must be selected so that the new simplex $\bar{s} = \bar{B} \cup \{\bar{v}\}$ be robust. But also we stipulate that as the iteration k gets large, the corresponding simplex, denoted s^k, gets arbitrarily small. In other words, the simplices can be large for a long while, but eventually they must get small. Explicitly, for large simplices require that:

$$l(s^k) \longrightarrow 0 \text{ as the iteration } k \longrightarrow \infty. \qquad (12.4.1)$$

When the simplices are large, the PL path generated might be far from an H^{-1} path. However, if the algorithm proceeds long enough, condition (12.4.1) implies that it will eventually latch on to an underlying path and follow it. Theorem 12.3.1 assures us of this. In the beginning the largeness and flexibility in the choice of simplices might be helpful to quickly home in on a point $x^1 \in H^{-1}(1)$. But as the iteration $k \longrightarrow \infty$ and the simplices get small, we know that the PL path must get close to and align itself with a path in H^{-1}.

Note that there is another difficulty. The PL path might run into a boundary of $D \times T$ while the simplices are still large. In other words, the PL path might terminate at a boundary of $D \times T$ before the simplices s^k are small enough to ensure that the PL path tracks an H^{-1} path.

Fortunately, it is easy to prevent the PL path from hitting a boundary before it follows an H^{-1} path. Just require the simplices to get arbitrarily small if a boundary of $D \times T$ is approached. In other words, choose the new vertex $\bar{v}$ in each iteration so that if the PL path approaches a boundary, the simplices get arbitrarily small. When the simplices are small, we know we must be following an underlying path. The possibility of the PL path hitting a boundary before the simplices are small enough to follow an H^{-1} path is thus eliminated.

We have just proved the following theorem.

THEOREM 12.4.1   Let a regular $H: D \times T \rightarrow R^n$, $H \in C^2$, be given where $D \subset R^n$ is the closure of a bounded open set. Suppose that the flex-simplicial algorithm generates robust simplices, and the G^{-1} path is regular. Let the size of the simplex s^k generated get arbitrarily small as:

1.   The iteration $k \rightarrow \infty$.

2.   The PL path approaches a boundary of $D \times T$.

Then the PL path generated eventually gets close to and follows an underlying path in H^{-1}. Moreover, the two paths are then similarly oriented, that is, $\dot{w}\dot{y} > 0$.

Proof.   The simplices generated must get small because either the PL path approaches a boundary or the iteration number gets large. But then, via Theorem 12.3.1, the PL path must follow a path in H^{-1}. The fact that the two paths are similarly oriented follows because if $y \in H^{-1}$ and $w \in G^{-1}$ are sufficiently close, $\dot{w}\dot{y} > 0$.   ☐

Figure 12.4.1 shows several instances of how a PL path can eventually track an H^{-1} path. Notice that if the PL path latches onto a loop like C, it can circle endlessly. If it attaches to a path like B, it can return to a point in $H^{-1}(0)$. To learn more about where the path may go, we must study orientation.

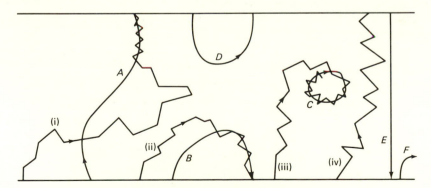

The piecewise linear path (i) eventually aligns itself to track underlying path A. Similarly, paths (ii) and (iii) eventually track underlying paths. Path (iv) runs into a boundary at $t = 1$. The hypotheses of Theorem 12.4.1 prevent (iv) from occurring. The arrows indicate the orientation of the path, that is, either $\dot{w}$ for the PL or $\dot{y}$ for the H^{-1} path.

Figure 12.4.1

The orientation of the two paths is quite important. Specifically, when a PL path and an underlying path are close,

$$\dot{w}\dot{y} > 0 \qquad\qquad (12.4.2)$$

must hold, and we must understand this inequality thoroughly. The vector $\dot{w}$ is the direction of increasing p for the PL path, while $\dot{y}$ is the direction of increasing p for the underlying path. The inequality requires that, if we increase the p for the PL path and if we increase the p for the underlying path, the two paths must move in similar directions. Similarly, if we decrease p on both paths, we must move in a similar direction. This is the key idea. Increasing (or decreasing) p along both paths takes us in a similar direction.

Note carefully, however, that the flex-simplicial algorithm does not necessarily move along the PL path in the direction of increasing p. Starting from a point at $t = 0$, we wish to increase t, but which way to change p depends on the path orientation. Given a starting point at $t = 0$, a path is called *positively oriented* if as p increases t increases. If p must be decreased to increase t from 0, the path is *negatively oriented*. In brief, if at $t = 0$ the path is positively oriented, we must increase p to increase t; if it is negatively oriented, p must be decreased to increase t.

To illustrate these concepts suppose that the negatively oriented PL path attached onto path A of Figure 12.4.1 or Figure 12.4.2. Notice that the path would not go to $t = 1$ but would go down to $t = 0$. This is because inequality (12.4.1) requires that decreasing p on both paths moves us in a similar direction. If we move in the direction of decreasing p on the PL path, we must move in the direction of decreasing p for the underlying path. That is why when it

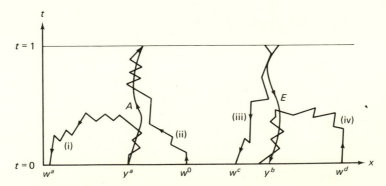

The arrows denote the direction of increasing p; that is, the arrow indicates the direction of $\dot{w}$ or $\dot{y}$. Note that starting from $t = 0$, we do not necessarily traverse the path in this direction. If the path is negatively oriented, we traverse it in the direction opposite the arrow.

Figure 12.4.2

latches onto A, a negatively oriented PL path will go to $t = 0$. (On the other hand, if a positively oriented PL latches onto A, it will go to $t = 1$).

Conceptually, when a PL path gets close to an underlying path, it can go in either of two directions along the underlying path. The direction is always selected so that by increasing both p's (or by decreasing both p's), the two paths move in a similar direction. This fact determines which of the two directions along the underlying path the algorithm will take.

Figure 12.4.2 explains this further. In it, paths (i), (iii) and E are negatively oriented; that is, at their start point (w^a, w^c, or y^b, respectively), decreasing p increases t. Paths A, (ii) and (iv) are positively oriented as, from their respective start points at $t = 0$, increasing p increases t. The algorithm moves along paths (i) and (iii) in the direction of decreasing p and along paths (ii) and (iv) in the direction of increasing p. The inequality $\dot{y}\dot{w} > 0$ must hold when a PL path and an underlying path are close. Thus the arrows, since they point in the direction of increasing p, must point in similar directions. For example, path (i) is traversed from its start w^a in the opposite direction of its arrow since it is negatively oriented. When it attaches onto path A, it then goes toward $t = 0$. Path (ii), on the other hand, goes toward $t = 1$. As path E is negatively oriented, path (iii) goes to $t = 1$ while path (iv) goes to $t = 0$.

These ideas are utilized in the next corollary.

COROLLARY 12.4.2   Suppose that the flex simplicial generates simplices that are always robust and the hypotheses of Theorem 12.4.1 hold. Also suppose that:

1. H^{-1} has no loops.

2. Given any $x^0 \in H^{-1}(0)$, it starts a path in H^{-1} such that:
    a. The path goes to a point $x^1 \in H^{-1}(1)$.
    b. The path is positively oriented; that is, increasing p takes us away from $y^0 = (x^0, 0)$ toward $y^1 = (x^1, 1)$.

Let the flex-simplicial algorithm starting from a point w^0 where $t = 0$ also be positively oriented. Then the flex simplicial will calculate a point approximating an $x^1 \in H^{-1}(1)$.

Proof.   Since there are no loops in H^{-1}, path C in Figure 12.4.1 cannot occur. Moreover, any path from $x^0 \in H^{-1}(0)$ must go to an $x^1 \in H^{-1}(1)$; thus paths such as B which return to $t = 0$, and F which runs off, cannot occur.

Path E is also excluded because it is oriented negatively; that is, increasing p moves us from $t = 1$ toward $(x^0, 0)$. Therefore, only paths such as A and D occur. If the PL path attaches to D, it goes to $t = 1$.

All paths such as A are oriented positively, so that if the PL path latches onto it, the PL path will follow it to $x^1 \in H^{-1}(1)$. This is because the PL path is also positively oriented, so will proceed in the same direction as A.   □

SPECIFICS ON ORIENTATION

Let us now provide some implementation details that utilize these orientation notions.

1. In practice, it might not be known which direction of a path heads to $t = 1$. That is, suppose that the PL path latches onto an H^{-1} path somewhere in the middle. It need not be clear which direction ends up at $t = 1$, and it might then be necessary to follow the path in both directions.

2. If desired, we can ensure that the G^{-1} path and a particular underlying path in H^{-1} are similarly oriented. Suppose that $y^0 = (x^0, 0)$ starts an underlying path. Select w^0, the PL path start, very near to $(x^0, 0)$ and use a small robust simplex. Then since the Jacobian G' will be similar to H', the BDE equations for the G^{-1} path will be similar to those for the H^{-1} path. The PL path will then inherit the same orientation as that particular underlying path.

This analysis helps us understand the flex-simplicial algorithm for large simplices, and we now summarize the overall procedure.

THE FLEX SIMPLICIAL
FOR LARGE SIMPLICES

On each iteration, assume that we can choose the new vertex so that the simplices generated:

1. are robust;

2. get arbitrarily small as either
    a. the iteration $k \to \infty$, or
    b. a simplex approaches the boundary of $D \times T$.

The flex simplicial will then eventually track a path in H^{-1}. Moreover, analysis of the orientation provides specifics on how to apply the algorithm in practice.

HUNTING BEHAVIOR

Understand, however, that it is not always trivial to generate simplices that satisfy both points 1 and 2 above. The complication is that the algorithm may not always be able to reduce the size of the simplex, that is, it might be difficult to make simplices small. There are circumstances (see Exercise 1) in which the simplices cannot be made smaller at all. Also, as the algorithm iterates, one vertex of a simplex might stick for a while, and prevent us from making the simplices smaller. That is, one vertex might keep staying in each of the new simplices generated for several iterations.

Figure 12.4.3 illustrates this sticking phenomenon. In such instances it might be necessary to generate different-size simplices for several iterations until the sticking vertex drops. After that occurs, we could then typically move out of that region and make the simplices smaller.

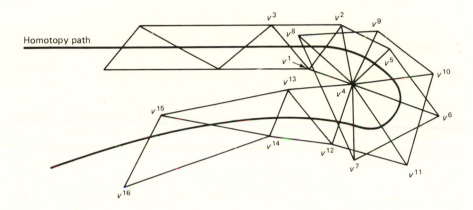

The initial simplex is $s^0 = \{v^1, v^2, v^3\}$. The new point v^4 has v^3 drop leaving simplex $s^1 = \{v^1, v^2, v^4\}$. The point v^5 has v^1 drop so that $s^2 = \{v^2, v^4, v^5\}$. . Ther vertex v^4 is sticking and stays until v^{13} is generated.

Here, $s^3 = \{v^4, v^5, v^6\}$, $s^4 = \{v^4, v^6, v^7\}$, $s^5 = \{v^4, v^7, v^8\}$, $s^6 = \{v^4, v^8, v^9\}$, $s^7 = \{v^4, v^9, v^{10}\}$, $s^8 = \{v^4, v^{10}, v^{11}\}$, $s^9 = \{v^4, v^{11}, v^{12}\}$, $s^{10} = \{v^4, v^{12}, v^{13}\}$, $s^{11} = \{v^{12}, v^{13}, v^{14}\}$, $s^{12} = \{v^{13}, v^{14}, v^{15}\}$, $s^{13} = \{v^{14}, v^{15}, v^{16}\}$.

Figure 12.4.3

This aspect of generating various-size simplices for a while is not uncommon. Sometimes the simplices cannot be made smaller just at will, but we might have to make them larger before making them smaller. The opposite also occurs and in order to get a sticking vertex to drop, the simplices might have to be made smaller for a while before they can be made larger.

In practice, often the algorithm will generate simplices that are larger, then smaller, then larger, then smaller, and so on. This indicates the algorithm is "hunting" for the solution. It uses large simplices to move rapidly and then small ones for a detailed search. Once it gets reasonably near a solution, the algorithm generates simplices that are continuously smaller and smaller as it homes in.

The algorithms in Chapter 13 are different from the flex simplicial because they choose the new vertex according to a fixed procedure. This fixed choice limits flexibility and does not let us pick the new vertex in a place that hopefully will help us converge quickly. However, the fixed choice might force a sticking vertex to drop more quickly. Thus the algorithms of Chapter 13 present a trade-off. They are not flexible but may more easily rid themselves of a sticking vertex and thus move to a new region of the space more rapidly. Finally, toward the end of Chapter 13 we introduce a hybrid algorithm that combines good features of both approaches.

SUMMARY

This chapter presented the flex-simplicial algorithm for path following. For small robust simplices it tracks an underlying path well. Larger simplices can also be used advantageously in an effort to move quickly or to rapidly home in on a solution. However, sometimes with large simplices the algorithm can stick somewhat. By generating larger and larger simplices for a while, the sticking can usually be overcome and the algorithm proceeds. Nevertheless, as sticking can at times be a difficulty, Chapter 13 presents a procedure to overcome it. Indeed, possibly the best algorithm combines features of both the flex simplicial and Chapter 13's algorithm, and that too will be discussed.

APPENDIX

JACOBIAN APPROXIMATION

Let us outline the proof that $G'(y)$ approximates $H'(y)$ if s is sufficiently small and robust.

Using the relation

$$\lambda_0 = 1 - \sum_{i=1}^{n+1} \lambda_i,$$

we obtain from (12.2.1) the equations

$$G(y) = H(v^0) + \sum_{i=1}^{n+1} \lambda_i(H(v^i) - H(v^0))$$

and

$$y = v^0 + \sum_{i=1}^{n+1} \lambda_i(v^i - v^0).$$

Letting $[v^i - v^0]$ be the $(n + 1) \times (n + 1)$ matrix with columns $v^i - v^0$, $i = 1, \ldots, n + 1$, $[H(v^i) - H(v^0)]$ be the $n \times (n + 1)$ matrix with columns $H(v^i) - H(v^0)$ and $\lambda = (\lambda_1, \ldots, \lambda_{n+1})$,

$$G(y) = H(v^0) + [H(v^i) - H(v^0)][v^i - v^0]^{-1}[y - v^0] \qquad \text{(A.12.1)}$$

since $[v^i - v^0]$ is invertible because the simplex $s = \{v^0, \ldots, v^{n+1}\}$ is composed of independent points.

Differentiating (A.12.1) yields

$$G'(y) = [H(v^i) - H(v^0)][v^i - v^0]^{-1}. \qquad \text{(A.12.2)}$$

By Taylor's expansion, since H is C^2,

$$[H(v^i) - H(v^0)] = H'(v^0)[v^i - v^0] + R. \qquad \text{(A.12.3)}$$

where the $n \times (n + 1)$ matrix $R = (R_{ij})$ is the remainder and where

$$|R_{ij}| \leq M \, ||v^i - v^0||^2 \leq Ml(s)^2 \qquad \text{(A.12.4)}$$

for some constant $M > 0$. Plugging (A.12.3) into (A.12.2), we obtain

$$G'(y) = H'(v^0) + R[v^i - v^0]^{-1}. \qquad \text{(A.12.5)}$$

Next, let us obtain a bound for the matrix $[v^i - v^0]^{-1}$. From linear algebra it can be shown that the absolute value of

$$\det \begin{bmatrix} v^0, \ldots, v^{n+1} \\ 1, \ldots, 1 \end{bmatrix} \qquad \text{(A.12.6)}$$

is the volume of the parallelepiped

$$\{y \in R^{n+1} \mid y = v^0 + \sum_{i=1}^{n+1} \lambda_i(v^i - v^0), \quad 0 \leq \lambda_i \leq 1, \quad \text{all } i\}.$$

But the volume of the simplex $s = \{v^0, \ldots, v^{n+1}\}$ is related to the volume of the parallelepiped by a constant L. Furthermore, subtracting $\begin{bmatrix} v^0 \\ 1 \end{bmatrix}$ from columns 2 through $n + 2$ of (A.12.6) and then expanding by its bottom row, the absolute value of (A.12.6) becomes the absolute value of

$$\det [v^i - v^0]. \qquad \text{(A.12.7)}$$

Consequently, since $r(s)$ is the volume of s,

$$r(s) = |\det [v^i - v^0]| \, L. \qquad \text{(A.12.8)}$$

Robustness assures us that for some $a > 0$,

$$\frac{r(s)}{(l(s))^{n+1}} > a. \tag{A.12.9}$$

Now express the (i, j)th element of $[v^i - v^0]^{-1}$, in magnitude, as

$$\frac{|A_{j,i}|}{|\det [v^i - v^0]|} \tag{A.12.10}$$

where $A_{i,j}$ is a cofactor of $[v^i - v^0]$.

But any cofactor is the sum of terms where each term is the product of n elements from $[v^i - v^0]$. This means, for some constant C,

$$|A_{j,i}| < Cl(s)^n.$$

Hence using (A.12.8), we obtain

$$\frac{|A_{j,i}|}{|\det [v^i - v^0]|} < \frac{L \cdot Cl(s)^n}{r(s)} < \frac{L \cdot Cl(s)^n}{a(l(s))^{n+1}} = \frac{L \cdot C}{al(s)}, \tag{A.12.11}$$

via (A.12.9). Since (A.12.11) holds for the (i, j)th element of $[v^i - v^0]^{-1}$, taking the matrix l_1-norm,

$$\| [v^i - v^0]^{-1} \| \leq \frac{(n+1)K}{l(s)} \tag{A.12.12}$$

for some constant K.

Calling on (A.12.4), (A.12.5), and (A.12.12), we obtain

$$\| G'(y) - H'(v^0) \| = \| R[v^i - v^0]^{-1} \| \leq \| R \| \| [v^i - v^0]^{-1} \|$$
$$\leq (n+1)^2 MKl(s). \tag{A.12.13}$$

Finally, using (A.12.13) and the continuity of H' yields

$$\| G'(y) - H'(y) \| \leq \| G'(y) - H'(v^0) \| + \| H'(v^0) - H'(y) \|$$
$$\leq (n+1)^2 MKl(s) + N \| v^0 - y \| \leq [(n+1)^2 MK + N]l(s) \tag{A.12.14}$$

for some constant $N > 0$. Thus we can make $G'(y)$ as close as we please to $H'(y)$ by selecting the simplex sufficiently small. Consequently, G' and H' approximate each other well.

EXERCISES/CHAPTER 12

1. Let D be the closure of an open bounded set. Suppose that there is no point $y \in H^{-1}$ in $D \times T$. That is, $H^{-1} \cap D \times T = \emptyset$.
    a. Show that by taking a large simplex it might be possible to start the flex simplicial in $D \times T$ from a starting point at $t = 0$.

    b. Suppose that the flex simplicial can be started at $t = 0$. Show that the simplices cannot be made arbitrarily small in $D \times T$ with robustness maintained.

    c. Discuss what happens to the PL path generated by the flex simplicial.

2. Explain where the consistency of the orientation of p as discussed in Chapter 2 was used in the proof of Theorem 12.3.1.

3. Let H be regular and G defined by (12.2.1). Show that for all sufficiently small and robust simplices, if s contains a solution to $G(w) = 0$, then matrix (A.11.1) has rank $n + 1$.

4. Let H be regular and G defined by (12.2.1). Let simplex s be sufficiently small and robust such that it contains a solution $G(w) = 0$ and such that matrix (A.11.1) has rank $n + 1$. If the line segment solution across s does not go through the interior of its facets, show that a wiggling of the vertices makes it do so.

5. Let H^{-1} consist of paths. Show that there is a $\delta > 0$ such that each δ-tube contains exactly one path of H^{-1}.

6. Let $v^0 = (0, 0)$, $v^1 = (1, 0)$, $v^k = (0, 1/k)$, $k = 2, 3, \ldots$ and $s^k = \{v^0, v^1, v^k\}$, $k = 2, 3, \ldots$. Show that for any $a > 0$, $r(s^k)/l(s^k)^2 < a$ for some $k = 2, 3, \ldots$.

7. In the flex-simplicial algorithm, suppose that $s = \bar{B} \cup \{v\}$ and the new vertex $\bar{v}$ is generated by the following rule:
    a. Determine the centroid g of $\bar{B}$.
    b. Let $\bar{v}$ be the unique point on the opposite side of $\bar{B}$, on the line determined by g and v, and such that the distance from g to $\bar{v}$ equals the distance from g to v. Determine the formula for computation of $\bar{v}$ given v and $\bar{B}$.

8. Let

$$H_1(x_1, x_2, t) = (1 - t)x_1 + t(x_1 - 5)x_2 = 0$$

$$H_2(x_1, x_2, t) = (1 - t)x_2 + t(x_2 - x_1 + 1)^2 = 0.$$

Starting with $v^0 = (-0.1, 0, 0)$, $v^1 = (0, -0.1, 0)$, $v^2 = (0.1, 0.1, 0)$, and $v^3 = (-0.1, 0, 0.1)$, do five iterations of the flex-simplicial algorithm using the replacement rule of Exercise 7.

9. Write a computer program for the flex-simplicial method using the replacement rule of Exercise 7. Run the program on Exercise 8.

10. In Exercise 8, calculate $\dot{y}$ at the point $x_1 = x_2 = t = 0$. Then calculate $\dot{w}$ at the initial simplex given in Exercise 8 and determine if $\dot{w}\dot{y} > 0$.

11. Let H be regular at $t = 0$ and $H(x^0, 0) = 0$. Show that the flex simplicial can be started at $t = 0$ in a neighborhood of $(x^0, 0)$.

NOTES

The flex-simplicial algorithm was introduced by Garcia and Zangwill [1980a].

13

Triangulation Algorithms

All simplicial algorithms have the same basic step, and differ only in their choice of the new vertex $\bar{v}$. In the flex-simplicial method, $\bar{v}$ could be chosen quite freely, almost randomly. Sometimes a sticking vertex may delay the flex simplicial, yet with astuteness and a little luck, it might be possible to choose $\bar{v}$ so that convergence is extremely rapid.

The triangulation algorithm introduced in this chapter forces $\bar{v}$ to be selected in a fixed, predescribed manner. The flexibility is virtually eliminated, but instead we obtain safety, steadfastness, and consistency. Like the tortoise in a race with the hare, we can be sure that it will get there. The flex simplicial on the other hand, is more like a hare, as it may get there very rapidly, yet has the habit of perhaps dawdling along the way. Both algorithms converge, of course, but which to bet on will become clearer in this chapter.

Conceptually, triangulation algorithms once examining a region of space never return, but always go to new regions. The flex simplicial, by contrast, might revisit which causes delay and can waste time.

Toward the end of the chapter we suggest a hybrid algorithm. It uses the flex simplicial much of the time for speed. But should the flex slow down and start revisiting a region of space, the hybrid switches to a triangulation approach.

13.1 THE TRIANGULATION

In the flex-simplicial method the simplices generated could be helter-skelter, even on top of each other, as shown in Figure 13.1.1. Indeed, the fact that they can be on top of each other is why the flex method can revisit a region of the space. The triangulation algorithm prevents revisiting and thereby gains speed, because the simplices are required to form a special pattern called a triangulation. The triangulation pattern eliminates any overlap and ensures, once a simplex is visited, that this portion of space is never revisited (see Figure 13.1.2).

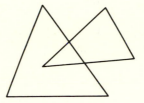

These two simplices overlap,
so this is not a triangulation.

Figure 13.1.1

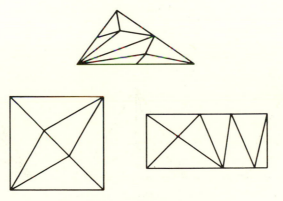

Three examples of triangulations.

Figure 13.1.2

Let us now be more precise. Define S to be a collection of $(n + 1)$-simplices, and let $D \subset R^n$ be the closure of an open set. The collection S is said to be a *triangulation* of $D \times [0, 1]$ if:

1. $D \times [0, 1]$ is contained in the union of the simplices of S.

2. If two simplices of S meet, they meet at a common face.

3.   Each bounded set in $D \times [0, 1]$ contains only a finite number of simplices of S.

The first condition above ensures that the whole set $D \times [0, 1]$ is triangulated, so that any point of $D \times [0, 1]$ is in some simplex of S. The second condition means that the simplices are carefully aligned. No simplex can overlap another's interior. If two simplices intersect, their intersection must comprise a face of both simplices (Figure 13.1.3). The third condition ensures that pathological cases such as that in Figure 13.1.4 will not arise. For instance, if there are an infinite number of smaller and smaller simplices, we could go from one to the next, to the next, and so on, and never really get anywhere.

The set $D \times [0, 1]$ can be triangulated in more than one way. In fact, it can be triangulated in an infinite number of ways. The only stipulation is that the three conditions stated above hold.

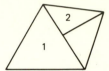 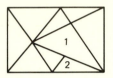

These examples are not triangulations
because simplices 1 and 2 intersect but
the intersection is not a face of simplex
1. The intersection must be a face of both
simplices.

Figure 13.1.3

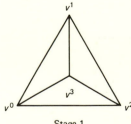 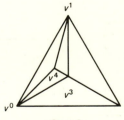 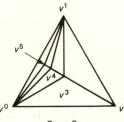

Stage 1   Stage 2   Stage 3

Let $\{v^0, v^1, v^2\}$ be a triangle in $R^1 \times [0, 1]$. In
stage 1, a point v^3 interior to the triangle is introduced
so as to subdivide the simplex into three triangles. In
stage 2, a point v^4 interior to $\{v^0, v^1, v^3\}$ is introduced,
subdividing the triangle. In stage 3, a point v^5 is introduced
and so on, ad infinitum, thereby creating an infinite collection of smaller and smaller triangles. The infinite collection
of triangles thus formed cannot be members of a triangulation S.

Figure 13.1.4

13.2 THE TRIANGULATION ALGORITHM

The triangulation algorithm is now easy to state. At each step choose the new vertex $\bar{v}$ so that the simplices thus generated form part of a triangulation. No longer can $\bar{v}$ be chosen freely, but the simplices must be such that they are part of some triangulation of $D \times [0, 1]$. The triangulation can be quite arbitrary. Also, presumably, the simplices actually generated by the algorithm would only be some of the simplices in the entire triangulation. Nevertheless, they must be part of a triangulation.

KINDS OF TRIANGULATIONS

Before we continue, let us remark on the kinds of triangulations that would be desirable. Since our objective is to have algorithms that are computer-implementable, it would be helpful if our triangulation were easy to construct. For example, we might want simplices to have similar shapes. (Thus, remembering one simplex would imply knowing all simplices.) Also, we might want the vertices to be at equally spaced distances from each other (Figure 13.2.1), as then it is easy to describe movement from simplex to simplex on the computer.

There are occasions, however, when we might want other forms of triangulations or to vary them. For instance, often one wishes to calculate a fixed point on a unit simplex. Then a pattern such as in Figure 13.2.2 might be utilized.

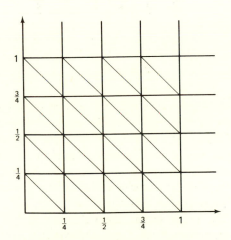

Each little square is diagonally divided.
This type of triangulation is straight-forward
to implement on a computer.

Figure 13.2.1

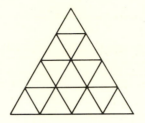

A triangulation of the unit simplex.

Figure 13.2.2

In many instances, the function values at $t = 0$ and $t = 1$ are of prime interest, and the other values of t, $0 < t < 1$, are of little or no interest. In these situations one might place vertices only at $t = 0$ or $t = 1$, as in Figure 13.2.3. Then, although $R^n \times [0, 1]$ is triangulated, the simplices are long and thin, and all vertices are in $R^n \times \{0\}$ or $R^n \times \{1\}$.

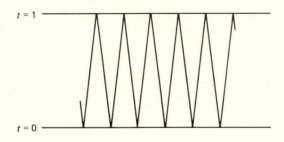

A triangulation of $R \times [0, 1]$ such that vertices are placed only at $t = 0$ or $t = 1$.

Figure 13.2.3

Finally, there is the important aspect of getting to an approximate solution quickly. We might not want to follow the H^{-1} path accurately in the beginning, as that might waste time. As t gets closer to 1 and the solution approaches, however, we generally would want greater precision. In this case triangulate $R^n \times [0, 1]$ so that the simplices in $R^n \times \{t\}$ become more and more dense as t approaches 1, as in Figure 13.2.4. Because of the speed gained by this approach, this triangulation is often used in practice.

In short, with the triangulation algorithm it is helpful for the simplices to form a fairly consistent pattern, as that will reduce computer effort. However, the particular pattern depends somewhat upon the problem under consideration. In most applications a pattern is chosen beforehand. At each step, as the algorithm moves from simplex to simplex, the computer generates the vertex which fits that specific pattern.

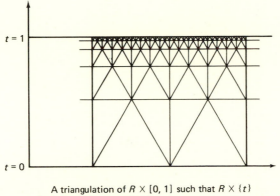

A triangulation of $R \times [0, 1]$ such that $R \times \{t\}$ is denser as t approaches 1.

Figure 13.2.4

FORMAL STATEMENT
OF THE TRIANGULATION ALGORITHM

With this background in triangulations we can provide a more formal statement of the triangulation algorithm. A homotopy function $H: D \times T \longrightarrow R^n$ will be given for $D \subset R^n$, the closure of an open set. An initial point w^0 and simplex s^0 will be given where $w^0 \in D \times \{0\}$ is in the facet B^0 of s^0. Generally, although not always, w^0 will be near a point $(x^0, 0)$ where $x^0 \in H^{-1}(0)$.

The algorithm will generate a piecewise-linear path in G^{-1} that approximates a path in H^{-1}. For the general step we will have a point $w \in B$, where B is a facet of a given simplex s. Then, letting G_s be the linear approximation of H on s, we will generate a linear path across s to a point $\bar{w} \in \bar{B}$ where $\bar{B}$ is another facet of s.

A new point $\bar{v}$ on the other side of $\bar{B}$ will be selected, forming the next simplex $\bar{s}$:

$$\bar{s} = \bar{B} \cup \{\bar{v}\}.$$

The process then continues from simplex to simplex in this manner. This is the basic step of all simplicial-type algorithms. The only difference for the triangulation algorithm is that all the simplices must be part of a triangulation of $D \times [0, 1]$.

Recall that G is regular if the points w and $\bar{w}$ generated are in the interior of their respective facets B and $\bar{B}$. Also, as discussed in Chapter 11, regularity almost always holds, and, if not, it is usually easy to obtain. Consequently, if G is regular, a triangulation algorithm will generate a path in G^{-1} that is PL (see Figure 13.2.5).

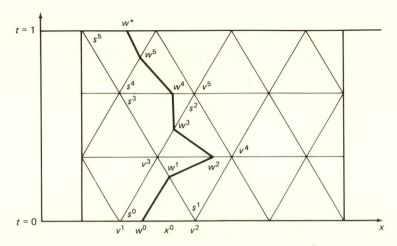

The triangulation is given beforehand. Starting at a point w^0 near $(x^0, 0)$, the triangulation algorthm generates the PL path G^{-1} from w^0 to w^*. A total of six simplices $s^0, s^1, \ldots, s^5$ of the triangulation were traversed. When the algorithm arrives at w^1, the vertex v^4 is added to form the new simplex $\{v^2, v^3, v^4\}$. At w^2 the new vertex v^5 is added, forming $\{v^3, v^4, v^5\}$, and so on.

Figure 13.2.5

13.3 COMPLEMENTARITY

We now come to complementarity. Complementarity is the premier theoretical concept underlying triangulation algorithms. Moreover, from a historical basis, this idea launched the entire field that this book is based upon. It may appear simple from our point of view today, but that is deceiving, because at the time of its discovery, it was revolutionary. Let us develop it.

Under regularity of G, a simplex s on the PL path has a single straight-line segment across it; and that line segment must go through the interior of s. This mean, as we will now show, that the PL path can never return to that simplex and can never cycle.

If the PL path were to return, some simplex on the path would have to have two different line segments of G^{-1} across it (see Figure 13.3.1). But under regularity only one line segment can exist, and it must cross the interior of the simplex.

Further, the simplices are part of a triangulation. With regularity we know that the path cannot revisit a simplex s. The triangulation ensures that no other simplex can overlap s. Thus the path can never return to that portion of space and can never cycle.

Note that both regularity and a triangulation are needed for this result. Regularity ensures one line segment of G^{-1} per simplex. It does not prevent two or more simplices from overlapping (see Figure 13.3.2). It is the triangulation

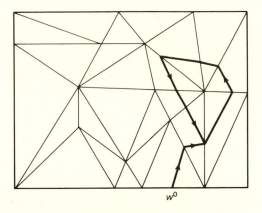

w^0

Notice how the PL path in the figure crosses a simplex on two separate line segments. Under regularity this cannot occur. The PL path G^{-1} thus cannot cycle.

Figure 13.3.1

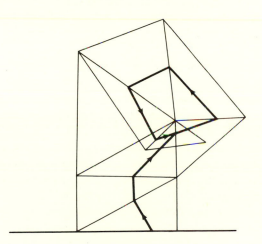

Without a triangulation the PL path G^{-1} might cycle. Here regularity holds, but there is no triangulation, so the simplices can overlap and cycling can occur.

Figure 13.3.2

aspect that prohibits the overlap. Thus both regularity and a triangulation are needed to prevent cycling.

This fact that the triangulation algorithm can never return to a simplex and can never cycle is termed the *principle of complementarity*. (Historically, the word "complementarity" arose from the linear complementarity problem, a very important problem that is discussed in Chapters 19 and 20.)

To summarize, we have the following theorem.

THEOREM 13.3.1 (COMPLEMENTARITY THEOREM).   Let $H: D \times T \longrightarrow R^n$ be a function where D is the closure of a bounded open set in R^n, and suppose that the triangulation algorithm generates simplices that are part of a triangulation of $D \times T$. Then, under regularity of G, the triangulation algorithm cannot cycle. If the PL path in G^{-1} starts at a boundary point of the triangulation, it must terminate at another different boundary point.

The proof follows from our previous discussion. Since the algorithm cannot cycle, it must always visit new simplices. Eventually, it will run out of new simplices to visit and must terminate at a boundary point. The terminal boundary point cannot be the initial point because the initial simplex is never revisited.

This theorem assures us that the triangulation algorithm can never cycle and thus is a highly viable addition to our techniques for solving actual problems.

TECHNICAL NOTE

For this theorem H need not be continuously differentiable. It need not even be continuous; it need merely be a function. The G is still defined by taking it as the linear approximation to H on each simplex generated.

The theorem, however, says nothing about how closely G^{-1} approximates H^{-1}, as that requires stronger assumptions. Theorems 12.2.1 and 12.3.1 state the conditions which ensure that G^{-1} is within a δ-tube of H^{-1} for each algorithm. For the flex simplicial method the simplices must be sufficiently small and robust and $H \in C^2$. For the triangulation method the simplices must be sufficiently small and H continuous.

Although these conditions are mathematically required for our proofs, from a practical viewpoint they are folderol and almost meaningless. All computer operations are discrete, which means that a computer cannot identify if a function is differentiable or continuous, or if the simplices are robust or not. Computationally, a differentiable function is usually easier to handle than a continuous one. However, in any particular instance and depending upon the shape of the functions, the opposite might be true. We must distinguish between conditions that are required for mathematical reasons and those that are required for computational considerations. They are different, yet both are essential. Indeed, both must be thoroughly understood to have an effective procedure.

We now continue to develop these ideas by studying the two algorithms in greater depth.

THE TRIANGULATION CHALLENGES
THE FLEX SIMPLICIAL

At this juncture we can make more precise comparisons between our two algorithmic approaches, the flex simplicial and the triangulation. First observe, under the hypotheses of Theorem 12.2.1, that as long as the simplices are sufficiently small, the G^{-1} path must be close to and follow an underlying H^{-1} path. This holds for both algorithms. Thus, whenever the simplices are small, both algorithms behave well and follow an H^{-1} path.

The problem, however, is with large simplices, as then the G^{-1} path may be far from an underlying H^{-1} path. It is here that the flex simplicial can cycle. Of course, when its simplices get sufficiently small the flex simplicial must follow an H^{-1} path. But note that if the flex simplicial attaches to a loop, it may cycle. The triangulation algorithm, as we have shown, will not cycle even with large simplices.

We see, therefore, a trade-off. The flex simplicial can choose its vertices freely, and that, by exploiting the problem structure, might speed convergence. On the other hand, large simplices may cause it to cycle. The triangulation algorithm, being restricted in its choice of vertices, is limited in exploiting any problem structure. But, of course, it does not cycle.

HYBRID ALGORITHM

These observations suggest the use of a hybrid or adaptive procedure. To get started, the triangulation algorithm with large simplices would hopefully move us rapidly away from $t = 0$. With t reasonably large we can change to the flex simplicial and attempt to exploit the problem structure and home in on a solution. However, if the flex simplicial is detected to cycle or otherwise not make progress, switch temporarily to the triangulation method.

EXERCISES/CHAPTER 13

1. Suppose that the plane is triangulated as in Figure E.13.1. Use the triangulation algorithm to solve

$$H(x, t) = (1 - t)(x - \tfrac{1}{2}) + t(x^2 - 4x - 21) = 0, \qquad 0 \le t \le 1,$$

starting at

$$s^0 = \{(0, 0), (1, 0), (1, 1)\}.$$

2. Prove Corollary 3.2.2 directly by use of the triangulation algorithm.

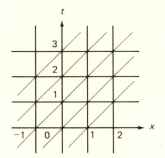

Figure E.13.1

3. Let an $(n + 1)$-simplex s be defined by a matrix

$$s = [v^0, v^1, \ldots, v^{n+1}], \tag{E.13.1}$$

where the columns of s are the vertices of the simplex. For $i = 1, 2, \ldots, n + 1$, let $\epsilon(i)$ be the $(n + 2)$-row vector with first i components equal to 0, and the remaining components equal to $\epsilon > 0$.

a. Suppose that

$$s = \begin{pmatrix} \epsilon(1) \\ \epsilon(2) \\ \cdot \\ \cdot \\ \cdot \\ \epsilon(n + 1) \end{pmatrix} = [v^0, v^1, \ldots, v^{n+1}]. \tag{E.13.2}$$

Show that $y \in s$ if and only if

$$\epsilon \geq y_1 \geq y_2 \cdots \geq y_{n+1} \geq 0.$$

b. Suppose that

$$s = \begin{bmatrix} \epsilon(q_1) \\ \epsilon(q_2) \\ \cdot \\ \cdot \\ \cdot \\ \epsilon(q_{n+1}) \end{bmatrix}, \tag{E.13.3}$$

where $\{q_1, \ldots, q_{n+1}\}$ is a permutation of $\{1, 2, \ldots, n + 1\}$. Show that $y \in s$ if and only if

$$\epsilon \geq y_{i_1} \geq y_{i_2} \geq \cdots \geq y_{i_{n+1}} \geq 0,$$

where $q_{i_j} = j$ for $j = 1, 2, \ldots, n + 1$.

4. Consider the set of simplices defined by (E.13.3), for all permutations $\{q_1, q_2, \ldots, q_{n+1}\}$ of $\{1, 2, \ldots, n + 1\}$. Prove that this set forms a triangulation of the $(n + 1)$-cube

$$\{y \in R^{n+1} \mid 0 \leq y_i \leq \epsilon, \quad \text{all } i\}.$$

5. a. Show how (E.13.3) is used to triangulate R^{n+1}. (*Hint:* Use Exercise 4.)
   b. Let s be a simplex of the form (E.13.3). Suppose that we are moving from s

to adjacent simplex $\bar{s}$ by removing the vertex v^k, $k = 0, 1, \ldots, n + 1$. Show how $\bar{s}$ of the form (E.13.3) is obtained from s.

6. Let $X = \{x \in R^n \mid \sum_1^n x_i = 1, \ x \geq 0\}$ denote the unit simplex. Describe a triangulation of X that is easily implementable on a computer.

7. Construct a two-dimensional example where the triangulation algorithm generates a PL path in G^{-1} that cycles.

8. Let $H(x, t) = 2x - e^{3t} = 0$. Suppose that the plane is triangulated as in Figure E.13.1. Sketch the PL paths in G^{-1}.

9. Let

$$F_1(x_1, x_2) = x_1$$

$$F_2(x_1, x_2) = (x_1^2 + x_2^2)^2 - 10(x_1^2 + x_2^2) + 9 \qquad F(x^0) = (-4, \quad 9109)$$

$$H(x, t) \equiv F(x) - (1 - t)F(x^0) = 0.$$

Using the triangulation constructed in Exercise 5, apply the triangulation algorithm on $H = 0$. Do at least five iterations.

10. Suppose that S is a triangulation of R^{n+1}. Prove that any n-simplex in S is a facet of exactly two $(n + 1)$-simplices of S.

11. Suppose that S is a triangulation of R^{n+1}. Prove that for any $x \in R^{n+1}$, where x is not a vertex of S, there is a unique k-simplex of S, $1 \leq k \leq n + 1$, which contains x in its relative interior.

12. Let $B = \{v^i\}_1^{n+1}$ be a facet of a triangulation of R^{n+1}. Let $s = B \cup \{v\}$ and $\bar{s} = B \cup \{\bar{v}\}$ be two nonintersecting simplices in R^{n+1}. Let

$$A = \begin{bmatrix} 1 & 1 & & 1 \\ & & \cdots & \\ v & v^1 & & v^{n+1} \end{bmatrix} \quad \text{and} \quad C = \begin{bmatrix} 1 & 1 & & 1 \\ & & \cdots & \\ \bar{v} & v^1 & & v^{n+1} \end{bmatrix}.$$

Show that $\det A$ and $\det C$ have opposite signs.

13. Prove that the simplices of the triangulation described in Exercise 5 are robust.

NOTES

13.1 Various triangulations are known for X and R^{n+1}. For a classical treatment, see Whitney [1957]. A well-known triangulation is "Kuhn's triangulation" (Freudenthal [1942], Kuhn [1960]). Eaves [1972], Shapley [1973], and Todd [1977], among others, have given other triangulations. Zangwill [1977] presented a triangulation algorithm that still maintains a degree of flexibility in the choice of the new vertex $\bar{v}$. The efficiency of triangulations was studied by Todd [1976c].

13.3 The principle of complementarity was developed by Lemke and Howson [1964].

14

Integer Labels

The algorithms presented in the preceding three chapters constructed a PL path in G^{-1} to approximate an H^{-1} path, and although these algorithms are quite effective and useful, all is not totally rosy. The calculation of the new facet $\bar{B}$ and a new point $\bar{w}$ can be lengthy. An $(n + 1) \times (n + 1)$ system of linear equations must be solved, and especially for large n, that can be prohibitively expensive.

We now state an integer procedure that overcomes this difficulty. Given a vertex v of a simplex, no longer is the full vector $H(v)$ utilized in the computation, but instead only an integer is required. With integers we throw away the $(n + 1) \times (n + 1)$ linear system altogether, and the movement from facet to facet becomes easy.

The experienced reader will no doubt recognize that a trade-off must lurk somewhere. The trade-off lies in our inability to approximate H^{-1} well. For G^{-1} to approximate H^{-1} closely, we would expect smaller simplices for the integer procedure than for those techniques previously described. In short, to solve a problem the integer algorithms generally require more simplices, but the effort per simplex is greatly reduced.

Far beyond its import in producing new algorithms, however, the integer concept has profound theoretical implications. Two famous results will be proved using it: Sperner's lemma and the Knaster–Kuratowski–Mazurkiewicz lemma. We even obtain the Brouwer fixed-point theorems again. Indeed, the integer technique of this chapter is fundamental both practically and theoretically, and may be considered a parallel approach to many of our previous results.

14.1 A TRIVIAL PIECEWISE-LINEAR MAP

For any $(n + 1)$-simplex $s = \{v^i\}_0^{n+1}$, we know that we can create a function $G_s: s \longrightarrow R^n$ which is linear on s by the formula

$$G(w) = \sum_{i=0}^{n+1} \lambda_i G(v^i)$$

$$w = \sum_{i=0}^{n+1} \lambda_i v^i$$

$$\sum_{i=0}^{n+1} \lambda_i = 1, \qquad \lambda_i \geq 0. \tag{14.1.1}$$

Given a homotopy $H: R^{n+1} \longrightarrow R^n$, our previous approach, recall, required that

$$G(v^i) = H(v^i), \qquad i = 0, \ldots, n + 1. \tag{14.1.2}$$

This definition of G was very useful because, as already shown, it ensured the following fact: If $G(w) = 0$, then w is an approximate solution to $H(y) = 0$.

However, many other ways of defining G on the vertices v^i exist besides (14.1.2), and different definitions of G lead to totally different algorithms. Indeed, there is only one property of G that must be preserved: that if w satisfies $G(w) = 0$, then w is an approximate solution to $H(y) = 0$. This property is essential, and we term it the *c-property*. Specifically, suppose that on a given facet $\{v^i\}_1^{n+1}$,

$$G(w) = \sum_{i=1}^{n+1} \lambda_i G(v^i) = 0, \tag{14.1.3}$$

where $w = \sum_1^n \lambda_i v^i$, $\sum_1^n \lambda_i = 1$, $\lambda_i \geq 0$. Then G satisfies the c-property if a solution w to $G(w) = 0$ approaches a solution y to $H(y) = 0$ as the facet becomes small. We reiterate, should G satisfy the c-property w is an approximate solution to $H(y) = 0$ (at least for sufficiently small facets).

A NEW G DEFINITION

If we define G by (14.1.2), then as we already know, G has the c-property. However, here is another G that also possesses the *c-property* and in fact will lead to our integer algorithm:

$$G(v) = \begin{cases} e^j & \text{if } H_j(v) > 0, \quad \text{some } j = 1, \ldots, n \\ -e & \text{if } H_j(v) \leq 0, \quad \text{all } j. \end{cases} \tag{14.1.4}$$

(Here e is a vector of all ones, and e^j is a vector of zeros except for a one in the jth position.) Ties can be broken arbitrarily.

To explain, suppose that we are given a facet $\{v^i\}_1^{n+1}$, and that we have evaluated G in each of the $n + 1$ vertices via (14.1.4). A set of $n + 1$ vectors $\{G(v^i)\}_1^{n+1}$ is obtained, where each $G(v^i)$ equals either $-e$ or e^j for some j, and $i \neq j$ is permissible.

For example, suppose that $H: R^3 \longrightarrow R^2$, so a facet will have three vertices. If $H_1(v^1) < 0$ and $H_2(v^1) > 0$, then $G(v^1) = e^2$. If $H_1(v^2) > 0$ and $H_2(v^2) > 0$, we may break the tie arbitrarily and let $G(v^2) = e^2$. If $H_1(v^3) \leq 0$ and $H_2(v^3) \leq 0$, then $G(v^3) = -e$. Hence $\{G(v^i)\} = \{e^2, e^2, -e\}$.

The set $\{G(v^i)\}_1^{n+1}$ that results by evaluating G at all of the $n + 1$ facet vertices is very important. In particular, we call the facet *complete* if

$$\{G(v^i)\}_1^{n+1} = \{e^1, e^2, \ldots, e^n, -e\}.$$

Verbally, if all e^i, $i = 1, \ldots, n$, and $-e$ appear in that set, the facet is complete. Notice that if the facet is complete, $G(v^i)$ must be different at each vertex of the facet. For instance, if $G(v^1) = e^3$ and $G(v^2) = e^3$, the facet cannot be complete.

COMPLETE FACETS

Complete facets will let us show that the G of (14.1.4) has the c-property.

Using (14.1.4), perceive that (14.1.3) has a solution if and only if the facet is complete. Specifically, if the facet $\{v^i\}_1^{n+1}$ is complete, system (14.1.3) then becomes

$$\sum_1^n \lambda_i e^i - \lambda_{n+1} e = 0$$

$$\sum_1^{n+1} \lambda_i = 1, \qquad \lambda_i \geq 0,$$

which has as its unique solution

$$\lambda_i = \frac{1}{n+1}, \qquad i = 1, \ldots, n + 1. \tag{14.1.5}$$

And from (14.1.1),

$$w = \sum_{i=1}^{n+1} \frac{1}{n+1} v^i,$$

which is the centroid of the facet. In words, if the facet is complete, $G(w) = 0$ where w is the centroid of the facet.

However, by reasoning backward it is easily seen that the converse is also true. Under (14.1.4), therefore, (14.1.3) has a solution w if and only if w is the facet centroid and the facet is complete. We emphasize this key implication of (14.1.4): a facet is complete if and only if $G(w) = 0$ has a solution.

THE C-PROPERTY HOLDS

We now can verify our claim that the G of (14.1.4) satisfies the c-property. Suppose that on a facet

$$G(w) = 0; \qquad (14.1.6)$$

then the facet must be complete. Assume, renumbering if necessary, that

$$G(v^i) = e^i, \qquad i = 1, \ldots, n$$
$$G(v^{n+1}) = -e.$$

Via (14.1.4), this means that for $i = 1, \ldots, n$,

$$H_i(v^i) > 0 \qquad \text{and} \qquad H_i(v^{n+1}) \leq 0. \qquad (14.1.7)$$

Assuming continuity, H_i must be equal to zero somewhere on the line segment connecting v^i and v^{n+1}. Yet this holds for all i; thus one might expect that for a y nearby, $H(y) = 0$.

Let us be even more precise. Suppose that there is a sequence of completely labeled facets

$$B^k = \{v^{i,k}\}_{i=1}^{n+1}, \qquad k = 1, 2, \ldots, \infty$$

and that this sequence converges to a point v^*. That is, as $k \longrightarrow \infty$ all vertices of all facets converge to v^*:

$$\lim_{k \to \infty} v^{i,k} = v^*, \qquad i = 1, \ldots, n+1. \qquad (14.1.8)$$

Employing (14.1.7) for $i = 1, \ldots, n$ and all k, we obtain

$$H_i(v^{i,k}) > 0 \qquad \text{and} \qquad H_i(v^{n+1,k}) \leq 0. \qquad (14.1.9)$$

Assuming that H is continuous, if we take limits of (14.1.9),

$$\lim_{k \to \infty} H_i(v^{i,k}) = H_i(v^*) \geq 0$$

$$\lim_{k \to \infty} H_i(v^{n+1,k}) = H_i(v^*) \leq 0;$$

thus

$$H(v^*) = 0. \qquad (14.1.10)$$

Given a sequence of complete facets, as the facets get small, the sequence converges to a solution to $H(y) = 0$. The c-property for the G of (14.1.4) is thus verified. Expressly, suppose that the facet is sufficiently small and $G(w) = 0$. Then w is near a point y where $H(y) = 0$. (The reader will obviously relate the word "complete" with the letter c of c-property.)

14.2 THE INTEGER ALGORITHM

We now develop the algorithm. First and most important, observe that the vectors of (14.1.4) can be replaced by integers. Simply identify e^j with the integer label j and $-e$ with the integer label $n + 1$. Now definition (14.1.4) is precisely equivalent to the following definition of G:

$$G(v) = \begin{cases} j & \text{if } H_j(v) > 0 \\ n + 1 & \text{if } H_j(v) \leq \text{all } j. \end{cases} \qquad (14.2.1)$$

There is no need to drag along the entire vector e^j or $-e$; just remember its integer label. Definitions (14.1.4) and (14.2.1) operate exactly the same way.

To illustrate, term a facet complete if its vertices contain all the labels $1, 2, \ldots, n + 1$. In other words, given a facet $\{v^i\}_1^{n+1}$, evaluate G via (14.2.1)

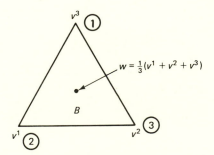

$$w = \tfrac{1}{3}(v^1 + v^2 + v^3)$$

Here the facet $B = \{v^1, v^2, v^3\}$, and the label of each vertex is indicated. Note the label number and vertex number can be totally different. The facet is completely labeled. Also, the centroid $w = \tfrac{1}{3}(v^1 + v^2 + v^3)$ is the unique point in $B \cap G^{-1}$.

Figure 14.2.1

on each of the vertices. If all labels $1, 2, \ldots, n + 1$ appear, the facet is complete, or completely labeled as it is often called (see Figure 14.2.1). Since the two definitions of complete are equivalent, we use them interchangeably. For instance, it still holds that for a completely labeled facet its centroid w

satisfies

$$G(w) = 0.$$

Also, for a sufficiently small facet, a point in H^{-1} is close by.

Integer labels are equivalent to the vector labels previously used, but are much easier to implement and produce quite simple algorithms.

THE ALGORITHMIC STEP

The operation of the algorithm hinges on one essential observation. Consider a simplex s which has a completely labeled facet B. Then s must have only one other completely labeled facet $\bar{B}$ (Figure 14.2.2a).

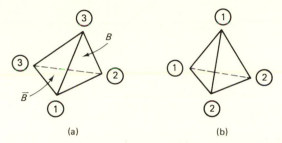

(a)   (b)

Simplex (a) has exactly two completely labeled facets. Simplex (b) has no completely labeled facet. An arbitrary simplex has either no completely labeled facet or exactly two completely labeled facets.

Figure 14.2.2

To see this, let $s = B \cup \{\bar{v}\}$, where $\bar{v}$ is the vertex of s that is not in B. Since B is completely labeled, it contains all labels $1, \ldots, n+1$. Now evaluate the label of $\bar{v}$ via (14.2.1). Suppose that label is l. But as B is completely labeled, one of the vertices of B also has the label l, say $G(v^l) = l$. Let $\bar{B}$ be the facet consisting of $\bar{v}$ and all the vertices of B except v^l. In other words, *drop* v^l and *replace* it by $\bar{v}$, forming the new facet $\bar{B}$ of the simplex s.

Clearly, the facet $\bar{B}$ is also completely labeled. This is because the label of v^l which was dropped and the label of $\bar{v}$ are the same. All the other vertices of B remain in $\bar{B}$. Most important, B and $\bar{B}$ are the only complete facets of s. Indeed, a simplex s has either no completely labeled facet, or exactly two of them.

This fact about the complete facets of s leads us to the basic step of the algorithm. In detail, suppose that we are given a completely labeled facet B and a vertex $\bar{v}$, where $s = B \cup \{\bar{v}\}$. Calculate the label of $\bar{v}$. Drop the vertex with the same label, and add $\bar{v}$ to form a new completely labeled facet $\bar{B}$. In this manner we move from one completely labeled facet to the next, and that is the main algorithmic step.

Moreover, let w be the centroid of B and $\bar{w}$ the centroid of $\bar{B}$, and consider the line segment across s from w to $\bar{w}$. Since w is an approximation to a point $y \in H^{-1}$ and $\bar{w}$ is an approximation to some point $\bar{y} \in H^{-1}$, the line segment from w to $\bar{w}$ will approximate a segment of a path in H^{-1}. Further, the line segment from w to $\bar{w}$ will comprise one portion of the G^{-1} path (see Figure 14.2.3). The algorithm is now easy to state.

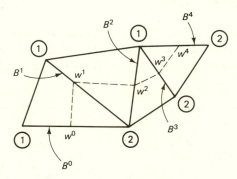

The label of each vertex is given. The dashed path is G^{-1}.

Figure 14.2.3

STATEMENT OF THE ALGORITHM

Let a triangulation of $D \times T$ be given where $H: D \times T \longrightarrow R^n$ and $D \subset R^n$ is the closure of an open set. An initial simplex s^0 with completely labeled facet B^0 will be given.

The general step of the algorithm is as previously described: A simplex $s = B \cup \{\bar{v}\}$ will be given with completely labeled facet B. Add $\bar{v}$ and drop the vertex of B with the same label of $\bar{v}$ to form another completely labeled facet $\bar{B}$ of s.

If we do not terminate, $\bar{B}$ will be a completely labeled facet of another simplex $\bar{s} = \bar{B} \cup \{\bar{v}\}$. The general step is then applied to $\bar{s}$ as before.

Starting from s^0 and B^0 and utilizing the general step in this manner each time, the algorithm will generate a sequence of simplices s^k, completely labeled facets B^k, and centroids w^k of B^k such that

$$B^k = s^{k-1} \cap s^k$$

and

$$G(w^k) = 0.$$

Note also all simplices must be from the triangulation. Furthermore, if we connect the centroids w^{k-1} and w^k for all k, the PL path created approximates a path in H^{-1}.

COMMENTS

The curious reader is encouraged to check that the vector labels of (14.1.4) with the usual pivoting generate the same path as the integer labels (14.2.1) (Exercise 1).

Because no pivoting in linear systems is required as in the preceding algorithms, computational burden per step is drastically cut by the integer algorithm. However, the G of (14.1.4) or (14.2.1) is not as accurate an approximation of H as is the G of (14.1.2). Thus we would expect the simplices of the integer approach to be much smaller to obtain the same accuracy.

Many integer labelings other than (14.2.1) are possible and some are given in the exercises. Indeed, a different labeling is used in the very next section and for a very different purpose.

14.3 SPERNER'S LEMMA

Integer labels were discovered prior to vector labels and were used to obtain certain powerful mathematical results, such as Sperner's lemma, the famous theorem from combinatorial topology. Years ago, Sperner's lemma was often employed to prove the existence of a Brouwer fixed point. Earlier in the book we showed something more powerful—not only the existence of, but how to calculate a fixed point. Our proof, moreover, was not based on Sperner's lemma but on path following. We now use path following to prove Sperner's lemma, thereby demonstrating that path following, this book's main concept, is a more general approach. The classical method of proving the Brouwer fixed-point theorem via Sperner's lemma is also obtained.

THE SPERNER LABELING

Consider the unit n-simplex

$$X = \left\{x \in R^{n+1} \,\middle|\, \sum_{1}^{n+1} x_i = 1, x_i \geq 0\right\} \tag{14.3.1}$$

and suppose that there is some triangulation S of X (see Figure 14.3.1). Also let $F: X \longrightarrow X$ be a function. By definition of X and of F, if $x = (x_i) \in X$, then

$$\sum_{i=1}^{n+1} x_i = 1, \qquad x_i \geq 0$$

$$\sum_{i=1}^{n+1} F_i(x) = 1, \qquad F_i(x) \geq 0. \tag{14.3.2}$$

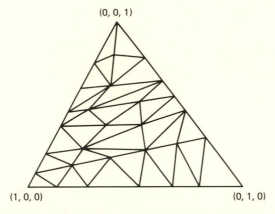

(0, 0, 1)

(1, 0, 0)   (0, 1, 0)

This figure is a triangulation of the unit 2-simplex. It is quite
uneven, in order to illustrate that the triangulation can be arbitrary.

Figure 14.3.1

Sperner's lemma employs a very special label function on X as follows:

$$G(x) = j \quad \text{if } x_j > 0 \quad \text{and} \quad F_j(x) \leq x_j, \qquad j = 1, \ldots, n + 1. \qquad (14.3.3)$$

In words, $x = (x_i) \in X$ is given the label j if both $x_j > 0$ and $F_j(x) \leq x_j$.
Ties are broken arbitrarily.

For example, suppose that for $n + 1 = 3$, $x = (\frac{1}{3}, \frac{1}{6}, \frac{1}{2})$ and $F(x) = (\frac{1}{4}, \frac{3}{8}, \frac{3}{8})$. Then

$$F_1(x) < x_1$$
$$F_2(x) > x_2$$
$$F_3(x) < x_3.$$

Thus the label is 1 or 3 and we may select either one.

The G of (14.3.3) is well defined. Note that $F_j(x) > x_j$ cannot hold for
all $x_j > 0$, since then $F_j(x) > x_j$ for all $x_j > 0$ and $F_j(x) \geq x_j$ for all $x_j = 0$
imply

$$\sum_1^{n+1} F_j(x) > \sum_1^{n+1} x_j = 1,$$

which violates (14.3.2). Thus any point $x \in X$ can be labeled. Also, any
n-simplex which has all labels $1 \ldots n + 1$ is called *completely labeled* (Figure
14.3.2).

PROPERTIES OF LABELING

This labeling function G has several properties which will enable us to
prove Sperner's lemma. First note that the vertices of the simplex X are e^i,
$i = 1, \ldots, n + 1$ (e^i is zeros except for a one in the ith position). Thus, via

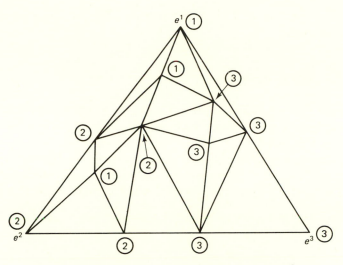

A labeling of a simplex via (14.3.3). Vertex e^i receives label ⓘ.
Also, by (14.3.3) the label ⓘ cannot appear on the facet opposite
e^i. That is, ① cannot appear on the face between e^2 and e^3, ②
cannot appear between e^1 and e^3, and ③ cannot appear between
e^1 and e^2.

Sperner's lemma proves that under the labeling (14.3.3), some
simplex of the triangulation must be completely labeled.

Figure 14.3.2

(14.3.3), the label of vertex e^i is i, which is rather convenient. Indeed, observe that the simplex $X = \{e^i\}$ is completely labeled since it has all labels $1, \ldots, n+1$.

Next examine a facet of X, say, the facet opposite the vertex e^1. For all $x = (x_i)$ on that facet the first component $x_1 = 0$. Therefore, $G(x)$ cannot be 1 for any x on that facet. Similarly, any point on the facet opposite vertex e^j cannot have a label j. We see that on a facet some label must be missing.

Sperner's lemma makes a brilliant application of the labeling function G. Given an arbitrary triangulation S of X, evaluate via G the label of each vertex in the triangulation S. Sperner's lemma proclaims that some simplex of the triangulation will be completely labeled. That is, suppose that in the triangulation we evaluate the labels of every vertex. Then some simplex must obtain all $n+1$ integers $1, 2, \ldots, n+1$; and thus is completely labeled.

This result is not intuitive. Given any triangulation S of X and any function $F: X \longrightarrow X$, consider an arbitrary simplex in the triangulation. Although we must label its vertices by (14.3.3), presumably its vertices could have any labels. Conceivably, they could all be 1, for example, or just odd integers, or anything else. Who would suspect that some simplex must have all the integers as its labels? Yet that is what Sperner proves. All that is necessary is to use the labeling function (14.3.3). And remember, this holds for any triangulation S

nd any function $F \colon X \longrightarrow X$. By Sperner's lemma, a completely labeled simplex must exist.

THE BROUWER THEOREM

Sperner's lemma establishes the existence of a completely labeled simplex s of the triangulation S. But first we must ask: Why all the interest in a completely labeled simplex?

Suppose that we have a sequence of completely labeled simplices $s^k = \{v^{i,k}\}_{i=1}^{n+1}$ which as $k \longrightarrow \infty$ get smaller and smaller and converge to a point $v^* = (v_j^*)$. As before, by convergence we mean all vertices of s^k converge to v^*, so that

$$v^{i,k} \longrightarrow v^* \qquad i = 1, \ldots, n+1 \quad \text{as } k \longrightarrow \infty. \tag{14.3.4}$$

Since $s^k = \{v^{i,k}\}$ is completely labeled, renumbering if necessary, we have

$$G(v^{i,k}) = i, \qquad i = 1, \ldots, n+1,$$

which means that for $v^{i,k} = (v_j^{i,k})$,

$$F_i(v^{i,k}) \leq v_i^{i,k}, \qquad \text{all } i.$$

Now suppose that F is continuous. As $k \longrightarrow \infty$ in the limit, (14.3.4) and the continuity of F provide

$$F_i(v^*) \leq v_i^*, \qquad i = 1, \ldots, n+1.$$

But (14.3.2) ensures that

$$F_i(v^*) = v_i^*, \qquad i = 1, \ldots, n+1, \tag{14.3.5}$$

so v^* is a fixed point.

We have hereby obtained the Sperner proof of the Brouwer theorem. In detail: Sperner's lemma ensures a completely labeled simplex s exists for any triangulation. By taking triangulations finer and finer, we can get a sequence of arbitrarily small, completely labeled simplices s^k. From compactness of X they must converge in the limit to a point v^* (at least on a subsequence). But as just observed, v^* must then be a fixed point. Indeed, we have established:

THEOREM 14.3.1 (BROUWER).   Let $F \colon X \longrightarrow X$ be continuous on the unit simplex X. Then F has a fixed point.

PROOF OF SPERNER'S LEMMA

The argument just given is the classical demonstration of the Brouwer theorem by Sperner's lemma, and we must now proceed to prove Sperner's lemma. Observe that it is not sufficient to merely verify the existence of a completely labeled simplex as Sperner's lemma does, however. To make the lemma practical, we must find the completely labeled simplex. Clearly, all simplices of the triangulation cannot be checked, as for even small n that could easily take a millenium on the fastest computer. This would be worse than finding a needle in a haystack.

Consequently, we will now do considerably more than Sperner's lemma, because we will not only prove that the completely labeled simplex exists but also find it.

First consider $X \times [0, 1)$, and let us triangulate it in a special way. A sequence of $t^i \in [0, 1)$ will be given where $t^0 = 0$, $t^i < t^{i+1}$, and $\lim_{i \to \infty} t^i = 1$. At any t^i require that we also have a triangulation of that level $X \times \{t^i\}$. Our triangulation of $X \times [0, 1)$ thus must create a triangulation of each level $X \times \{t^i\}$. Under this property, therefore, an n-simplex of any level $X \times \{t^i\}$ will be a facet of an $(n + 1)$-simplex of $X \times [0, 1)$. (Study Figure 14.3.3.)

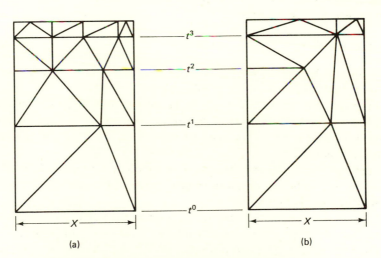

Both figures illustrate different triangulations of $X \times [0, 1)$; Notice in (b) that although the set $X \times [0, 1)$ itself is triangulated, its triangulation does not induce a triangulation of level t^2. Thus the triangulation of (b) does not fit our criteria. However, the triangulation of (a) does, because that triangulation of $X \times [0, 1)$ creates a triangulation of each level $X \times \{t^i\}$.

$X \times [0, 1)$ is triangulated by $(n + 1)$-simplices. Any level is triangulated by facets of $(n + 1)$-simplices (i.e., n-simplices). That is, each level $X \times \{t^i\}$ is triangulated and the triangulation of any level is "induced" by the triangulation of $X \times [0, 1)$.

Figure 14.3.3

The starting level $B^0 = X \times \{0\}$ is required to be a single n-simplex, namely X itself. But levels $X \times \{t^i\}$ will have finer and finer triangulations as $t^i \rightarrow 1$. Also as $t^i \rightarrow 1$, the triangulation must get arbitrarily fine; that is, the simplices must shrink to zero.

One other stipulation is required. Sperner's lemma holds for an arbitrary triangulation S of X. Require that for some $t^* = t^i$, $0 < t^* < 1$, the triangulation of level $X \times \{t^*\}$ be the triangulation S. Thus when we reach level t^*, we have exactly the triangulation S of Sperner's lemma.

To review, triangulate $X \times [0, 1)$, so that at each level $X \times \{t^i\}$ we also get a triangulation of that level. For $t^0 = 0$, $B^0 = X \times \{t^0\}$ will be the untriangulated set X. As t^i increases the triangulation gets finer, until at the t^* level, $X \times \{t^*\}$ has the triangulation S of X. For $t^i > t^*$ the triangulation is even finer than S. Indeed, as t^i approaches 1 the triangulation will get arbitrarily fine.

Notice that $X \times [0, 1)$ is triangulated by $(n + 1)$-simplices. Each level is triangulated by facets of $(n + 1)$-simplices.

LABELING

The labeling function is identical to that of (14.3.3) except that we must redefine it slightly because now our points are $y = (x, t) \in X \times [0, 1)$, whereas before we had $x \in X$. Specifically,

$$G(y) = j \quad \text{if } x_j > 0 \quad \text{and} \quad F_j(x) \leq x_j, \qquad j = 1, \ldots, n + 1. \qquad (14.3.6)$$

Clearly, this is identical to the previous labeling as the label is independent of t. A point x gets the same label for all levels t.

First for $t^0 = 0$ select $X \times \{t^0\}$ to be the initial completely labeled facet B^0. By construction the vertices of B^0 are the vertices of X. Consequently, as previously observed, the vertices of B^0 are $(e^i, 0)$, $i = 1, \ldots, n + 1$, and B^0 is completely labeled.

Starting from B^0 we can then generate a path of completely labeled facets. This is done just as before by adding a new vertex and dropping the vertex with the duplicated label (Figure 14.3.4.).

Second, observe there cannot be a completely labeled facet (n-simplex) on the boundary $\partial X \times [0, 1)$. As noted with (14.3.3), if we are on a facet of X, one label must be missing. Thus no n-simplex on $\partial X \times [0, 1)$ can be complete (see Figure 14.3.5). The path, of course, only goes through complete facets. Consequently, the path of completely labeled facets cannot go to the boundary $\partial X \times [0, 1)$. Restated, the path cannot penetrate the boundary.

Overall, then, observe these key properties of the path. At $t^0 = 0$, B^0 is the unique completely labeled facet. The path can therefore never return to $t^0 = 0$. Also the path cannot penetrate a boundary. Thus the path from $t^0 = 0$ must eventually go through all levels t^i. Furthermore, it passes through each level t^i at a completely labeled n-simplex (facet).

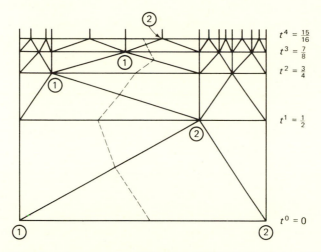

$$t^4 = \tfrac{15}{16}$$
$$t^3 = \tfrac{7}{8}$$
$$t^2 = \tfrac{3}{4}$$

$$t^1 = \tfrac{1}{2}$$

$$t^0 = 0$$

There is a unique, completely labeled facet B^0 on level $t^0 = 0$. As t^i increases, the triangulation on any level $X \times \{t^i\}$ gets finer and finer. As $t^i \to 1$, it gets arbitrarily fine. Note how the triangulation of $X \times [0, 1)$ also center a triangulation of each level $X \times \{t^i\}$. At some t^* the triangulation level of $X \times \{t^*\}$ would, by construction, be the triangulation S of X in Sperner's lemma. The dashed line traces out the path of completely labeled facets.

Figure 14.3.4

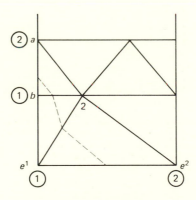

This path cannot occur. Here X is the line segment between e^1 and e^2. By (14.3.6), point a is mislabeled and should be a ①. For any $x \in X$, all points "above" it receive the same label, because (14.3.6) is independent of t.

Since the algorithm operates from completely labeled facet to completely labeled facet, it can never go to the boundary $\partial X \times [0, 1)$.

Figure 14.3.5

Sperner's lemma follows immediately. The path eventually passes through level t^* and it must do so at a completely labeled simplex of the triangulation S, thereby proving the lemma. Not only is the existence of a completely labeled simplex of S demonstrated, but by path following we actually determine it. We have proved:

LEMMA 14.3.2 (SPERNER). Let S be an arbitrary triangulation of the unit simplex X. Suppose that each vertex of S is labeled via (14.3.3). Then some simplex of S is completely labeled.

ODD SOLUTIONS

Examining Figure 14.3.6, note that there can be three types of paths. The path from $t^0 = 0$ to $t = 1$, loops, and paths from $t = 1$ that return to $t = 1$. (Technically, $t \longrightarrow 1$, but it is convenient to consider $t = 1$ here.) The fact that only these three types of paths exist proves that there must be an *odd* number of completely labeled simplices on any level t^i.

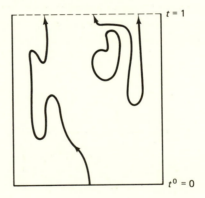

The possible paths of completely labeled facets. There is one path from B^0 by construction, and no path can end in the boundary $\partial X \times [0, 1)$. Note that the paths can reverse themselves and go up and down for a while. However, the path from t^0 must eventually approach $t = 1$. The only other forms of paths could be in the form of a loop, or a path that "starts" at $t = 1$ and "returns" to $t = 1$.

Figure 14.3.6

FIXED POINTS

Finally, by following the path as $t \longrightarrow 1$, we determine a fixed point of F. All the simplices the path goes through will be completely labeled. Also, for $t^i \longrightarrow 1$ they will be getting arbitrarily small and, in fact, converging to some limit v^*. As observed in (14.3.5), that limit will be a fixed point of F. Essentially, the path generates the simplices that satisfy (14.3.4) and thus produces a fixed point.

Overall, we see the advantages of path following via integer labels to prove Sperner's lemma. We not only verify Sperner's lemma that a completely labeled simplex exists, but we calculate that simplex. Moreover, by letting the path continue so $t \longrightarrow 1$, we also obtain a fixed point.

14.4 THE KNASTER–KURATOWSKI– MAZURKIEWICZ LEMMA

We continue to show the power of integer labels by proving the Knaster–Kuratowski–Mazurkiewicz (KKM) lemma. The KKM lemma, Sperner's lemma, and the Brouwer theorem are a sort of mathematical trinity. All are extremely important and, although seemingly different, are in a deep sense equivalent. Having demonstrated the Brouwer and Sperner already, we now take on the KKM.

The proof via integer labels is rather easy, as it follows the proof of Sperner's lemma almost to the letter.

As before, let X be the unit simplex

$$X = \left\{ x \in R^{n+1} \,\Big|\, \sum_{1}^{n+1} x_i = 1, \quad x_i \geq 0 \right\}.$$

Define J as any subset of the integers $1, 2, \ldots, n + 1$. Thus

$$J \subset M,$$

where for convenience we let $M = \{1, 2, \ldots, n + 1\}$. Now specify

$$X_J = \{x = (x_i) \in X \,|\, x_i \geq 0, \quad i \in J, \quad x_i = 0, \quad i \notin J\}. \qquad (14.4.1)$$

For X_J, the components $x_i \geq 0$ for $i \in J$, but $x_i = 0$ for those i not in J.

To understand X_J, suppose that $J = M$. Then since $x_i \geq 0$ for all i, $X_J = X$. If J is strictly smaller than M, X_J will be a face of X. For example, if $J = \{2, \ldots, n + 1\}$, then $x_1 = 0$, so X_J is the facet opposite vertex e^1 of X. Depending upon J, X_J will be all of X or a face of X, and the J tells us which face.

In short, simply remember that "X_J is the face of J"—that is, the face generated by the vertices e^i, $i \in J$. Also, if $x \in X_J$, then $x_i \geq 0$, $i \in J$, but we must have $x_i = 0$, $i \notin J$.

The KKM lemma utilizes a collection of closed sets A_i, $i = 1, \ldots, n + 1$, which satisfy the condition

$$X_J \subset \bigcup_{i \in J} A_i, \qquad \text{all } J \subset M. \qquad (14.4.2)$$

First note by setting $J = M$ that the sets A_i must cover the entire simplex X. If $J = \{1, 2\}$, X_J is the face for which only components x_1 or x_2 can be positive. Then the union of A_1 and A_2 must cover X_J. More generally, any face X_J must be in the union of those A_i for $i \in J$. Succinctly, the A_i, $i \in J$, cover face X_J.

The KKM condition (14.4.2) tells us only about the unions of the A_i. The surprising conclusion is that their intersection is not empty (see Figure 14.4.1).

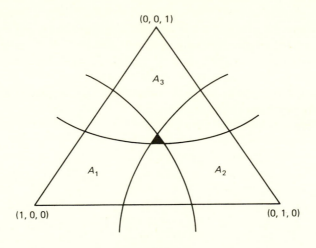

The KKM lemma.

Figure 14.4.1

THEOREM 14.4.1 (KKM) Consider the unit simplex X and the sets X_J defined by (14.4.1). Suppose that there are closed sets A_i such that

$$X_J \subset \bigcup_{i \in J} A_i, \qquad \text{all } J \subset M. \qquad \text{(KKM)}$$

Then

$$\emptyset \neq X \cap (\bigcap_{\text{all } i} A_i).$$

The proof uses a triangulation of $X \times [0, 1)$ similar to that in Sperner's lemma. Let a sequence $t^0 = 0$, $t^i < t^{i+1}$, $\lim t^i = 1$ be given. Starting with

$B^0 = X \times \{t^0\}$ untriangulated, triangulate $X \times [0, 1)$ so that each level $X \times \{t^i\}$ is also triangulated. Moreover, as $t^i \rightarrow 1$ require the size of the simplices in the $X \times \{t^i\}$ triangulations to get arbitrarily small.

This triangulation of $X \times [0, 1)$ is like that of Sperner's lemma except here we do not require a particular triangulation at level $X \times \{t^*\}$. That is dispensed with.

As for the labeling function, given $y = (x, t) \in X \times [0, 1)$, define

$$G(y) = j \quad \text{if } x_j > 0 \text{ and } x \in A_j. \tag{14.4.3}$$

Ties are broken arbitrarily.

Under the KKM condition, this labeling is always defined. For example, suppose that given $y = (x, t)$, $x_1 > 0, x_2 > 0$, and $x_3 > 0$. Then $x \in X_J$ for $J = \{1, 2, 3\}$ and by the KKM condition

$$x \in A_1 \cup A_2 \cup A_3. \tag{14.4.4}$$

Consequently, x is in at least one of the $A_i, i = 1, 2, 3$—say, $x \in A_3$. But $x_3 > 0$. Thus $G(y) = 3$, and we see the labeling is well defined.

This labeling has the properties we need to prove the theorem. First notice the initial simplex $B^0 = X \times \{t^0\}$ is complete. Of course, $X = \{e^i\}_{i=1}^{n+1}$. If $x = e^i, x_i = 1 > 0$, so for $J = \{i\}, x \in X_J \subset A_i$. Therefore, e^i gets the label i and X is completely labeled.

Second, no facet of X can contain all the labels. On a facet some component $x_i = 0$, so the label i cannot appear. For the triangulation of $X \times [0, 1)$, therefore, a completely labeled facet (n-simplex) cannot be in the boundary $\partial X \times [0, 1)$.

Start a path from $B^0 = X \times \{t^0\}$. Since at $t^0 = 0$ the completely labeled simplex is unique, and since the path cannot penetrate a boundary, it must cross all levels $X \times \{t^i\}$ for all t^i. The path therefore generates a sequence of completely labeled simplices that converge as $t^i \rightarrow 1$ to some limit $v^* \in X$.

Let $s^k = \{v^{i,k}\}_{i=1}^{n+1}$ be that sequence so that

$$\lim_{k \to \infty} v^{i,k} = v^*, \quad \text{all } i. \tag{14.4.5}$$

As s^k is complete, by labeling (14.4.3),

$$v^{i,k} \in A_i, \quad i = 1, \ldots, n+1 \quad \text{and all } k,$$

where we have renumbered if necessary so that $G(v^{i,k}) = i$. Applying the limit (14.4.5) and since A_i is closed,

$$v^* \in A_i, \quad \text{all } i,$$

Consequently,

$$X \cap (\bigcap_{\text{all } i} A_i) \neq \varnothing.$$

The KKM lemma thus is a direct consequence of the integer labels. The proof is virtually the same as the Sperner lemma except that a different label function was employed.

SUMMARY

We previously presented path following via vector labels, and in this chapter introduced path following by integer labels. The integer labels historically predate the vector label, and yield algorithms that are extremely easy to implement. The trade-off is that with integer labels smaller simplices are usually required for the same degree of accuracy.

Integer labels are basic not only to algorithms but to theory and yield proofs of the Sperner lemma, KKM lemma, and, via Sperner, the Brouwer fixed-point theorem. These theorems underpin many powerful results and are in a mathematical sense equivalent (Exercises 10 and 11). On both the practical and theoretical levels, therefore, integer labels provide a convenient and effective extension of our path-following concept.

EXERCISES/CHAPTER 14

1. Check that the vector labels of (14.1.4) and the triangulation algorithm generate the same path as the integer labels (14.2.1).

2. Find a solution to the function $F(x) = x^2 - 1$:
   a. By the integer label algorithm.
   b. By the vector label algorithm in which $G(v^i) = H(v^i)$ on the simplex vertices. Use the homotopy $H(x, t) = F(x) - (1 - t)F(\frac{1}{4}) = 0$, and triangulation (E.13.1) with mesh size $\epsilon = \frac{1}{2}$.

3. Suppose we are given the set

$$Y = \left\{ x = (x_i) \in R^n \,\middle|\, \sum_1^n x_i \leq 1, \quad x_i \geq 0 \right\}$$

   and a continuous $F: Y \longrightarrow Y$. How can we modify F to $\bar{F}$ so that $\bar{F}: X \longrightarrow X$, where X is the unit simplex (14.3.1), thereby establishing that F has a fixed point?

4. Suppose that a continuous $F: D \longrightarrow D$ is given where D is a nonempty compact convex set. Show how to revise the function F into $\bar{F}: X \longrightarrow X$, and thereby prove that F has a fixed point in D. (*Hint:* Embed D in a simplex X that contains it.)

5. The Sperner lemma required a certain triangulation of $X \times [0, 1)$. In particular, the level $X \times \{t^*\}$ had to be triangulated in the same manner as a given triangulation S of X. Show that this can be done. That is, given an arbitrary triangulation S of X, show that for some t^*, $0 < t^* < 1$, level $X \times \{t^*\}$ can be triangulated like S.

6. Suppose that the integer labeling is defined by

$$G(v^i) = \begin{cases} j & \text{if } H_j(v^i) < 0 \\ n + 1 & \text{if } H_j(v^i) \geq 0, \end{cases} \quad \text{all } j.$$

   Show that G satisfies the c-property.

7. Suppose that the integer labeling is defined by

$$G(v^i) = \begin{cases} 1 & \text{if } H_1(v^i) > 0 \\ j & \text{if } H_j(v^i) > 0 \text{ and } H_{j-1}(v^i) \leq 0, \quad j = 2, 3, \dots, n \\ n + 1 & \text{if } H_j(v^i) \leq 0, \quad \text{all } j. \end{cases}$$

   Show that G satisfies the c-property.

8. Let $H(x, t) = (1 - t)(x + 1) + t(3x - 12) = 0$. Using the integer labeling (14.2.1), label all points on the $x - t$ plane. Show the path H^{-1} and explain why the integer algorithm will follow that path.

9. Let $f: X \longrightarrow X$ be defined by

$$f(x_1, x_2) = ((x_1 - \tfrac{1}{2})^2, 1 - (x_1 - \tfrac{1}{2})^2).$$

   Using the triangulation in Figure E.14.1, draw the path followed by the integer algorithm up to $t = \frac{31}{32}$. To which point will the path converge?

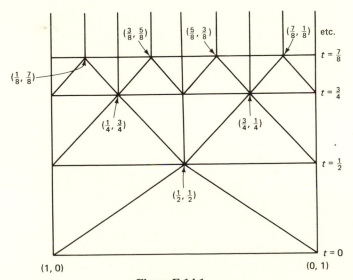

Figure E.14.1

10. In Section 14.3 we showed that the Brouwer theorem follows from Sperner's lemma. Prove Sperner's lemma by using the Brouwer theorem.

11. Prove the Brouwer theorem directly from the KKM lemma.

NOTES

14.3. The first constructive proof of Sperner's lemma (Sperner [1928]) is due to Cohen [1967]. The proof of the lemma given here follows that given by Eaves [1972]. Kuhn [1969], Garcia, Lemke, and Lüthi [1973], and Shapley [1973] generalize the lemma to the case where vertices on a facet can be assigned arbitrary labels. For an interesting application of integer labels, see Gale [1979].

14.4. The KKM lemma is proved in Knaster, Kuratowski, and Mazurkiewicz [1929]. Freidenfelds [1971] shows a constructive proof of it.

15

Differential Equations

We hereby commence the first of two chapters on differential equation methods for path following. Interestingly, although the simplicial methods of Chapters 11 through 14 are a recent development, they were applied to fixed-point calculations first. Differential equation approaches are centuries old, but only lately was their value in fixed-point and homotopy calculations emphasized. Nevertheless, differential equations have been studied extensively, numerous approaches for solving them exist, and they are readily applicable. They indeed provide a potent means for path following.

This chapter starts with the simplest differential equation solution procedure, the Euler method. It will be applied both to the BDE and to a new system which is often easier to implement than the BDE, the homotopy differential equations. The Euler method, unfortunately, tends to drift off the path, so to overcome that, other methods including a special restart procedure, are introduced. Finally, a technique is presented that does not follow the path but instead tries to leap ahead to a solution. This is the famed Newton method.

Chapter 16 suggests techniques for getting back on the path called predictor–corrector methods, and will provide even deeper insights into this fundamental approach of differential equations.

15.1 EULER'S METHOD

Given a continuous function $f: R^{l+1} \longrightarrow R^l$, suppose we are to find the solution path $w(u)$ satisfying

$$\frac{dw}{du} = f(w, u), \qquad w \in R^l, \quad u \in R^1, \tag{15.1.1}$$

$$w(0) \text{ given.}$$

This problem is termed an *initial value problem* because an initial value $w(0)$ is given and the path starts from that point. (Here we are deliberately using a different notation to emphasize the generality of the approach. Later we will identify (15.1.1) with our path following situation.)

If f can be integrated explicitly, then the solution $w(u)$ is

$$w(u) = w(0) + \int_0^u f(w, \theta) \, d\theta.$$

More often than not, however, f cannot be integrated to obtain a solution, and a numerical method for approximation $w(u)$ is then normally employed.

Many numerical solution methods exist, some quite complicated, but the simplest is due to Euler. It approximates the derivative by its corresponding difference, so that

$$\frac{w^{k+1} - w^k}{u^{k+1} - u^k} = f(w^k, u^k).$$

Then, given $0 < u^1 < u^2 < \ldots < u^k < u^{k+1}$, let

$$w^0 = w(0)$$

and specify

$$w^{k+1} = w^k + f(w^k, u^k)(u^{k+1} - u^k). \tag{15.1.2}$$

This is the Euler formula (see Figure 15.1.1).

Suppose that $w^k = w(u^k)$ is on the path. The function $f(w^k, u^k)$ is then the tangent to the path at w^k. Also $d^k = u^{k+1} - u^k$ is called the *step length*.

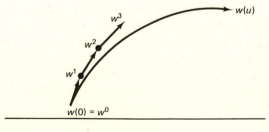

The Euler method:

$$w^{k+1} = w^k + f(w^k, u^k)(u^{k+1} - u^k).$$

Figure 15.1.1

Conceptually, Euler's method takes a linear approximation to the path at w^k and moves a distance $d^k f(w^k, u^k)$ along the linear approximation. Then it takes a new linear approximation at w^{k+1} and repeats the process. If the step lengths d^k are small, then presumably these approximations should not be too far off the path. (However, this will be discussed more in Section 15.3.)

APPLICATION TO BDE

Euler's method immediately applies to the BDE. Recall that $y \in R^{n+1}$, $(y, p) \in R^{n+2}$, $\dot{y}_i = dy_i/dp$, and the BDE stated that

$$\dot{y}_i = (-1)^i \det H'_{-i}(y), \qquad i = 1, \ldots, n+1, \qquad (15.1.3)$$

where $y(0) = y^0$ is given. Since this is of the form (15.1.1), we may apply Euler's method and obtain

$$y^0 = y(0)$$
$$y_i^{k+1} = y_i^k + (p^{k+1} - p^k)(-1)^i \det H'_{-i}(y^k), \qquad i = 1, \ldots, n+1,$$

where $0 < p^1 < p^2 < \ldots < p^k < p^{k+1} < \ldots$. We thus have determined a differential equation method of following the path.

15.2   THE HOMOTOPY DIFFERENTIAL EQUATIONS

Although the BDE can follow a path in H^{-1}, a related differential equation system exists which also follows a path in H^{-1}, yet is usually easier to implement.

To develop this new system, recollect that in Chapter 1, x was written in terms of the parameter t. In Chapter 2 we revised the notation so that $y = (x, t)$ depended upon the parameter p. To distinguish the two approaches notationally, let $x[t]$ indicate x as a function of t, whereas for y as a function of p, of course, $y(p) = (x(p), t(p))$.

Assuming that H is C^1, the implicit function theorem states conditions under which each of these two notations is valid. Specifically, if H'_x is of full rank for all (x, t) then $x[t]$ is defined and is a continuously differentiable function. If H is regular (i.e., if H' is of rank n), then $y(p) = (x(p), t(p))$ is defined and is continuously differentiable. In short, if H is regular, we can use the parameter p and the notation $(x(p), t(p))$. However, to write $x[t]$ requires the slightly stronger condition that H'_x be of full rank. (Note that H'_x of full rank implies that H is regular.)

Recall the BDE were developed using the parameter p. That is, the BDE were obtained by differentiating

$$H(x(p), t(p)) = 0$$

with respect to p. The new system, called the *homotopy differential equations* (HDE), utilizes instead the parameter t. Explicitly, the HDE is obtained by differentiating

$$H(x[t], t) = 0 \qquad (15.2.1)$$

with respect to t.

Carrying out the differentiation, where H_t' is the tth column of H' and the prime indicates the derivative,

$$H_x' \frac{dx}{dt} + H_t' = 0 \qquad (15.2.2)$$

or, using the inverse,

$$\frac{dx}{dt} = -[H_x']^{-1} H_t'. \qquad (15.2.3)$$

System (15.2.3) holds whenever H_x' is of full rank and is termed the homotopy differential equations. Solving the HDE will give a path in H^{-1} since the HDE were obtained directly from the homotopy equation (15.2.1) itself. The proof of this follows immediately from the differentiation equations theory result stated in Chapter 2, and we have:

THEOREM 15.1.1   Let $H \in C^2$, $H_x'(y)$ be of full rank for all $y \in H^{-1}$, and y^0 be given where $H(y^0) = 0$. Then solving the HDE

$$\frac{dx}{dt} = -[H_x']^{-1} H_t',$$

from $y(0) = y^0$, yields a path in H^{-1}.

In conclusion, if H is regular we may write $y = (x, t)$ as a function of p and obtain the BDE. Under the slightly stronger assumption that H_x' is of full rank we may write x as a function of t and obtain the HDE.

VERIFICATION OF PATH EQUIVALENCE

Two alternative differential equation systems have been developed, the BDE and the HDE, and since they were derived from the same homotopy, we know they must generate the same path. However, let us verify this in more detail to be doubly sure.

Suppose that H is continuously differentiable and H'_x is of full rank. Also suppose that an initial point is given,

$$x^0 = x[t^0] = x(p^0), \tag{15.2.4}$$

where

$$t^0 = t(p^0).$$

From this same initial point let us generate an HDE path and a BDE path. That is, from the same point we will generate $x[t]$ from the HDE and $y(p)$ $= (x(p), t(p))$ from the BDE. We must prove that $x[t] = x(p)$ or, since $t = t(p)$, that

$$x[t(p)] = x(p). \tag{15.2.5}$$

This will then verify that the paths generated by the two different differential equation systems are identical.

From the HDE

$$H'_x \frac{dx}{dt} = -H'_t.$$

Using Cramer's rule, the ith component dx_i/dt becomes

$$\frac{dx_i}{dt} = \frac{\det [H'_1, H'_2, \ldots, H'_{i-1}, -H'_t, H'_{i+1}, \ldots, H'_n]}{\det H'_x}$$

$$= \frac{(-1)^{n-(i-1)} \det [H'_1, \ldots, H'_{i-1}, H'_{i+1}, \ldots, H'_n, H'_t]}{\det H'_x}$$

$$= \frac{(-1)^i \det H'_{-i}}{(-1)^{n+1} \det H'_x}.$$

Applying the BDE to the right-hand side yields

$$\frac{dx_i}{dt} = \frac{dx_i/dp}{dt/dp}. \tag{15.2.6}$$

But notice this expression is an identity because it is simply $x[t(p)] = x(p)$ differentiated by the chain rule. The two paths must, therefore, be the same.

If H is regular, we know that the BDE always hold whether H'_x is of full rank or not. The HDE, however, requires that H'_x be of full rank and they can fail otherwise. Interestingly, the BDE pinpoint exactly where the HDE can fail. Since

$$i = (-1)^{n+1} \det H'_x,$$

if H'_x is singular, then $i = 0$. The HDE therefore are seen to fail at exactly the points where the path reverses itself in the t dimension (see Figure 15.2.1). (An inflection point may also exist, but that is of less concern.)

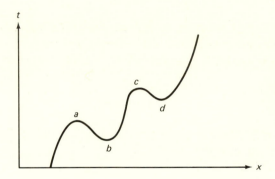

At points a, b, c, and d, $i = 0$ and the HDE cannot be written as a function of t. The BDE hold throughout the path, however.

Figure 15.2.1

To continue this thinking, suppose that after using the HDE for a while we come to a point where $i = 0$, so the HDE do not hold. What do we do then? Assuming regularity, the BDE still operate even at $i = 0$, so we can switch to them. However, a totally different approach also exists because a new HDE can be established with a new parameter other than t. More exactly, even if H'_x is singular, assuming H regular, some partial Jacobian H'_{-i} must be of full rank. We can then reparameterize with variable x_i. The implicit function theorem permits this because as long as any H'_{-i} is of full rank, the other variables can be expressed as a function of the corresponding x_i.

For example, suppose that at $i = 0$, H'_{-1} is of full rank. Then parameterize the variables $x_2, \ldots, x_n$ and t using x_1. Letting $t = x_{n+1}$, we obtain $x_i(x_1)$, $i = 2, \ldots, n + 1$. The homotopy equations then become

$$H(x_1, x_2(x_1), \ldots, x_{n+1}(x_1)) = 0.$$

Differentiating with respect to x_1,

$$H_1' + H_{-1}' \frac{d\tilde{x}}{dx_1} = 0,$$

where $\tilde{x} = (x_2, \ldots, x_{n+1})$. Since H_{-1}' is invertible,

$$\frac{d\tilde{x}}{dx_1} = -(H_{-1}')^{-1}H_1'. \qquad (15.2.7)$$

System (15.2.7) is the same as the HDE except for the different parameter.

In short, suppose that at some point H_x' becomes singular, so the HDE parameterized by t fail. To get through that point we can either use the BDE or reparameterize the HDE using a new variable.

COMPUTATIONAL CONSIDERATIONS

The HDE are suggested as an alternative to the BDE because whenever there is difficulty in evaluating the determinants the HDE may be easier to use. On the other hand, evaluation of the inverse $[H_x']^{-1}$ in the HDE might also be computationally expensive. To overcome this in practice, rarely is the inverse evaluated at each iteration. Rather, it might be calculated only once every several iterations. Also, to save computer effort, the n derivatives in H_t' and the n^2 derivatives in H_x' may not be calculated at all, but difference approximations might be utilized. These and other modifications are well known in the numerical analysis literature and are designed to enhance computational speed. The HDE, replete with their modifications, represent a very widespread and commonly used means of path following.

15.3 TROUBLE WITH EULER
AND POSSIBLE ALTERNATIVES

The BDE, HDE, or in fact any other differential equation can be solved by the Euler approach. The Euler has previously been stated for the BDE, and let us do the same for the HDE. Explicitly, the Euler formula for the HDE is

$$x^{k+1} = x^k - (t^{k+1} - t^k)H_x'(x^k, t^k)^{-1}H_t'(x^k, t^k).$$

Despite its value the Euler has a major difficulty in that it can drift off the path, as shown in Figure 15.3.1. The repeated linear approximations tend to get farther and farther from a path in H^{-1}, and since we often must follow the path accurately, this drift can be a serious problem.

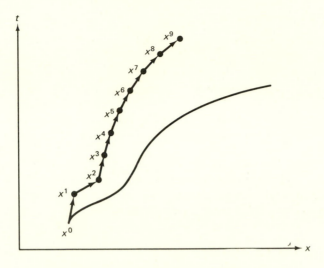

The Euler method tends to drift farther and farther from the underlying path.

Figure 15.3.1

Linear approximations, as in the Euler, are, of course, the simplest technique, but other methods utilize more sophisticated approximations in an effort to better follow the path. We will omit the details, as they can be found in standard numerical analysis texts, yet let us understand the concepts. [Since these methods are useful for arbitrary differential equations, they are presented in their general form (15.1.1). The reader should have no difficulty adapting them to the BDE or HDE].

For specificity, recall the Euler method

$$w^{k+1} = w^k + d^k \dot{w}^k,$$

where we define $\dot{w}$ as

$$\dot{w} = \frac{dw}{du} = f(w, u)$$

and

$$\dot{w}^k = f(w^k, u^k). \tag{15.3.1}$$

Here $d^k = u^{k+1} - u^k$ is the step length. A well-known method that attempts to overcome much of the drifting of the Euler is the *Adams–Bashforth*, which uses the formula

$$w^{k+1} = (w^k + d^k \dot{w}^k) + \frac{(d^k)^2}{d^{k-1}} \frac{(\dot{w}^k - \dot{w}^{k-1})}{2}. \tag{15.3.2}$$

Note that the Euler formula, which is the first term, is "corrected" by the second term. The second term utilizes information from the previous linear approximation $\dot{w}^{k-1}$, thereby "smoothing" the direction. Other correction procedures extend this notion by using weighted averages of even earlier linear approximations, such as $\dot{w}^{k-2}$, $\dot{w}^{k-3}$, and so on. Rather than moving in the linear approximation direction, the idea is to use information gained in previous iterations to move in a more sophisticated and accurate direction.

Although there are innumerable variations and extensions of the Euler, recall the two main means for improving the Euler: (1) increase the accuracy of the approximation using additional information, and (2) speed computation by use of approximation formulas. These concepts are quite powerful, and much research has been conducted on them. In the next section, however, we present a totally different approach.

15.4   A RESTART METHOD

This section presents another procedure for handling the drift away from the path. It does not attempt to follow the old path at all. Instead, it provides a novel use of the homotopy because it restarts on an entirely new path.

For perspective recall that (15.1.1) is an initial value problem. In other words, given a point w^0, the corresponding differential equations would generate a path from that point. The initial value approach is quite useful in physical, social, or economic systems. In these systems, w^0 represents the state of the system today and the differential equations

$$\frac{dw}{du} = f(w, u)$$

describe how the system evolves in the future.

Our problem, however, is often of a different sort, namely a *terminal value problem*. That is, we are interested in calculating $x^* \in H^{-1}(1)$. The point x^* is the terminal value. Exactly where we start $x^0 \in H^{-1}(0)$ is usually of less concern. In fact, in many applications x^0 can be chosen rather freely. It is obtaining x^* that counts. For these problems there is no need to stay on the precise path from x^0. All that is necessary is to end up at x^*, the terminal value.

This analysis immediately suggests a new approach to overcome the fact that we may drift off the path. Should we drift off, do not attempt to get back on the original path. Instead, restart the entire procedure from the current point. Then proceed to follow the path (actually a new path) from the current point to the terminal point x^*. No longer do we follow the path from x^0 to x^*; instead, we follow a different path to x^*, namely the path from wherever we are to x^*.

Referring to our physical system once more, suppose x^0 is the state of the system on the first month, and that due to imperfect foresight we have drifted off the path after a few months. Then, to attempt to return to the original path is to be inflexible to new changes in the state of the world. The original path reflects old data and assumptions (it predicted that today we would be on the old path, and clearly we are not) and to try to recapture it would be naive. Rather, we should look toward the future, and keeping our objective x^* in focus, proceed to follow a new path from the current state to x^*. Indeed, we see that the restart method is the method used by the renaissance person who is flexible yet always knows and pursues the ultimate goal x^*.

Let us discuss the method in detail.

RESTART HOMOTOPY

Although the notation is slightly cumbersome, define the *restart homotopy:*

$$\bar{H}(y; y^k) = H(y) - \frac{(1 - t)H(y^k)}{1 - t^k}, \tag{15.4.1}$$

where $t^k < 1$. The point $y^k = (x^k, t^k)$ is fixed. Observe that at $t = 1$, both H and $\bar{H}$ have the same solution. That is, where here $y = (x, 1)$,

$$\bar{H}((x, 1); y^k) = H(x, 1). \tag{15.4.2}$$

Also at $t = t^k$,

$$\bar{H}(y^k; y^k) = 0. \tag{15.4.3}$$

Suppose that, given an initial point $y^0 = (x^0, 0)$, we generate via a differential equation method the points $y^1, y^2, \ldots, y^k$. However, suppose that y^k has drifted off the path, so that

$$H(y^k) \neq 0.$$

It is here, from y^k, that we restart the procedure. But we use the new homotopy $\bar{H}$. That is, from y^k restart the differential equations. But use y^k as the new initial point and use the restart homotopy $\bar{H}$ instead of H to form the differential equations. Also, start t from $t = t^k$.

Conceptually, with the new homotopy we are creating a new path that starts from y^k. If y^k is not too far from the original path, the new path, under reasonable assumptions, will end up at the same terminal point. This is because of (15.4.1). In short, using the restart homotopy $\bar{H}$ we create a new path that goes from y^k to the same terminal point (see Figure 15.4.1).

Certainly, after solving the new differential equations restarted from y^k for a while, the points generated may drift off the new path. Then start again

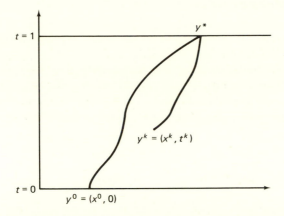

Suppose that y^k is not too far from the path from y^0 to y^*.
Then the new path using the restart homotopy $\bar{H}(y; y^k)$ will go
from y^k to y^*.

Figure 15.4.1

from a new point. Specifically, suppose that a point $y^{k'}$ is generated from the restart y^k where $y^{k'}$ is off the new path. Then restart from $y^{k'}$ and use the restart homotopy

$$\bar{H}(y; y^{k'}) = 0.$$

Clearly, we can restart as many times as we wish, should the points generated drift sufficiently off the path. Just be sure whenever you restart from a point, say $y^{\bar{k}}$, that the restart homotopy $\bar{H}(y; y^{\bar{k}})$ is used to form the differential equations and also that you start anew from $y^{\bar{k}} = (x^{\bar{k}}, t^{\bar{k}})$.

OTHER RESTART HOMOTOPIES

The restart concept is quite powerful and forms of restart homotopy other than (15.4.1) are often useful. For example, in many circumstances $t^k = 1$, yet (15.4.1) requires that $t^k < 1$, so (15.4.1) cannot be used.

One approach given $y^k = (x^k, t^k)$, where $t^k = 1$, is to restart from the current point x^k but to reset t^k so that $t^k = 0$. The appropriate restart homotopy for this is

$$\bar{H}(y; y^k) = H(y) - (1 - t)H(x^k, 0). \qquad (15.4.4)$$

Notice that at $t = 1$

$$\bar{H}((x, 1); y^k) = H(x, 1).$$

Also, at a restart $(x^k, 0)$ we have $\bar{H}((x^k, 0); y^k) = H(x^k, 0) - H(x^k, 0) = 0$. Thus from (x^k, t^k), we may restart with $t^k = 0$ at $(x^k, 0)$ and still end up at a solution

299

to $H(y) = 0$ when $t = 1$. Notice also that the restart homotopy (15.4.4) may be used not just for $t^k = 1$ but $0 \leq t^k \leq 1$.

Clearly, the application of (15.4.4) is analogous to (15.4.1). To summarize its use, suppose that a point $y^k = (x^k, t^k)$ has been generated which is off the path. Then restart from $(x^k, 0)$ and use (15.4.4) as the new homotopy. Although other restart homotopies are also useful, (15.4.1) and (15.4.4) are the most common.

INITIAL VERSUS TERMINAL

A terminal value problem, which is the type we are often interested in, is usually easier to solve than an initial value problem (15.1.1). The reason is that the initial value problem (15.1.1) uses only differential equation information. However, the terminal value problem employs both the homotopy H and the differential equations (the BDE or HDE). In essence, the homotopy permits us to correct and restart. [Observe that H was needed to form the restart homotopy $\bar{H}$ in (15.4.1) or (15.4.4).] The initial value problem (15.1.1) uses only the differential equations and no homotopy information, so has no way to correct itself. The homotopy, which is used for the terminal value but not the initial value problem, provides a great deal of additional data and permits the correction.

It should be noted that although we often only seek a terminal point x^*, for some applications the entire path in H^{-1} must be followed closely. The dynamic nonlinear programming and dynamic equilibrium programming problems of Chapters 4 and 5 were of this sort. Then the restart procedures of this section cannot be utilized, as the entire original path from x^0 must be determined. Having the homotopy can still help, however, and in Chapter 16 we employ it to develop predictor–corrector methods that accurately follow the entire path from x^0.

15.5 NEWTON'S SOLUTION METHOD

This section presents Newton's solution method for solving a system of nonlinear equations. Like the Euler, it operates by taking linear approximations. However, it takes larger steps, uses a special homotopy, and is focused on solving nonlinear equations instead of differential equations. After presenting it we show how it relates to our previous ideas. Indeed, our emphasis here is on how Newton's method fits into our more general concepts and may be considered a large-step path-following approach with restarts.

NEWTON'S METHOD FOR SOLVING EQUATIONS

Centuries ago, Sir Isaac Newton suggested an approach to find equation solutions. It is remarkable that even today it stands as one of the most powerful and widely used procedures for solving nonlinear equations. These days, when most ideas are transitory and are spent in but a brief moment, Newton's method stands in direct contrast. It, in the truest sense, has passed the test of time.

Newton's method is, of course, only a minor one of his towering achievements. His physical laws provided the foundation of our understanding of the universe. Everything in the world followed his laws—until that other towering genius, Einstein, revised our perception of the universe with his concept of relativity. But even though it is a bit more mundane, let us return now to Newton's method for solving equations.

Given a continuously differentiable $F: R^n \longrightarrow R^n$, consider the problem of finding a solution x^* to

$$F(x) = 0.$$

Let x^0 be a guess at a solution and make a linear approximation to F via a Taylor expansion,

$$F(x) \doteq F(x^0) + F'(x^0)(x - x^0).$$

(Recall that $\doteq$ means approximately equal.) If F were actually linear, we would find a solution x^1 where

$$F(x^1) = 0$$

by solving

$$0 = F(x^1) = F(x^0) + F'(x^0)(x^1 - x^0).$$

Assuming that the Jacobian F' is invertible at x^0,

$$x^1 = x^0 - F'(x^0)^{-1}F(x^0). \tag{15.5.1}$$

The point x^1 is a new guess at a solution and in fact if F were linear, it would be a solution. Since in general x^1 will not be a solution, we repeat the process from x^1. That is, make a linear approximation to F at x^1 and then solve that for x^2. The resulting formula for x^1, by analogy to (15.5.1), is

$$x^2 = x^1 - F'(x^1)^{-1}F(x^1).$$

The process is then continued in like manner from x^2.

This, then, is Newton's method for solving equations. Specifically, given x^0, a sequence x^k is generated iteratively by the recursion

$$x^{k+1} = x^k - F'(x^k)^{-1}F(x^k). \qquad (15.5.2)$$

Newton's method thus operates by taking repeated linear approximations to the function F. As we will discuss shortly, a limit point $x^* = \lim_{k \to \infty} x^k$, under certain assumptions, solves the equations.

RELATION TO PATH FOLLOWING

To provide an overview about Newton's solution method, let us consider how Newton's method is related to path following. Specify the homotopy to be the *Newton homotopy*,

$$H(x, t) = F(x) - (1 - t)F(x^0). \qquad (15.5.3)$$

Assuming that $F'(x)$ is invertible, the HDE become

$$\frac{dx}{dt} = -F'(x)^{-1}F(x^0).$$

The first Euler step would then be

$$x^1 = x^0 - (t^1 - t^0)F'(x^0)^{-1}F(x^0).$$

At the start $t^0 = 0$, so

$$x^1 = x^0 - t^1 F'(x^0)^{-1}F(x^0).$$

Examining (15.5.1) the first Newton point is obtained by setting the step length $d = 1$, where $d = t^1 - t^0$. The Newton step is thus like the Euler, but larger. Euler would select d small and thus take a small step. Newton, by setting $d = 1$, attempts to leap immediately to a solution at $t = 1$.

Suppose we set $d = 1$ and take a Newton step to x^1. The large step would no doubt take us off the path, so let us use the restart homotopy at x^1. However, $t^1 = 1$, so we must apply the restart homotopy (15.4.4), which is

$$\bar{H}(y; y^k) = H(y) - (1 - t)H(x^1, 0).$$

In terms of the Newton homotopy (15.5.3), this becomes

$$\begin{aligned}
\bar{H}(y; y^1) &= F(x) - (1 - t)F(x^0) - (1 - t)(F(x^1) - F(x^0)) \\
&= F(x) - (1 - t)F(x^1).
\end{aligned}$$

For this restart homotopy the HDE, assuming that $F'(x)$ is invertible, is

$$\frac{dx}{dt} = -F'(x)^{-1}F(x^1).$$

The Euler step then is

$$x^2 = x^1 - (t^1 - t^0)F'(x^1)^{-1}F(x^1).$$

Since we are restarting the homotopy, $t^0 = 0$. Notice that the Newton step at x^1 is obtained by again setting $t^1 = 1$.

The pattern of the Newton solution method should be clear. At each point x^k it uses the restart homotopy (15.4.4) and t^k is reset to zero. The Newton then uses a large Euler step; that is, the Newton step length is 1.

Conceptually, the Euler takes small steps and attempts to inch along the path toward a solution. Newton's method takes great leaps by guessing at a solution for $t = 1$ and then jumping to that guess. However, because the guess is usually wrong, it must restart.

CONVERGENCE

Newton's solution method takes large steps and does not follow a path closely; thus the question arises: Will it converge? If the starting point x^0 is sufficiently close to the solution x^* and if $F'(x^*)$ is invertible, the answer is yes. This we validate in the theorem below. Let us stress, however, that Newton's method is assured to converge only if the start x^0 is sufficiently close to x^*, that is, within a certain ball $B(x^*)$ centered at x^*. (The ball centered at x^* of radius r is the set of all points less than a distance r from x^*.) Other methods, such as Euler's, can converge even if x^0 is far from x^*. The difference is that the other methods take small steps and follow a path into x^*. They thus can start far from x^* and still converge as long as they stick to the path, because the path leads to x^*. The Newton method does not path-follow, so in general must start closer to x^*. However, because it takes large steps, when it does converge the convergence is usually extremely rapid. To reiterate, Euler-type methods succeed even from a distance because they inch carefully along a path to a solution. The Newton method takes great leaps, may be wrong, and generally must start close to a solution. Yet when convergence occurs, it is usually very rapid.

CONVERGENCE PROOF

In the convergence theorem we will use the notation

$$G(x) = x - F'(x)^{-1}F(x). \tag{15.5.4}$$

Newton's method then can be stated succinctly as

$$x^{k+1} = G(x^k).$$ (15.5.5)

Also, since $F(x^*) = 0$,

$$G(x^*) = x^*.$$ (15.5.6)

Let us now differentiate (15.5.4). First simplify and assume that $n = 1$, so that x, F, and G are all scalars. Then differentiating (15.5.4) yields

$$\frac{dG(x)}{dx} = 1 + \left[\left(\frac{dF}{dx}\right)^{-1}\right]^2 \frac{d^2F}{dx^2} F(x) - \left(\frac{dF}{dx}\right)^{-1} \frac{dF}{dx}.$$ (15.5.7)

But $(dF/dx)^{-1}(dF/dx) = 1$, and also at x^*, $F(x^*) = 0$, so at x^*

$$\frac{dG(x^*)}{dx} = 0.$$ (15.5.8)

Letting G' denote the derivative, (15.5.8) can be rewritten as

$$G'(x^*) = 0.$$ (15.5.9)

We have verified (15.5.9) for $n = 1$, but it also holds for arbitrary n. The proof for general n is similar, although technical, and is given in the chapter appendix.

The convergence of Newton's method now follows easily.

THEOREM 15.5.1   Let $F: R^n \longrightarrow R^n$ be C^2. Let x^* be a solution to $F(x) = 0$ with $F'(x^*)$ nonsingular. Then there is an open ball $B(x^*)$ such that for any $x^0 \in B(x^*)$, the sequence $\{x^k\}_0^\infty$ generated by the Newton formula (15.5.2) converges to x^*.

Proof.   Define $G(x) = x - F'(x)^{-1}F(x)$ as above. By Taylor's expansion of $G(x)$ about x^* and using (15.5.9),

$$G(x) = G(x^*) + R(\|x - x^*\|),$$ (15.5.10)

where R is the remainder and

$$\lim_{x \to x^*} \frac{R(\|x - x^*\|)}{\|x - x^*\|} = 0.$$

Let $0 < \alpha < 1$ and let $B(x^*)$ be an open ball such that

$$\|R(\|x - x^*\|)\| \leq \alpha \|x - x^*\|$$ (15.5.11)

for all $x \in B(x^*)$. For any $x \in B(x^*)$, using (15.5.6), (15.5.10), and

(15.5.11),

$$\| G(x) - x^* \| = \| G(x) - G(x^*) \| = \| R(\| x - x^* \|) \|$$

$$\leq \alpha \| x - x^* \|. \qquad (15.5.12)$$

Moreover, since $\| G(x) - x^* \| < \| x - x^* \|$,

$$G(x) \in B(x^*). \qquad (15.5.13)$$

Now let $x^0 \in B(x^*)$; since $x^1 = G(x^0)$,

$$\| x^1 - x^* \| = \| G(x^0) - x^* \| \leq \alpha \| x^0 - x^* \| \qquad (15.5.14)$$

from (15.5.12).

Also by (15.5.13), $G(x^0) = x^1 \in B(x^*)$, so we may do the same operation on x^1. Explicitly,

$$x^2 = G(x^1) = G(G(x^0))$$

and

$$\| x^2 - x^* \| = \| G(G(x^0)) - x^* \| \leq \alpha \| G(x^0) - x^* \| \leq \alpha^2 \| x^0 - x^* \|$$

via (15.5.12) and (15.5.14). Continuing the induction,

$$\| x^k - x^* \| \leq \alpha^k \| x^0 - x^* \|.$$

Because $\alpha < 1$, we have

$$\lim_{k \to \infty} x^k = x^*. \qquad \square$$

The Newton method will converge quickly if it is started sufficiently close to a solution. If it is started too far out, it may not converge and the points generated will jump around, cycle, or diverge to infinity. From far out we must path-follow.

In practice a dual approach is often employed. First a path-following procedure is used to get fairly close. Then for final convergence we switch to the Newton method because of its great speed.

MODIFICATIONS

Notice that the Newton solution method employs (1) the Newton homotopy, (2) the unit step length in the Euler formula, and (3) the restart homotopy at each step. Clearly, any of these features can be altered. Indeed, changing any of these three features leads to related approaches usually called modifications of Newton's method.

One common modification, for example, is that the inverse may be recalculated only periodically, since determination of $F'(x^k)^{-1}$ at each step is expensive. Also, to avoid any explicit inverse calculation, the point x^{k+1} is often found by solving the system of linear equations

$$F'(x^k)x^{k+1} = F'(x^k)x^k - F(x^k).$$

In fact, rarely is the pure form of the Newton method utilized, but in standard usage at least some modification is employed.

SUMMARY

This chapter introduced the idea of path following by differential equations. The Euler method is basic, as it uses linear approximations. Yet because it tends to drift off the path, we were led to more subtle methods and to the idea of restarting. Finally, the very important Newton solution method was presented and shown to be like the Euler method but with large steps and restarts. Moreover, in this chapter we developed the homotopy differential equations as alternatives to the BDE, and which, in actual practice, are usually easier to utilize than the BDE. This chapter thereby presented a variety of useful techniques, all of which are valuable for actual computational implementation by the practitioner.

APPENDIX

For arbitrary n we will obtain equation (15.5.9), but because we are now differentiating vector matrix products, we must be very careful.

Rewriting (15.5.4) yields

$$F'(x)(G(x) - x) = -F(x)$$

or in terms of components,

$$\sum_{j=1}^{n} \frac{\partial F_i}{\partial x_j}[G_j(x) - x_j] = -F_i(x), \qquad i = 1, \ldots, n.$$

Differentiating with respect to x_k, we obtain

$$\sum_j \frac{\partial^2 F_i}{\partial x_j \partial x_k}[G_j(x) - x_j] + \sum_j \frac{\partial F_i}{\partial x_j}\left[\frac{\partial G_j(x)}{\partial x_k} - \delta_{jk}\right] = -\frac{\partial F_i}{\partial x_k} \qquad \begin{matrix} i = 1, \ldots, n \\ k = 1, \ldots, n, \end{matrix}$$

$$\text{(A.15.1)}$$

where

$$\delta_{jk} = \begin{cases} 0 & \text{if } j \neq k \\ 1 & \text{if } j = k. \end{cases}$$

Now define A_i as the Hessian matrix for F_i. That is,

$$A_i = \begin{bmatrix} \dfrac{\partial^2 F_i}{\partial x_1\, \partial x_1} & \cdots & \dfrac{\partial^2 F_i}{\partial x_1\, \partial x_n} \\[2ex] \cdot & & \cdot \\ \cdot & & \cdot \\ \cdot & & \cdot \\[1ex] \dfrac{\partial^2 F_i}{\partial x_n\, \partial x_1} & \cdots & \dfrac{\partial^2 F_i}{\partial x_n\, \partial x_n} \end{bmatrix}.$$

Also observe that where G' is the Jacobian matrix and I the identity matrix,

$$G' - I = \left[\frac{\partial G_j}{\partial x_k} - \delta_{jk}\right].$$

Moreover, differentiating F_i yields the row vector

$$F_i' = \left(\frac{\partial F_i}{\partial x_1}, \cdots, \frac{\partial F_i}{\partial x_n}\right) = \left(\frac{\partial F_i}{\partial x_j}\right) = \left(\frac{\partial F_i}{\partial x_k}\right).$$

We may now rewrite (A.15.1) in matrix notation as

$$[G(x) - x]^{\mathrm{T}} A_i + F_i'(G'(x) - I) = -F_i', \qquad i = 1, \ldots, n. \qquad \text{(A.15.2)}$$

But from (15.5.6) at x^*,

$$G(x^*) - x^* = 0,$$

so (A.15.2) can be written at x^* as

$$F_i'(x^*)(G'(x^*) - I) = -F_i'(x^*), \qquad i = 1, \ldots, n. \qquad \text{(A.15.3)}$$

Since the full Jacobian matrix of F can be expressed as

$$F' = \begin{pmatrix} F_1' \\ \cdot \\ \cdot \\ \cdot \\ F_n' \end{pmatrix},$$

(A.15.3) becomes

$$F'(x^*)[G'(x^*) - I] = -F'(x^*)$$

or, equivalently,

$$G'(x^*) = I - F'(x^*)^{-1} F'(x^*).$$

We conclude that

$$G'(x^*) = 0,$$

just as we surmised.

EXERCISES/CHAPTER 15

1. Consider the equation of the circle

$$x^2 + t^2 = 1. \tag{E.15.1}$$

Differentiating with respect to t, we have

$$x \frac{dx}{dt} + t = 0, \tag{E.15.2}$$

while differentiating with respect to p yields

$$x \frac{dx}{dp} + t \frac{dt}{dp} = 0. \tag{E.15.3}$$

Given the initial condition $(x^0, t^0) = (1, 0)$, explain why (E.15.1) is a solution to (E.15.3) but is not a solution to (E.15.2).

2. Let

$$H_1(x_1, x_2, t) = x_1 - e^{2t} = 0$$

$$H_2(x_1, x_2, t) = x_2 - e^{-t/2} = 0.$$

Graph the Euler direction for $(x_1, x_2) = (-1, 0)$, $(0, 1)$ and $(1, 1)$ on the $x_1 - x_2$ plane.

3. Consider the system

$$\dot{x} = 5x + 3t, \qquad x(0) = 1, \quad t(0) = 1$$

$$\dot{t} = -6x - 4t.$$

Find the solution $(x(p), t(p))$. [*Hint:* Make a change of coordinates (y_1, y_2), where

$$y_1 = 2x + t$$

$$y_2 = x + t.$$

4. Let $H(x, t) = (1 - t)Ax + t(Ax - b) = 0$, where A is a nonsingular $n \times n$ matrix.
   a. Set up the BDE for $H = 0$.
   b. Show that the Euler directions at all points (x, t) are parallel.

5. Let

$$H_1(x, t) = x_1 - 3t = 0$$

$$H_2(x, t) = (x_1^2 + x_2^2)^2 - 10(x_1^2 + x_2^2) - 3t = 0.$$

   a. Set up the HDE for $H = 0$.
   b. Set up the BDE for $H = 0$.
   c. For what values of (x_1, x_2, t) are the HDE well defined?

6. a. Consider the differential equation (E.15.3), where $(x(0), t(0)) = (1, 0)$. Show that the Euler method will drift off the path.
    b. Consider the equation $H = 0$ defined in Exercise 4. Show that the Euler method, when started at any point on the path, will not drift off the path.

7. Let $H(x, t) = x - t^2 = 0$.
    a. Use the Euler method until $t = 1$ is reached or pierced, starting from $(x, t) = (\frac{1}{16}, \frac{1}{4})$ and using a step size of $d = \frac{1}{4}$.
    b. Use the restart method, restarting each step until $t = 1$ is reached or pierced, starting from $(x, t) = (\frac{1}{16}, \frac{1}{4})$ and using a step size of $d = \frac{1}{4}$.
    c. Determine which of the two approaches yields a better approximation upon termination of the method.

8. Let $F(x) = x - w^2 = 0$, $w > 0$ a real number.
    a. Suppose that $H(x, t) = (1 - t)(x - x^0) + tF(x) = 0$. Show that for any $x^0 \neq -w$, the path H^{-1} cannot go to $x = -w$.
    b. Suppose that $H(x, t) = (1 - t)(x^0 - x) + tF(x) = 0$. Describe where the path H^{-1} goes for $x^0 \neq \pm w$.
    c. Suppose that $H(x, t) = (1 - t)F'(x^0)(x - x^0) + tF(x) = 0$. Describe where the path H^{-1} goes if $x^0 \neq \pm w$.

9. Consider the problem

    $$F_1(x_1, x_2) = x_1 - 2x_2 + 5(x_2)^2 - (x_2)^3 - 13 = 0$$
    $$F_2(x_1, x_2) = x_1 - 14x_2 + (x_2)^2 + (x_2)^3 - 29 = 0.$$

    Use Newton's method, starting from

    $$x^0 = (15, -2),$$

    and verify that it fails to converge.

10. Use Euler's method for Exercise 9, starting at $x^0 = (15, -2)$, a Newton homotopy, and a step size $d = \frac{1}{4}$. Do at least five iterations. Will the method converge to a solution for small enough step size? Explain your answer.

NOTES

For a rigorous treatment on differential equations, see Birkhoff and Rota [1962], Brauer and Nohel [1967], and Shampine and Gordon [1975]. For variations and modifications on Newton's method and Euler's method, see, for example, Ortega and Rheinboldt [1970].

Kellogg, Li, and Yorke [1976] recognized that differential equations can also be used for finding Brouwer fixed points. Alexander and Yorke [1978], Chow, Mallet-Paret, and Yorke [1978], Garcia and Gould [1978, 1979], Garcia and Zangwill [1979b, d, 1980b], Gould and Schmidt [1980], Hirsch and Smale [1979], and Smale [1976] use differential equations for solving Leray–Schauder fixed points, Newton homotopies, polynomial equations, and so on. Historically, these methods were developed after the simplicial methods described in Chapters 11 through 14.

16

Predictor-Corrector Methods

Two basic techniques were introduced in Chapter 15: the Euler for following a path and the Newton for solving equations. In this chapter we unite these two approaches, thereby creating *predictor–corrector methods* of path following. The Euler, or related method, serves as the predictor. It predicts where the path is going, and we move in that direction. Yet as previously noted, after several steps the Euler can drift off the path. The Newton method is then called upon as a corrector to get us back on the path. This sequence is followed iteratively. We predict for a while, which moves us ahead on the path, yet when the drift is sufficiently large, we correct. This predictor–corrector approach, as we will see, enables us to path-follow quite accurately.

16.1 THE BASIC IDEA OF PREDICTOR–CORRECTOR METHODS

Let us now specify the predictor–corrector method in detail. For convenience we present it using the BDE, although it is equally applicable to the HDE. Given an initial point $y^0 = (x^0, 0)$, suppose that on iteration k, we are at a point y^k that is near a point $y(p)$ on the path [i.e., $y(p) \in H^{-1}$].

PREDICTOR

From y^k we take a *predictor* step. That is, the next point y^{k+1} is a predictor of a point farther along the path. A wide variety of predictor steps exist, although a common one is the Euler step:

$$y^{k+1} = y^k + d^k \dot{y}^k, \qquad (16.1.1)$$

where d^k is the step length and

$$\dot{y}_i^k = (-1)^i \det H'_{-i}(x^k \; t^k), \qquad i = 1, \ldots, n+1.$$

A different and especially convenient predictor is called the *elevator step*. If $y^k = (x^k, t^k)$ and $H'_x(y^k)$ is nonsingular, then let

$$y^{k+1} = (x^k, t^k + d^k) \qquad (16.1.2)$$

for some step size d^k. Note that only t^k changes. Assuming that the t axis is vertical, as in Figure 16.1.1, the elevator step takes us straight up.

Although the Euler, using a straight-line approximation, is more accurate, the elevator has the advantage of extreme simplicity. Nevertheless, both are quite useful and both illustrate the predictor idea of moving us in a direction like that of the path.

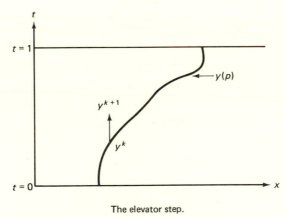

The elevator step.

Figure 16.1.1

CORRECTOR

As observed in Chapter 15, a sequence of predictor steps, be they Euler, elevator, or something else, can drift away from the path. In detail, if

$$H(y^k) = 0,$$

then y^k is on the path. If

$$H(y^k) \neq 0,$$

then y^k is off the path, and as the drift off the path increases,

$$\| H(y^k) \|$$

gets larger. After a number of predictor steps, when the drift gets sufficiently large, we must correct.

The *corrector step* is a sequence of Newton method iterations designed to get us back onto or very close to the path. To develop it, let y^k be the current point. Also for $b \in R^{n+1}$, consider a hyperplane $b(y - y^k) = 0$ (recall b is normal to the hyperplane). Suppose that for some selection of b, the hyperplane intersects the path at a unique point $y(p^k)$, where $y(p^k)$ is farther along the path (see Figure 16.1.2).

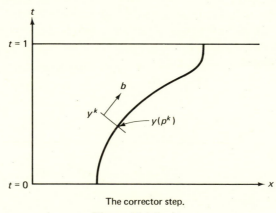

The corrector step.

Figure 16.1.2

Now examine very carefully the function $F: R^{n+1} \longrightarrow R^{n+1}$:

$$F(y) = \begin{pmatrix} H(y) \\ b(y - y^k) \end{pmatrix} = 0. \qquad (16.1.3)$$

This is the function Newton's method will be applied to and it has exactly the properties we want; namely, it will take us from y^k to the point on the path $y(p^k)$.

To see this, by choice, $y(p^k)$ is on the path, so

$$H(y(p^k)) = 0.$$

Also, $y(p^k)$ is on the hyperplane, so

$$b(y(p^k) - y^k) = 0.$$

Thus

$$F(y(p^k)) = 0. \tag{16.1.4}$$

Further, assume that $F'(y(p^k))$ is nonsingular and y^k is sufficiently close to $y(p^k)$. Because $y(p^k)$ solves (16.1.3), the Newton method starting from the point y^k will then converge to $y(p^k)$. In short, suppose that y^k is close to $y(p^k)$. Then applying Newton's method to (16.1.3) will take us from y^k to $y(p^k)$.

Letting $y^k = y^{k,0}$ be the starting point, and $y^{k,l}$, $l = 0, 1, \ldots$, be the Newton points, Newton's method applied to (16.1.3) becomes

$$y^{k,l+1} = y^{k,l} - \begin{pmatrix} H'(y^{k,l}) \\ b \end{pmatrix}^{-1} \begin{pmatrix} H(y^{k,l}) \\ b(y^{k,l} - y^k) \end{pmatrix}. \tag{16.1.5}$$

Since the initial point $y^k = y^{k,0}$ is already on the hyperplane, it is an easy induction (Exercise 1) to show that all points generated by (16.1.5) are on the hyperplane. Explicitly,

$$b(y^{k,l} - y^k) = 0, \qquad \text{for all } l. \tag{16.1.6}$$

Thus (16.1.5) reduces to

$$y^{k,l+1} = y^{k,l} - \begin{pmatrix} H'(y^{k,l}) \\ b \end{pmatrix}^{-1} \begin{pmatrix} H(y^{k,l}) \\ 0 \end{pmatrix}. \tag{16.1.7}$$

This is the formula for the corrector step. Note that the iterations are over l for k fixed.

In brief, for the corrector step we are given a point y^k near but not on the path. Also, a hyperplane $b(y - y^k) = 0$ goes through y^k and intersects the path at $y(p^k)$. Starting from y^k, formula (16.1.7) generates points that in the limit converge to $y(p^k)$. In actual practice, of course, we would not converge to $y(p^k)$, as that would take too long. Instead, a finite number of corrector steps would be taken, enough, say, so that

$$\| H(y^{k,l}) \| < \epsilon.$$

PREDICTOR–CORRECTOR

The predictor–corrector approach should now be clear. Given a point near the path, a sequence of predictor steps moves us in a pathlike direction. Because drift occurs, a sequence of Newton steps (16.1.7) corrects and takes us back close to the path. Repetition of this process will path-follow quite well. Here now is the formal statement of the method.

Initialization

Let $\epsilon_1 > \epsilon_2 > 0$ be given, and set $y^0 = y(0)$, $k = 0$.

Predictor

Given y^k, take a predictor step from y^k to obtain y^{k+1}. Replace k by $k + 1$. Repeat the predictor step if $\| H(y^k) \| < \epsilon_1$. Otherwise, go to the corrector step.

Corrector

Let $b \in R^{n+1}$ describe a hyperplane

$$b(y - y^k) = 0,$$

which intersects the path $y(p)$ at a nearby point $y(p^k)$ farther along the path. Setting $y^k = y^{k,0}$ compute

$$y^{k,l+1} = y^{k,l} - \begin{pmatrix} H'(y^{k,l}) \\ b \end{pmatrix}^{-1} \begin{pmatrix} H(y^{k,l}) \\ 0 \end{pmatrix}, \qquad l = 0, 1, \ldots. \qquad (16.1.8)$$

Stop when $\| H(y^{k,l+1}) \| < \epsilon_2$.

If $y^{k,l+1}$ is near $t = 1$, terminate. Otherwise, let $y^{k+1} = y^{k,l+1}$ and go to the predictor step with $k + 1$ replacing k.

OBSERVATIONS

Although it is assumed that p increases along the path, the analogous process holds if p decreases.

Also, observe that the computation of the inverse

$$\begin{pmatrix} H'(y^{k,l}) \\ b \end{pmatrix}^{-1} \qquad (16.1.9)$$

is expensive. It may be more desirable to replace it by

$$\begin{pmatrix} H'(y^{k,0}) \\ b \end{pmatrix}^{-1} \qquad (16.1.10)$$

or calculate $y^{k,l+1}$ as the solution to the system of linear equations

$$\begin{pmatrix} H' \\ b \end{pmatrix} y^{k,l+1} = \begin{pmatrix} H' \\ b \end{pmatrix} y^{k,l} - \begin{pmatrix} H(y^{k,l}) \\ 0 \end{pmatrix},$$

where H' is selected as in (16.1.9) or (16.1.10).

16.2 HORIZONTAL CORRECTOR

The vector b, the normal to the hyperplane, can often be chosen astutely. A particularly convenient choice is to select b parallel to the t-axis. This, as we will see, creates the *horizontal* corrector (see Figure 16.2.1).

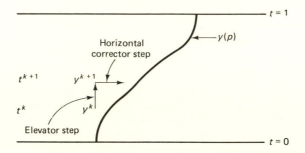

The elevator predictor–horizontal corrector method.

Figure 16.2.1

Since $b \in R^{n+1}$ and $y = (x, t) \in R^{n+1}$, b will point in the t direction if

$$b = (0, 0, \ldots, 0, 1). \qquad (16.2.1)$$

Here b is all zeros except for the $(n + 1)$st position. The hyperplane

$$b(y - y^k)$$

is parallel to the x axes, and also, the value of t is constant in this hyperplane. Furthermore, formula (16.1.7) reduces to

$$x^{k, l+1} = x^{k, l} - H'_x(x^{k, l}, t^k)^{-1}H(x^{k, l}, t^k), \qquad (16.2.2)$$

where $t^k = t^{k, l}$ all l. (Exercise 2). This formula is very simple which, naturally, is a prime advantage.

If we consider t as vertical and the x as horizontal, formula (16.2.2) moves horizontally only—hence its name, horizontal corrector. Also note that (16.2.2) is simply Newton's method on the system $H(x, t^k) = 0$, where t^k is fixed.

ELEVATOR APPLICATION

If we select the elevator as the predictor, and use it with the horizontal corrector, a nice geometrical interpretation results. The elevator step

$$y^{k+1} = (x^k, t^k + d^k)$$

takes us up vertically. The corrector step moves us horizontally. Thus we take the elevator up a floor. We then get off and via the corrector move closer to the path on that floor. There we take the elevator up to the next floor, get off, move closer to the path on that floor, and so on. The elevator predictor and horizontal corrector combine very well.

EULER PREDICTOR

The Euler step can also be used with the horizontal corrector (16.2.2). Moreover, Euler tracks the path better than the elevator. Instead of moving directly up, Euler proceeds more in the direction of path movement. Thus, given y^k the Euler point is y^{k+1}, where

$$y^{k+1} = y^k + d^k \dot{y}^k,$$

should be closer to the path than the corresponding elevator point. The corrector (16.2.2) still operates horizontally, of course, in that given y^{k+1}, it will generate points with the same t value.

The Euler predictor takes more computational effort per step than the elevator but is more accurate. However, either can be used nicely with the horizontal corrector.

CONVERGENCE OF HORIZONTAL CORRECTOR

Convergence of the predictor–corrector method using the horizontal corrector and the elevator predictor is straightforward and will now be discussed. (Convergence with the Euler is similar and is left as Exercise 4.) Of course, the step sizes in the elevator must be kept sufficiently small so that it does not stray too far off the path $y(p)$. Also, there must be enough corrector iterations to move us back close to the path.

Since t^k is fixed, (16.2.2) generates a sequence $x^{k,l}$ which converges to $x(p^k)$. Via Theorem 15.5.1, for this to occur we know that $H'_x(x(p^k), t^k)$ must be of full rank. Furthermore, if the start point x^k is not too far from $x(p^k)$, then the points can be generated arbitrarily close to $x(p^k)$. The following theorem immediately results.

THEOREM 16.2.1   For a homotopy $H \in C^2$ at any $y \in H^{-1}$, let $H'_x(y)$ be of full rank. Also, for $y^0 \in H^{-1}$, let the path be of finite length. Now suppose that a predictor–corrector method uses the horizontal corrector and the elevator predictor.

Given $\epsilon_1 > \epsilon_2 > 0$ sufficiently small, if for all iterations k the predictor step length d^k is sufficiently small, then the method will follow the entire path length to any degree of accuracy.

Proof. The details of the proof are left as Exercise 3, as it follows readily from our previous observations, but here are the key ideas: (1) since H'_x is of full rank, the elevator moves us along the path, ensuring that $t^k < t^{k+1}$; (2) small step size guarantees that the predictor does not stray far from the path; (3) the corrector gets us back close to the path again, but without decreasing t, thus t keeps increasing monotonically; and (4) a compactness argument (because the path is of finite length) provides that all this can be done uniformly along the path. Thus by taking d^k sufficiently small, we can follow the entire path to any given degree of accuracy. $\square$

16.3 FAILURE OF THE HORIZONTAL CORRECTOR

Theorem 16.2.1 requires that the matrix $H'_x(y)$ be of full rank for all y on the path. From the BDE this means that the path must be monotonic in t; that is, it cannot reverse but must strictly increase in t. However, if H'_x is singular at some $\bar{y}$, we could have trouble. At such a $\bar{y}$ the BDE tell us that

$$\dot{t} = (-1)^{n+1} \det H'_x(\bar{y}) = 0,$$

so the path, assuming that $\bar{y}$ is not an inflection point, may reverse itself at $\bar{y}$. Notice that the hyperplane

$$b(y - y^k) = 0$$

then need not intersect the path at all. See Figure 16.3.1, where y^k is a point

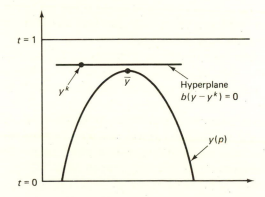

The horizontal hyperplane at y^k misses the path because the path reverses itself at $\bar{y}$.

Figure 16.3.1

generated by the predictor. The path ducks under the hyperplane at $\bar{y}$, so there is no path point $y(p^k)$ on the hyperplane. Clearly, the theorem fails here. In practice, moreover, paths are very likely to reverse like this; hence a means is needed to handle this difficulty.

CHANGING THE DIRECTION
OF THE HYPERPLANE

One technique to overcome the difficulty with $i = 0$ or even i near zero is to switch variables. Assuming regularity, some partial Jacobian is of full rank at $i = 0$. Then use the corresponding variable instead of t.

To illustrate, suppose that at $i = 0$, H'_{-1} is of full rank, so the corresponding variable is x_1. Then change b so that it points along the x_1 axis. Specifically, in the corrector formula (16.1.7) employ

$$b = (1, 0, \ldots, 0)$$

(1 in the first position and the rest are zeros). Also, if the elevator corrector were used before, do not use it. Instead, replace it by a step that increases the component x_1 only; thus where $y^k = (x_1^k, \ldots, x_n^k, t^k)$, let the corrector be

$$y^{k+1} = (x_1^k + d^k, x_2^k, x_3^k, \ldots, x_n^k, t^k). \tag{16.3.1}$$

The direction of change (increase or decrease) of x_1 is determined from the BDE

$$\dot{x}_1 = (-1)^1 \det H'_{-1}(x, t).$$

Explicitly, since we want to change x_1 to move us along the path, the sign of $\dot{x}_1$ tells us whether to increase or decrease x_1.

Everthing is conceptually similar to the preceding discussion except that a new direction has been chosen as "up" instead of t. The predictor moves parallel to the x_1 axis, while the corrector operates in the hyperplane perpendicular to that axis.

Of course, after using x_1 for a while, it might be necessary to change variables again should H'_{-1} become nearly singular. For instance, Figure 16.3.2 illustrates that variables might be changed several times. This changing of variables technique overcomes the requirement that $H'_x(y)$ be nonsingular throughout the entire path. In sum, under regularity some partial Jacobian matrix is nonsingular. If the variable we are operating with has its corresponding matrix singular, we can switch to another variable. Of course, the b in the elevator step must also be revised.

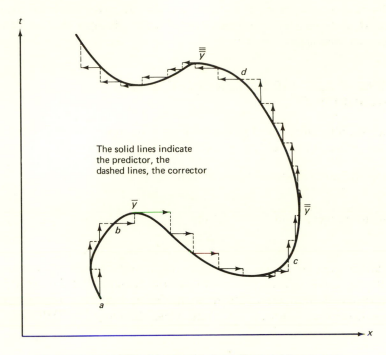

The solid lines indicate the predictor, the dashed lines, the corrector

From a to b the variable t is increased as the predictor, and the corrector is horizontal. However, at $\overline{y}$, $H'_x(\overline{y})$ is singular. Therefore, at b, we switch to increasing x for the predictor, and the corrector is now vertical. At c, we switch back to increasing t as the predictor because at $\overline{\overline{y}}$, H'_{-1} is singular but H'_x is not. At d, we must switch back to changing x, but now x must be decreased.

Figure 16.3.2

COMMENTS

It is clear that Theorem 16.2.1 can be used to prove that this new horizontal corrector with variable switching converges. Essentially, the path can be subdivided into a finite number of subpaths, where on each subpath, $H'_{-i}(y)$ is nonsingular for all y in the subpath. Then with y_i as the independent variable, Theorem 16.2.1 tells us that the whole path can be traversed by the elevator predictor and horizontal corrector (with variable switching).

In practice, we would use a variable, say y_i, until $|\dot{y}_i|$ by the BDE became close to zero. That would indicate that H'_{-i} was becoming singular. Then we would switch to a new variable y_j, where

$$|\dot{y}_j| = \max_{i=1,\ldots,n+1} |\dot{y}_i|.$$

This new variable y_j under regularity would have its matrix H'_{-j} of full rank and would be chosen the "up" direction for the algorithm.

The next section presents an approach that also overcomes the requirement that H'_x be nonsingular, but requires no variable changes at all and handles everything automatically.

16.4   THE EULER PREDICTOR–CORRECTOR ALGORITHM

To avoid the difficulties with variable switching, we now state the Euler predictor–corrector method. It:

1. uses the Euler step as a predictor, that is,

$$y^{k+1} = y^k + d^k \dot{y}^k. \qquad (16.4.1)$$

2. for the corrector step at y^{k+1} selects

$$b = (\dot{y}^k)^\tau. \qquad (16.4.2)$$

The predictor step utilizes the linear approximation $\dot{y}^k$, and the corrector step uses that very same linear approximation $\dot{y}^k$ for the normal to the hyperplane. The hyperplane is thus perpendicular to the Euler step which was just taken as the predictor (see Figure 16.4.1). Here $\dot{y}^k$ is, of course, the BDE evaluated at y^k.

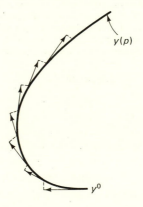

The Euler predictor-corrector algorithm. The corrector steps, shown by dashes, are perpendicular to the predictor steps.

Figure 16.4.1

Explicitly, the Euler predictor–corrector algorithm is:

Initialization

Given $\epsilon_1 > \epsilon_2 > 0$, set $y^0 = y(0)$, and $k = 0$.

Predictor

At y^k, let

$$y^{k+1} = y^k + d^k \dot{y}^k. \tag{16.4.3}$$

Replace k by $k + 1$. Repeat the predictor step if $\|H(y^k)\| < \epsilon_1$. Otherwise, go to the corrector step.

Corrector

Let $y^k = y^{k,0}$ and $b = (\dot{y}^{k-1})^\tau$. Compute

$$y^{k,l+1} = y^{k,l} - \begin{bmatrix} H'(y^{k,l}) \\ b \end{bmatrix}^{-1} \begin{bmatrix} H(y^{k,l}) \\ 0 \end{bmatrix}, \qquad l = 0, 1, \dots . \tag{16.4.4}$$

Stop when $\|H(y^{k,l+1})\| < \epsilon_2$. If $y^{k,l+1}$ is near $t = 1$, terminate. Otherwise, let $y^{k+1} = y^{k,l+1}$ and go to the predictor step with $k + 1$ replacing k.

Note carefully that the corrector starts from y^k but uses $b = (\dot{y}^{k-1})^\tau$. This is because the corrector uses the same direction $\dot{y}^{k-1}$ as the Euler, yet starts from the point y^k, which is the result of the Euler step.

ANALYSIS

Clearly, the Euler predictor–corrector requires no switching of variables. Also, the predictor step, since it is a fairly good approximation to the path, will propel us along the path nicely.

As for the corrector step, Theorem 15.5.1 requires for these iterations to converge that the matrix

$$\begin{bmatrix} H'(y(p^k)) \\ (\dot{y}^{k-1})^\tau \end{bmatrix} \tag{16.4.5}$$

be of full rank. To verify this, recall that for all points $y(p) \in H^{-1}$,

$$H(y(p)) = 0$$

and differentiating, where $\dot{y} = \dot{y}(p)$,

$$H'(y(p))\dot{y} = 0. \tag{16.4.6}$$

Thus $\dot{y}$ is perpendicular to all rows of $H'(y(p))$.

If H is regular, the rows of H' are linearly independent, and then by (16.4.6) the matrix

$$\begin{bmatrix} H'(y(p)) \\ \dot{y}(p)^\tau \end{bmatrix}$$

must be of full rank. In the algorithm $\dot{y}^{k-1}$ will be very close to $\dot{y}(p^k)$; therefore, the matrix in (16.4.5) will also be of full rank. The Newton iterates (16.4.4) will then converge if started sufficiently close.

16.5 GENERAL DISCUSSION

LARGE PREDICTOR STEPS

Although the algorithm stated in this chapter will follow a path closely, that could be time consuming since the predictor steps d^k must be small and the number of corrector iterations large. Often, we only seek a terminal point y^* $= (x^*, 1) \in H^{-1}(1)$ and not the entire path $y(p) \in H^{-1}$. It might then be desirable to keep the step sizes d^k large and the ϵ_1 and ϵ_2 large, at least until near $t = 1$. This way, we proceed to level $t = 1$, perhaps quickly.

Some potential dangers can occur in doing this, however. For example, the algorithm may latch onto a different path; it may cycle; or it might even attach itself to a loop in H^{-1} (see Figure 16.5.1). Should such difficulties occur,

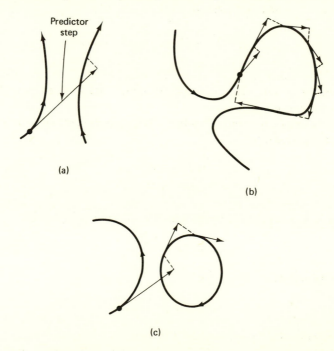

Predictor
step

(a)

(b)

(c)

By taking large steps, various problems can occur. In (a) the algorithm goes to a different path; in (b) it cycles; and in (c) it attaches itself to a loop in H^{-1} and then proceeds around the loop endlessly, even if small steps are taken henceforth.

Figure 16.5.1

one must back up to where the problem first occurred if possible, and use small step sizes, at least until safely out of the area of difficulty. The restarting procedure discussed in Chapter 15 might also be helpful.

COMPARISON BETWEEN THE PREDICTOR–CORRECTOR AND THE FLEX SIMPLICIAL

Two general ways for closely tracking a homotopy path $y(p) \in H^{-1}$ have been presented in this book, the flex simplicial and the predictor–corrector. The flex simplicial will track the path as long as robustness and small simplices are used. At each iteration, recall, $n + 2$ points are being carried along. The predictor–corrector method will also track the path, provided that small predictor steps are used and enough corrector steps are used to return close to the path.

Intuitively, the flex simplicial is like an animal with $n + 2$ feet that must drag each foot one at a time as it moves along, whereas the predictor–corrector shoots like an arrow. However, although the flex simplicial need never worry about drifting away from the path, the predictor–corrector is like a child that must regularly be disciplined for falling out of line. The reason is that the predictor step contains only the path direction information of the BDE and hence, even though it can scout speedily across regions, it cannot know if the homotopy equations are being violated. It is the corrector step that contains the homotopy equations information, and tells how far the point is from the path. The flex simplicial, on the other hand, contains both the BDE information and the homotopy equations information in the vectors $H(v^i)$ given on the vertices v^i of the current simplex. Hence the flex simplicial knows at every time where to go next, and how near the path is.

In short, the flex simplicial simultaneously maintains information of the path direction and of how close the path is. The predictor–corrector must do that in two separate parts, the predictor part and the corrector part.

SUMMARY

This chapter presented predictor–corrector path-following methodology. A predictor step, the Euler or elevator, took us along the path but drifted somewhat. The corrector step, the Newton method in a hyperplane, corrected for the drift and returned us back closer to the path. Also, by adjusting the hyperplane we obtained several useful versions of the corrector step. Indeed, different hyperplane directions combined with the different predictor steps create a potent variety of predictor–corrector methods to follow a path accurately.

1. Prove that (16.1.6) holds for the Newton points $y^{k,l}$ of the corrector steps.

2. Show that formula (16.1.7) reduces to (16.2.2) for the horizontal corrector.

3. Prove Theorem 16.2.1.

4. Let $H \in C^2$ be regular and let $y(p) \in H^{-1}$ be a given path of finite length. Prove that the predictor–corrector with the Euler predictor and horizontal corrector will follow the path for step size and $\epsilon_1 > \epsilon_2 > 0$ sufficiently small.

5. Let $H \in C^2$ be regular and $y(p) \in H^{-1}$ a given path of finite length. Prove that the Euler predictor–corrector method of Section 16.4 follows the path for step size and $\epsilon_1 > \epsilon_2 > 0$ sufficiently small.

6. Apply the predictor–corrector method to the problem

$$H(x, t) = 3x - t^2 - 3t - 5 = 0, \qquad -5 \leq t \leq 5.$$

Start from $(x(0), t(0)) = (5, -5)$; use the elevator predictor with step size $d = 0.5$, and a horizontal corrector to try to follow the path H^{-1} up to $t = 5$.

7. Suppose that $F: R^n \longrightarrow R^n$ is C^2 and that $F'(x)$ is nonsingular for all $x \in R^n$. Consider the Newton homotopy

$$H(x, t) = F(x) - (1 - t)F(x^0) = 0. \qquad \text{(E.16.1)}$$

Show that the process

$$x^{k+1} = x^k - F'(x^k)^{-1}\left[F(x^k) + \left(\frac{k}{N} - 1\right)F(x^0)\right], \qquad k = 0, 1, \ldots, N - 1$$

$$x^{k+1} = x^k - F'(x^k)^{-1}F(x^k), \qquad k = N, N + 1, \ldots$$

$$\text{(E.16.2)}$$

is a special case of the predictor–corrector applied on H.

8. Suppose that $F: R^n \longrightarrow R^n$ is C^2, $F'(x)$ is nonsingular for all $x \in R^n$, and $\|F(x)\| \longrightarrow \infty$ as $\|x\| \longrightarrow \infty$. Show that for any $x^0 \in R^n$, there is an integer $N_0 \geq 1$ such that for any $N \geq N_0$, the process (E.16.2) converges to the unique solution x^* of $F(x) = 0$ in R^n.

9. Consider the process

$$x^{k+1} = x^k - hF'(x^k)^{-1}F(x^0), \qquad k = 0, 1, \ldots, N - 1, \quad h = 1/N$$

$$x^{k+1} = x^k - F'(x^k)^{-1}F(x^k), \qquad k = N, N + 1, \ldots. \qquad \text{(E.16.3)}$$

Suppose that $F: R^n \longrightarrow R^n$ is C^2 and satisfies $\|F'(x)^{-1}\| \leq A$ for all $x \in R^n$. Show that for any $x^0 \in R^n$, there exists an $N_0 \geq 1$ such that for any $N \geq N_0$, the process (E.16.3) converges to the unique solution x^* of $F(x) = 0$ in R^n.

10. Construct examples of mappings $H: R^2 \longrightarrow R^1$ that exhibit each of the three cases in Figure 16.5.1.

11. Let $H: R^{n+1} \longrightarrow R^n$ be a given C^2 and regular mapping and let $y(p^k)$ be a point on the path $y(p) \in H^{-1}$. Show that for any b such that $b\dot{y}(p^k) \neq 0, \|b\| = 1$, there exists an open ball $B(y(p^k), \delta)$ with center $y(p^k)$ and radius δ such that for

any

$$y^k = y^{k,0} \in B(y(p^k), \delta),$$

the corrector (16.1.7) converges to a point in $y(p)$.

12. In Exercise 11, prove that the point of convergence of the corrector (16.1.7) is the unique solution of $F(y) = 0$ [as defined by (16.1.3)] in an open ball $B(y^k, \delta_1)$ for some $\delta_1 > 0$.

13. Suppose that $F: R^n \longrightarrow R^n$ is C^2, $F'(x)$ is nonsingular for all $x \in R^n$, and $\| F(x) \| \longrightarrow \infty$ as $\| x \| \longrightarrow \infty$. Given $x^0 \in R^n$, assume the Newton homotopy

$$H(x, t) = F(x) - (1 - t)F(x^0) = 0$$

is regular. Show that H^{-1} is a unique path from $(x^0, 0)$ to the point $(x^*, 1)$ where x^* is the unique solution of $F(x) = 0$.

NOTES

The predictor–corrector method is a globalization of the method of Davidenko [1953b]. Proofs of convergence of the method are furnished by Menzel and Schwetlick [1978] and Garcia and Li [1979]. Variations and modifications are given by Schmidt [1979].

17

Special Functions:
Contraction and Separability

Often in practical application the function employed has a special form that permits us to obtain considerably more powerful results than for an arbitrary function. This chapter explores two such special forms, the contraction property and separability, and states the sometimes striking efficiencies that can be gained when either of these two forms hold.

The first section of the chapter analyzes the contraction property and proves the contraction or Banach fixed-point theorem. It is different from the Brouwer theorem, and when contraction holds, creates results of exceptional power. Formal algorithms, for instance, can be dispensed with, as the function itself becomes the algorithmic step.

Next we explore separability, a property that occurs whenever the homotopy does not have cross-product terms. Under separability, simplices are no longer needed, as our algorithm can instead operate on rectangles. Rectangles are so easy to state, use, and manipulate that the algorithms obtained become markedly more efficient.

Overall, this chapter establishes results that, when either the contraction or the separability property holds, are far more potent than our previous, more general results.

17.1 CONTRACTION

In many applications the function involved has a property called contraction. Specifically $F: R^n \rightarrow R^n$ is *contractive* if for some $0 < \alpha < 1$ and $x, y \in R^n$,

$$\| F(x) - F(y) \| \leq \alpha \| x - y \|. \tag{17.1.1}$$

Because $\alpha < 1$, $F(x)$ and $F(y)$ are closer together than x and y. Thus F in some sense shrinks distances in space.

It is easy to verify from (17.1.1) that F is continuous (Exercise 1). Also, with more effort F can be shown to operate on a compact convex set D so that $F: D \rightarrow D$ (Exercise 2). The Brouwer theorem then assures us that F has a fixed point. We will now demonstrate far more than that however. In particular, our proof:

1. automatically provides the algorithm.

2. gives a bound on where the fixed point is.

3. establishes that the fixed point is unique.

Not one of these three properties holds in the Brouwer case, which clearly demonstrates how much more powerful these contraction results are. For example, the algorithm given x^0 is

$$x^{k+1} = F(x^k). \tag{17.1.2}$$

The algorithm applies the function itself repeatedly and calculates the fixed point. This type of algorithm, note, will not in general work for the Brouwer, as that requires the considerably more complicated procedures of homotopies and path-following algorithms to obtain the fixed point.

For the proof, recall two technical facts. Given $0 < \alpha < 1$,

$$\sum_{i=0}^{l-1} \alpha^i < \sum_{i=0}^{\infty} \alpha^i = \frac{1}{1 - \alpha} \tag{17.1.3}$$

and

$$\| x^{k+l} - x^k \| = \| x^{k+l} - x^{k+l-1} + x^{k+l-1} \cdots - x^{k+1} + x^{k+1} - x^k \|$$
$$\leq \sum_{i=0}^{l-1} \| x^{k+i+1} - x^{k+i} \|. \tag{17.1.4}$$

We now prove the Banach fixed-point theorem, also called the contraction fixed-point theorem.

THEOREM 17.1.1 Let F be contractive. Then given x^0, the sequence $\{x^k\}$ generated by

$$x^{k+1} = F(x^k)$$

converges in the limit to a unique fixed point x^*. Also,

$$\|x^k - x^*\| \leq \frac{\alpha}{1-\alpha} \|x^k - x^{k-1}\|, \qquad k = 1, 2, \ldots. \qquad (17.1.5)$$

Proof.

$$\|x^2 - x^1\| = \|F(x^1) - F(x^0)\| \leq \alpha \|x^1 - x^0\|$$

by the contraction property. Iterating,

$$\|x^{k+1} - x^k\| \leq \alpha^k \|x^1 - x^0\|$$

and

$$\|x^{k+i+1} - x^{k+i}\| \leq \alpha^i \|x^{k+1} - x^k\|.$$

Then using (17.1.3) and (17.1.4), we get

$$
\begin{aligned}
\|x^{k+l} - x^k\| &\leq \sum_{i=0}^{l-1} \|x^{k+i+1} - x^{k+i}\| \\
&\leq \sum_{i=0}^{l-1} \alpha^i \|x^{k+1} - x^k\| \\
&\leq \alpha^k \sum_{i=0}^{l-1} \alpha^i \|x^1 - x^0\| \\
&\leq \frac{\alpha^k}{1-\alpha} \|x^1 - x^0\|.
\end{aligned}
\qquad (17.1.6)
$$

Thus

$$\|x^{k+l} - x^k\| \leq \frac{\alpha^k}{1-\alpha} \|x^1 - x^0\|. \qquad (17.1.7)$$

As $\alpha < 1$, the sequence $\{x^k\}$ is a *Cauchy sequence* and therefore has a limit x^* (Exercise 3). Using the continuity of F, we obtain

$$x^* = \lim x^{k+1} = \lim F(x^k) = F(x^*),$$

so x^* is the fixed point.

The fixed point is unique because suppose that there were two of them, $\bar{x}$ and x^*. Then because $\bar{x}$ and x^* are fixed points,

$$\|\bar{x} - x^*\| = \|F(\bar{x}) - F(x^*)\| \leq \alpha \|\bar{x} - x^*\| < \|\bar{x} - x^*\|,$$

which is a contradiction.

As for the bound, from (17.1.6),

$$\|x^{k+l} - x^k\| \leq \sum_{i=0}^{l-1} \alpha^i \|x^{k+1} - x^k\| \leq \alpha \sum_{i=0}^{l-1} \alpha^i \|x^k - x^{k-1}\|$$

$$\leq \frac{\alpha}{1-\alpha} \|x^k - x^{k-1}\|.$$

Since this holds for all l, take limits as $l \to \infty$ and k fixed:

$$\| x^* - x^k \| \le \frac{\alpha}{1 - \alpha} \| x^k - x^{k-1} \|. \qquad \Box$$

If F is contractive, considerably more information is obtained via this theorem than from the Brouwer theorem alone. Important is the fact that repeated iteration of (17.1.2) yields the fixed point. For instance, F may represent an economic or physical operation. Then by doing the same operation repeatedly, we obtain the fixed point. Further (17.1.5) provides a computable error estimate: that is, if the contraction constant is known, the distance from x^k to the fixed point x^* can be bounded in terms of the last step, $x^k - x^{k-1}$.

Historically, since only recently was it discovered how to calculate a Brouwer fixed point, (17.1.2), when it could be applied, was the standard means of calculating a fixed point. Unfortunately, the contraction assumption is quite restrictive and frequently holds only near the fixed point (e.g., see Exercise 11). Thus most often in practice the Brouwer theorem and homotopy methods for calculating the fixed point must be employed. Nevertheless, whenever its assumptions hold, the Banach theorem is extremely valuable. Indeed, note that we have already seen that the Newton iterates (15.5.5) are exactly of the form (17.1.2). In fact, Newton's method is powerful precisely because the map G of (15.5.4) was shown to be contractive in Theorem 15.5.1.

17.2   SEPARABLE HOMOTOPIES

Often the homotopy $H: R^{n+1} \to R^n$ has a particular structure called separable that can greatly speed the efficiency of simplicial-type algorithms. Interestingly, the algorithms will no longer operate on simplices but on rectangles. Rectangles are very nice geometric figures with excellent properties, and our modified algorithms will move very rapidly from rectangle to rectangle as they track a path in H^{-1}.

A homotopy H is called *separable* if

$$H(y) = \sum_{i=1}^{n+1} h^i(y_i)$$

where

$$h^i: R^1 \longrightarrow R^n, \qquad i = 1, \ldots, n+1.$$

Notice that h^i depends on the component y_i only. Also, recall that $y = (x_1, \ldots, x_n, t)$.

For example, for $H: R^3 \to R^2$, let

$$\begin{aligned}
H_1(y) &= (x_1)^2 + 3x_1 + 2(x_2)^2 + 4t = 0 \\
H_2(y) &= \phantom{(x_1)^2 + 3x_1 + {}} 4x_1 - x_2 \phantom{{} + 2(x_2)^2} + 10t = 0.
\end{aligned}$$

Then H is separable and

$$h^1(x_1) = \begin{pmatrix} (x_1)^2 + 3x_1 \\ 4x_1 \end{pmatrix}$$

$$h^2(x_2) = \begin{pmatrix} 2(x_2)^2 \\ -x_2 \end{pmatrix}$$

$$h^3(t) = \begin{pmatrix} 4t \\ 10t \end{pmatrix}.$$

For H to be separable, no cross-product terms are allowed. For instance,

$$H_1(y) = x_1 x_2 + x_2 - t$$
$$H_2(y) = \quad x_1 - 4x_2 + 3t$$

would not be separable.

RECTANGULAR SUBDIVISION

To efficiently exploit the separability of H, we subdivide $D \subset R^{n+1}$ into $(n + 1)$-dimensional rectangles. Precisely, we require D to be subdivided so that:

1. D is contained in the union of the rectangles.
2. Two adjacent rectangles intersect only at a common face.
3. Every bounded region has only a finite number of rectangles.

Call this a *rectangular subdivision*. Clearly, all the properties of a rectangular subdivision are identical to a triangulation except that rectangles are used instead of simplices.

For notation, let

$$a = (a_1, \ldots, a_{n+1}) \quad \text{and} \quad b = (b_1, \ldots, b_{n+1})$$

be such that

$$S = \{y \in R^{n+1} \mid a_i \leq y_i \leq b_i\}$$

represents a typical rectangle. Intuitively, in three dimensions a is the corner nearest and to the lower left and b is on the far side and to the upper right. Normally, a_i is an integer multiple of a given $\epsilon > 0$ and

$$b_i = a_i + \epsilon. \tag{17.2.1}$$

For example, if for all i, $a_i = 0$ and $b_i = \epsilon$, then the rectangle is a cube starting at the origin and going into the positive orthant with sides of length ϵ.

Rectangle S has 2^n vertices of the form $v = (v_i)$, where each v_i is either a_i or b_i. Also, any point $y = (y_i) \in S$ is expressible as

$$y_i = (1 - \lambda_i)a_i + \lambda_i b_i, \quad 0 \le \lambda_i \le 1, \quad i = 1, \ldots, n+1$$

for some unique $\lambda = (\lambda_1, \ldots, \lambda_{n+1})$.

THE PL APPROXIMATION

Given a separable H, let us now define a linear approximation G of H on a rectangle S. Given $y \in S$, we know that

$$y_i = (1 - \lambda_i)a_i + \lambda_i b_i, \quad 0 \le \lambda_i \le 1.$$

Let

$$g^i(y_i) = (1 - \lambda_i)h^i(a_i) + \lambda_i h^i(b_i). \tag{17.2.2}$$

Then define the linear approximation G by

$$G(y) = \sum_1^{n+1} g^i(y_i) \tag{17.2.3}$$

$$= \sum_1^{n+1} [(1 - \lambda_i)h^i(a_i) + \lambda_i h^i(b_i)]. \tag{17.2.4}$$

Clearly, G is linear on S (Exercise 4). Also, for any vertex $v \in S$,

$$G(v) = H(v).$$

For example, suppose that we are at a vertex $v = (a_1, a_2, \ldots, a_k, b_{k+1}, \ldots, b_{n+1})$. Then $\lambda_i = 0, i = 1, \ldots, k, \lambda_i = 1, i = k+1, \ldots, n+1$ and

$$G(v) = \sum_1^k h^i(a_i) + \sum_{k+1}^{n+1} h^i(b_i) = H(v).$$

In short, we have created a linear approximation G of H on the rectangle S which equals H on the 2^n vertices of the rectangle. (See Figure 17.2.1).

THE ALGORITHMIC PROCESS

The operation of the algorithm will be virtually identical to triangulation algorithms. A facet (actually an n-rectangle) of an $(n+1)$-rectangle will be given to start. More precisely, an initial point w^0 in a facet B^0 will be given so that

$$G(w^0) = 0,$$

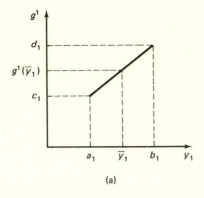

(a)

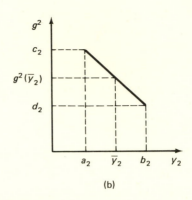

(b)

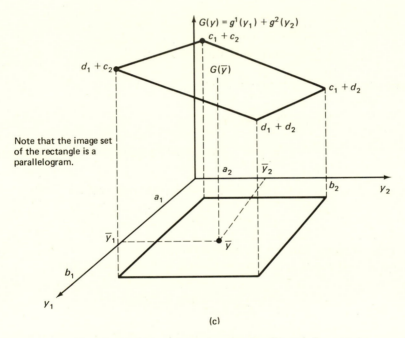

Note that the image set of the rectangle is a parallelogram.

(c)

For $n = 1$, diagram (c) shows what $G(y) = g^1(y_1) + g^2(y_2)$ looks like for the g^1 given in (a) and the g^2 given in (b).

Figure 17.2.1

where G is formed by using (17.2.3) on the initial $(n + 1)$-rectangle S^0. A straight-line segment across S^0 will then be generated by solving

$$G(w) = 0,$$

thereby yielding a point w^1 on another facet B^1 of S^0.

If we do not terminate, there will be another $(n + 1)$-rectangle S^1 on the other side of the facet B^1 and the process will continue on S^1. That is, we will construct the approximation G of H on S^1 and generate the line segment from w^1 to a point w^2 on another facet B^2 of S^1. Assuming regularity, that

$$\text{each } w^k \text{ is interior to a facet } B^k \tag{17.2.5}$$

for all k, a PL path in G^{-1} will be created. Also, if the rectangles are sufficiently small, G^{-1} will approximate H^{-1} well. The process is identical to that for triangulation algorithms except for the use of rectangles instead of simplices.

SPECIFICS

Let us examine in more detail the process of generating the line segment across a rectangle S. Given $\bar{w}$ in a facet $\bar{B}$ of S, we must solve

$$G(w) = 0$$

to generate $\bar{\bar{w}}$ in another facet $\bar{\bar{B}}$ of S. Using the formula for G given by equation (17.2.4) yields

$$G(w) = \sum_{1}^{n+1} [(1 - \lambda_i)h^i(a_i) + \lambda_i h^i(b_i)] = 0$$

or

$$\sum_{1}^{n+1} \lambda_i (h^i(a_i) - h^i(b_i)) = \sum_{1}^{n+1} h^i(a_i), \tag{17.2.6}$$

where

$$w_i = (1 - \lambda_i)a_i + \lambda_i b_i, \qquad 0 \le \lambda_i \le 1, i = 1, \dots, n + 1.$$

The system (17.2.6) has $n + 1$ variables and n equations and under regularity will generate a straight-line segment across S.

System (17.2.6) must be solved to find the new point $\bar{\bar{w}}$. The complication in solving (17.2.6), however, is the inequality constraints, $0 \le \lambda_i \le 1$. There are a variety of ways of handling this. For example, the upper bounds $\lambda_i \le 1$ could be eliminated by adding slack variables

$$u_i = 1 - \lambda_i. \qquad u_i \ge 0.$$

System (17.2.6) would then become $2n + 1$ equations in $2n + 2$ unknowns, a much larger system. Our approach, however, is different because we will show that via the upper bounding techniques of linear programming, (17.2.6) can be solved directly and easily.

Under regularity $\bar{w} = (\bar{w}_i)$ will be interior to facet $\bar{B}$. Then for some k,

$$\bar{w}_k = a_k \quad \text{or} \quad b_k \qquad \text{and} \qquad \bar{\lambda}_k = 0 \quad \text{or} \quad \bar{\lambda}_k = 1, \tag{17.2.7}$$

while for all $i \neq k$,

$$\bar{w}_i = (1 - \bar{\lambda}_i)a_i + \bar{\lambda}_i b_i, \qquad \text{where } 0 < \bar{\lambda}_i < 1.$$

This follows because $\bar{\lambda}_k = 0$ or 1 will force $\bar{w}$ to be on a facet. Also, if $\bar{\lambda}_k = 0$, then $\bar{w}_k = a_k$, whereas if $\bar{\lambda}_k = 1$, $\bar{w}_k = b_k$. Since $\bar{w}$ is interior to that facet, all other $\bar{\lambda}_i$ must be $0 < \bar{\lambda}_i < 1$.

Similarly, as $\bar{\bar{w}}$ is interior to a different facet $\bar{\bar{B}}$,

$$\bar{\bar{w}}_i = (1 - \bar{\bar{\lambda}}_i)a_i + \bar{\bar{\lambda}}_i b_i, \tag{17.2.8}$$

where for some l,

$$0 < \bar{\bar{\lambda}}_i < 1, \quad i \neq l \qquad \text{and} \qquad \bar{\bar{\lambda}}_l = 0 \quad \text{or} \quad 1.$$

Given $\bar{w}$, let us now solve (17.2.6) to determine $\bar{\bar{w}}$. First suppose that $\bar{\lambda}_k = 0$; then we increase λ_k in (17.2.6) until either:

1. some λ_i, $i \neq k$ becomes 0 or 1, or
2. $\lambda_k = 1$,

whichever occurs first. The corresponding λ is $\bar{\bar{\lambda}}$, from which $\bar{\bar{w}}$ is determined by (17.2.8).

If $\bar{\lambda}_k = 1$, decrease λ_k in (17.2.6) until either $\lambda_k = 0$, or some λ_i, $i \neq k$, becomes zero or 1, whichever occurs first. The corresponding λ yields $\bar{\bar{\lambda}}$ for $\bar{\bar{w}}$.

In the upper-bounding terminology of linear programming, suppose that $\bar{\lambda}_k = 0$. Then λ_k is at its lower bound and λ_k must be increased until some λ_i hits its upper or lower bound. If $\bar{\lambda}_k = 1$, λ_k is at its upper bound. Then λ_k must be decreased until some λ_i hits its upper or lower bound.

Let the index of the variable that hits its bound be l. Thus at $\bar{\bar{w}}$ some variable $\bar{\bar{\lambda}}_l = 0$ or 1, and all other variables $\bar{\bar{\lambda}}_i$ are between zero and 1. Suppose that $k = l$, so that λ_l has increased from 0 to 1 (or decreased from 1 to 0). Then facet $\bar{\bar{B}}$ is parallel to $\bar{B}$; that is, the two facets are on opposite sides of the rectangle. If $k \neq l$, then facet $\bar{\bar{B}}$ is adjacent to $\bar{B}$.

CONTINUING ON

At $\bar{\bar{w}}$, assuming there is no termination, we would now look at a rectangle S' on the other side of $\bar{\bar{B}}$. For S' the system (17.2.6) will be similar to that for S with all components the same except those with subscript l. This is because if $\bar{B}$ is an upper facet of S, it will be a lower facet of S', and conversely. The details are straightforward and are left as Exercise 6. In brief, and assuming all rectangles are of the form (17.2.1), suppose that for S, point $\bar{w}$ has $\lambda_l = 0$. Then to form S', a_l must be decreased to $a_l - \epsilon$. Also in S', $\bar{w}$ will have $\lambda_l = 1$. If in S we have $\lambda_l = 1$, then to form S', a_l is increased to $a_l + \epsilon$. Also in S', point $\bar{w}$ will have $\lambda_l = 0$. The following example will clarify this process.

17.3   EXAMPLE

Let us now illustrate the algorithm with an example. Suppose that $n = 1$ and $G(y) = g^1(y_1) + g^2(y_2)$, where g^1 and g^2 are as depicted in Figure 17.3.1. Here the g^i have already been calculated using $g^i(a_i) = h^i(a_i)$, $g^i(b_i) = h^i(b_i)$. The path generated by the algorithm is also given in that figure.

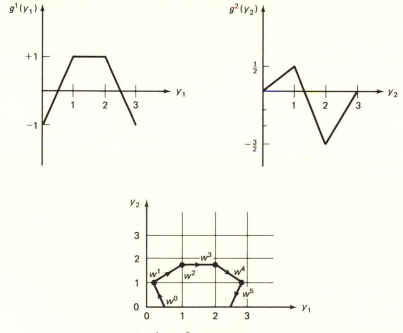

The g^1 and g^2 functions and the PL path.

Figure 17.3.1

The rectangles will be of unit size, so that always $b = a + (1, 1)$. The initial rectangle S^0 has $a = (a_1, a_2) = (0, 0)$; hence $b = (b_1, b_2) = (1, 1)$ and

$$S^0 = \{(y_1, y_2) \mid 0 \le y_1 \le 1, 0 \le y_2 \le 1\}.$$

For this rectangle

$$g^1(a_1) = g^1(0) = -1 \qquad g^1(b_1) = g^1(1) = +1$$
$$g^2(a_2) = g^2(0) = 0 \qquad g^2(b_2) = g^2(1) = \tfrac{1}{2}.$$

System (17.2.6) becomes, recalling that $g^i(a_i) = h^i(a_i)$ and $g^i(b_i) = h^i(b_i)$,

$$\lambda_1(g^1(a_1) - g^1(b_1)) + \lambda_2(g^2(a_2) - g^2(b_2)) = g^1(a_1) + g^2(a_2). \qquad (17.3.1)$$

Thus

$$\lambda_1(-2) + \lambda_2(-\tfrac{1}{2}) = -1$$

or

$$\lambda_1 + \tfrac{1}{4}\lambda_2 = \tfrac{1}{2}. \qquad (17.3.2)$$

The initial facet is $B^0 = \{(y_1, 0) \,|\, 0 \le y_1 \le 1\}$. Also, since for $w = (w_1, w_2)$,

$$w_1 = (1 - \lambda_1)a_1 + \lambda_1 b_1$$
$$w_2 = (1 - \lambda_2)a_2 + \lambda_2 b_2 \qquad (17.3.3)$$
$$w = (w_1, w_2) \in B^0 \qquad \text{if } 0 \le w_1 \le 1 \quad \text{and} \quad w_2 = 0$$

causing $\lambda_2 = 0$. From (17.3.2), $\lambda_1 = \tfrac{1}{2}$. Thus the starting point is $w^0 = (\tfrac{1}{2}, 0)$, where $\lambda_1 = \tfrac{1}{2}, \lambda_2 = 0$.

To determine w^1 we increase λ_2 in (17.3.2) until λ_1 becomes 0 or 1, or λ_2 becomes 1. This produces

$$\lambda_1 = \tfrac{1}{4}, \qquad \lambda_2 = 1 \qquad (17.3.4)$$

and from (17.3.3),

$$w^1 = (\tfrac{1}{4}, 1).$$

The new facet is

$$B^1 = \{(y_1, 1) \,|\, 0 \le y_1 \le 1\}.$$

To form rectangle S^1, since $\lambda_2 = 1$, we must increase a_2 by 1.

RECTANGLE S^1

The next rectangle has $a = (a_1, a_2) = (0, 1)$ because a_2 must be increased. Also $b = (1, 2)$ and

$$S^1 = \{(y_1, y_2) | 0 \le y_1 \le 1, 1 \le y_2 \le 2\}.$$

For this rectangle notice that from (17.3.4), $w^1 = (\frac{1}{4}, 1)$ has

$$\begin{aligned}
\tfrac{1}{4} &= w_1^1 = (1 - \lambda_1)a_1 + \lambda_1 b_1 = (1 - \lambda_1)0 + \lambda_1 \\
1 &= w_2^1 = (1 - \lambda_2)a_2 + \lambda_2 b_2 = (1 - \lambda_2)1 + 2\lambda_2,
\end{aligned}$$
$$(17.3.5)$$

so $\lambda_1 = \frac{1}{4}$ and $\lambda_2 = 0$, which is different from (17.3.4) because the a and b of S^1 is different from S^0. (Always, if $\lambda_i = 1$ in the previous rectangle, $\lambda_i = 0$ in this new one, and if $\lambda_i = 0$ in the previous rectangle, $\lambda_i = 1$ in the new one.) Here

$$\begin{aligned}
g^1(a_1) &= g^1(0) = -1 & g^1(b_1) &= g^1(1) = +1 \\
g^2(a_2) &= g^2(1) = \tfrac{1}{2} & g^2(b_2) &= g^2(2) = -\tfrac{3}{2}.
\end{aligned}$$

Then for S^1 (17.3.1) becomes

$$\lambda_1(-2) + \lambda_2(2) = -\tfrac{1}{2}. \tag{17.3.6}$$

At w^1 since $\lambda_1 = \frac{1}{4}$, $\lambda_2 = 0$, increase λ_2 in (17.3.6) until λ_1 becomes 0 or 1, or $\lambda_2 = 1$. Here $\lambda_1 = 1$ first, so

$$\lambda_1 = 1, \lambda_2 = \tfrac{3}{4}$$

and, via (17.3.3),

$$w^2 = (1, \tfrac{7}{4}).$$

Because $\lambda_1 = 1$, to form S^2 we increase a_1. (Also, in the new rectangle, $\lambda_1 = 0$.)

RECTANGLE S^2

The rectangle S^2 has $a = (1, 1)$ and $b = (2, 2)$; also, $w^2 = (1, \frac{7}{4})$ has $\lambda_1 = 0$ and $\lambda_2 = \frac{3}{4}$ for this rectangle, and

$$\begin{aligned}
g^1(a_1) &= +1 & g^1(b_1) &= +1 \\
g^2(a_2) &= \tfrac{1}{2} & g^2(b_2) &= -\tfrac{3}{2}.
\end{aligned}$$

Thus (17.3.1) becomes

$$\lambda_1(0) + \lambda_2(2) = \tfrac{3}{2}.$$

Increasing λ_1 from 0, λ_1 becomes 1, so $\lambda_1 = 1$, $\lambda_2 = \frac{3}{4}$ and

$$w^3 = (2, \tfrac{7}{4}).$$

Since $\lambda_1 = 1$, we increase a_1 to form S^3; also in S^3, $\lambda_1 = 0$.

RECTANGLE S³

For rectangle S^3, $a = (2, 1)$, $b = (3, 2)$, and for $w^3 = (2, \frac{7}{4})$, $\lambda_1 = 0$ and $\lambda_2 = \frac{3}{4}$. Also,

$$g^1(a_1) = +1 \qquad g^1(b_1) = -1$$
$$g^2(a_2) = \tfrac{1}{2} \qquad g^2(b_2) = -\tfrac{3}{2}$$

and (17.3.1) becomes

$$\lambda_1(2) + \lambda_2(2) = \tfrac{3}{2}.$$

We increase λ_1 from 0, obtaining $\lambda_1 = \frac{3}{4}$, $\lambda_2 = 0$, and $w^4 = (\frac{11}{4}, 1)$. Also, since $\lambda_2 = 0$, we decrease a_2 to form S^4.

RECTANGLE S⁴

At S^4, $a = (2, 0)$, $b = (3, 1)$, and for $w^4 = (\frac{11}{4}, 1)$, $\lambda_1 = \frac{3}{4}$ and $\lambda_2 = 1$. Also,

$$g^1(a_1) = +1 \qquad g^1(b_1) = -1$$
$$g^2(a_2) = 0 \qquad g^2(b_2) = \tfrac{1}{2}.$$

Equation (17.3.1) is

$$\lambda_1(2) + \lambda_2(-\tfrac{1}{2}) = +1.$$

Decrease λ_2 from 1, yielding $\lambda_1 = \frac{1}{2}$, $\lambda_2 = 0$,

$$w^5 = (\tfrac{5}{2}, 0).$$

And so on.

Matrix Pivoting

We have discussed the algorithmic moves in great detail to make it clear. In actual computer operation the movement from $\bar{w}$ to $\bar{w}$ across a rectangle would be done using matrix operations (Exercise 13).

SUMMARY

This chapter presented both the classical Banach fixed-point theorem and separability. The Banach theorem yields powerful existence and uniqueness results, but because the hypothesis is strong, the theorem is useful only in limited circumstances. Separability is a valuable property of functions, and if the homotopy is separable, the PL algorithm can be substantially speeded up by operating on rectangles instead of simplices. Although these two properties do not hold in general, they occur with surprising frequency in applications and thus are quite beneficial to the practitioner.

EXERCISES/CHAPTER 17

1. Prove that if $F: R^n \longrightarrow R^n$ is contractive, then F is continuous.

2. Prove that if $F: R^n \longrightarrow R^n$ is contractive, then there is a nonempty compact convex set $D \subset R^n$ such that $F(D) \subset D$.

3. Prove that if a sequence $\{x^n\}$ is a Cauchy sequence, that is, if given $\epsilon > 0$ there is an N such that for all $n \geq N$ and all $m \geq N$ we have
    $$\|x^n - x^m\| < \epsilon,$$
    then the sequence converges to a limit point x^*.

4. Let H be separable on a rectangle S. Suppose that G is defined by (17.2.3) on the set S. Prove that G is linear on S.

5. Let H be separable on a rectangle S and let G be defined by (17.2.3) on the set S. Suppose that S is triangulated into simplices without introducing new vertices and that on any simplex $s = \{v^i\}_0^{n+1}$ of S, the map
    $$G_s(y) = \sum_{i=0}^{n+1} \lambda_i H(v^i), \qquad \text{where } y = \sum_{i=0}^{n+1} \lambda_i v^i, \quad \sum_{i=0}^{n+1} \lambda_i = 1, \quad \lambda_i \geq 0.$$
    Prove that $G_s(y) = G(y)$.
    (In other words, the piecewise linear approximation is the same for both the triangulation and the rectangular subdivision.)

6. In (17.2.8), suppose that $\bar{w} \in \bar{B}$ and $\bar{B}$ is a facet of rectangles S and S'. Prove that:
    a. If $\bar{w}$ has $\lambda_l = 0$ in S, then to form S', a_l must be decreased to $a_l - \epsilon$. Also, $\bar{w}$ has $\lambda_l = 1$ in S'.
    b. If $\bar{w}$ has $\lambda_l = 1$ in S, then to form S', a_l must be increased to $a_l + \epsilon$. Also, $\bar{w}$ has $\lambda_l = 0$ in S'.

7. Let $n = 1$ and suppose that the vertices of the rectangular subdivision on the plane are the set of integers. Suppose that
    $$H(y) = g^1(y_1) + g^2(y_2),$$
    where g^1 and g^2 have the following values:

$y_i, i = 1, 2$	-3	-2	-1	0	1	2
$g^1(y_1)$	-1	2	1	1	-1	-1
$g^2(y_2)$	0	$\frac{3}{2}$	$-\frac{1}{2}$	0	$\frac{1}{2}$	$-\frac{3}{2}$

Starting from $w^0 = (\frac{1}{2}, 0)$, generate the PL path containing w^0 and lying in the region $\{(y_1, y_2) \mid -3 \leq y_i \leq 2, i = 1, 2\}$.

8. Suppose that F is a mapping from R^n onto R^n. Suppose that for some $\alpha > 1$,

$$\|F(x) - F(y)\| > \alpha \|x - y\|, \qquad \text{all } x \text{ and } y.$$

(Note, therefore, that F is one-to-one.) Prove or disprove: F has a unique fixed point.

9. Consider the linear system

$$x = Ax, \qquad A \text{ an } n \times n \text{ matrix and real.} \qquad \text{(E.17.1)}$$

Prove that if the matrix norm $\|A\| < 1$, then (E.17.1) has a unique solution.

10. Let $G = I - F$, where $F: R^n \longrightarrow R^n$ is contractive. Prove that G is a homeomorphism from R^n onto R^n, that is, that G is continuous, one-to-one and onto, and that the inverse of G exists and is continuous.

11. Suppose that $F: R^n \longrightarrow R^n$ is C^1 and $\|F'(x^0)^{-1}\| \leq \beta$. Assume that $\|F'(y) - F'(x)\| \leq \gamma$, all x, y and that $2\beta\gamma < 1$. Show that the Newton iterates

$$x^{k+1} = x^k - F'(x^k)^{-1}F(x^k), \qquad k = 0, 1, \ldots$$

converge to a unique solution of $F(x) = 0$.

12. To solve (17.2.6), the text also suggested using slack variables

$$u_i = 1 - \lambda_i, \qquad u_i \geq 0.$$

Discuss the advantages and disadvantages of this approach.

13. Discuss how to use matrix pivot and update operations to generate the PL path in the separable algorithm.

14. Why is the regularity assumption (17.2.5) made? What would happen if it does not hold?

NOTES

17.1. The contraction fixed-point theorem is also valid in arbitrary Banach spaces. The theorem was first formulated and proved in the famous dissertation of Banach [1922]. Extensions have been made by Browder [1965], Browder and Petryshyn [1967], Belluce and Kirk [1969], and Kirk [1965]. Applications of the Banach theorem are many. For example, see Collatz [1964], Kantorovich and Akilov [1959], and Keller [1968].

17.2. The approach for separable homotopies was described by Kojima [1978a]. Todd [1980a, b] builds on this idea further.

IV

FUNDAMENTAL CONCEPTS
AND EXTENSIONS

18

All Solutions

The ability to calculate an equation solution, equilibrium, or a fixed point was indeed an achievement, and thus far in the book numerous techniques have been presented for such a calculation. Yet all of these techniques can guarantee to obtain only one such point, and this is despite the fact that most nonlinear problems have not one but many solutions. Since numerous applications require all or most solutions and the previous techniques guarantee only one, the previous procedures are inadequate.

This chapter overcomes these limitations and presents means to obtain all solutions to a system (or, equivalently, to obtain all fixed points or all equilibria). The key is to use a different form of homotopy, the all-solutions homotopy. It has many different start points in $H^{-1}(0)$, each of which starts a different path. Following each of these paths leads us to each of the different solutions.

The ability to obtain all solutions opens up totally new and exciting areas of research and application. Previously, we were limited to finding a local optimum for a maximization problem, for example. Now we can in many circumstances determine the global optimum. The idea is simple. Use the all-solutions approach of this chapter to determine all stationary points. Assuming that this number is not too large, sort among them to determine the one that yields the highest function value. That will be the global optimum. Indeed, the all-solutions approach permits us to solve many problems that previously were unapproachable, and we now proceed to examine it.

18.1 COMPLEX SPACES

The first thing to notice is that even in the simplest of nonlinear equations, the quadratic, we must use complex variables. Even

$$x^2 - x + 1 = 0$$

is seen to have two complex solutions. If we are to have any hope of obtaining all solutions, then, we must of necessity deal with complex spaces, and let us now quickly review some essential features of complex spaces.

To begin specify $\mathbb{C}^1$ as the complex plane and consider a function F: $\mathbb{C}^m \longrightarrow \mathbb{C}^m$. Since F is defined on a complex space, it is a function of the complex vector $z = (z_i) \in \mathbb{C}^m$, where each complex component z_i can be written

$$z_i = u_i + \hat{i}v_i, \qquad i = 1, \ldots, m.$$

Here u_i is the real part and v_i the imaginary part, where $u = (u_i)$, $v = (v_i)$, and $\hat{i} = \sqrt{-1}$. For example, the complex space $\mathbb{C}^m$ becomes

$$\mathbb{C}^m = \{z = (z_i) \,|\, z_i = u_i + \hat{i}v_i, \quad i = 1, \ldots, m\}.$$

Elementary complex variables teach us that each z_i can also be written as

$$z_i = r_i e^{i\theta_i}, \tag{18.1.1}$$

where

$$e^{i\theta_i} = \cos\theta_i + \hat{i}\sin\theta_i,$$

so that

$$u_i = r_i \cos\theta_i \qquad \text{and} \qquad v_i = r_i \sin\theta_i.$$

Clearly,

$$e^{i2\pi k} = 1 \qquad \text{for any nonnegative integer } k. \tag{18.1.2}$$

To focus our ideas and because it will be utilized later, let $z \in \mathbb{C}^1$, and suppose that we wish to solve

$$z^q = 1 \tag{18.1.3}$$

for q a positive integer. From (18.1.1) and (18.1.2) we can rewrite this as

$$(re^{i\theta})^q = 1e^{i2\pi k}$$

or

$$r^q e^{i\theta q} = 1 e^{i 2\pi k},$$

so

$$r^q = 1 \quad \text{and} \quad \theta = \frac{2\pi k}{q}.$$

Solving yields

$$r = 1 \quad \text{and} \quad \theta = \frac{2\pi k}{q}, \qquad k = 0, 1, 2, \ldots, q - 1. \qquad (18.1.4)$$

Note that (18.1.3) has q solutions. If we plot these on a complex plane, they are equally spaced around the unit circle.

As a particular example,

$$z^3 = 1$$

has three solutions:

$$1, \quad -\frac{1}{2} + \frac{\hat{i}\sqrt{3}}{2}, \quad -\frac{1}{2} - \frac{\hat{i}\sqrt{3}}{2}.$$

Another example is shown in Figure 18.1.1 for the equation $z^6 = 1$.

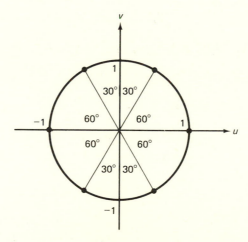

Solutions for $z^6 = 1$.

Figure 18.1.1

EXPANDING F

A valuable feature we will use is that a complex function $F = (F_i)$ is expressible in terms of its real parts F^r and imaginary parts F^s. Thus

$$F_i(z) = F_i^r(u, v) + \hat{i}F_i^s(u, v), \qquad i = 1, \ldots, m.$$

This permits us to define a new function $\hat{F}: R^{2m} \longrightarrow R^{2m}$, precisely equivalent to F, by

$$
\begin{aligned}
\hat{F}_1 &= F_1^r \\
\hat{F}_2 &= F_1^s \\
\hat{F}_3 &= F_2^r \\
\hat{F}_4 &= F_2^s \\
&\cdot \qquad \cdot \\
&\cdot \qquad \cdot \\
&\cdot \qquad \cdot \\
\hat{F}_{2m-1} &= F_m^r \\
\hat{F}_{2m} &= F_m^s
\end{aligned}
\tag{18.1.5}
$$

Note that

$$F(z) = 0 \tag{18.1.6}$$

if and only if

$$\hat{F}(u, v) = 0 \tag{18.1.7}$$

and z solves (18.1.6) if and only if (u, v) solves (18.1.7), where

$$z_i = u_i + \hat{i}v_i.$$

For example, suppose that

$$
\begin{aligned}
F_1(z) &= z_1^3 + 2z_2^2 + 1 \\
F_2(z) &= z_1 z_2.
\end{aligned}
$$

Then

$$
\begin{aligned}
F_1(z) &= (u_1 + \hat{i}v_1)^3 + 2(u_2 + \hat{i}v_2)^2 + 1 \\
&= u_1^3 - 3u_1 v_1^2 + 2u_2^2 - 2v_2^2 + 1 + \hat{i}(3u_1^2 v_1 - v_1^3 + 4u_2 v_2) \\
F_2(z) &= (u_1 + \hat{i}v_1)(u_2 + \hat{i}v_2) \\
&= u_1 u_2 - v_1 v_2 + \hat{i}(u_1 v_2 + v_1 u_2),
\end{aligned}
$$

and we obtain

$$\hat{F}_1 = F_1^r = u_1^3 - 3u_1v_1^2 + 2u_2^2 - 2v_2^2 + 1$$
$$\hat{F}_2 = F_1^s = 3u_1^2v_1 - v_1^3 + 4u_2v_2$$
$$\hat{F}_3 = F_2^r = u_1u_2 - v_1v_2$$
$$\hat{F}_4 = F_2^s = u_1v_2 + v_1u_2.$$

The function $\hat{F}$ will often be more convenient to work with, and in fact define $n = 2m$ and $w = (u, v)$ so that $\hat{F}: R^n \rightarrow R^n$. We term $\hat{F}$ the *expansion* of F. Note especially that since the two functions are precisely equivalent, we will often use them interchangeably.

The expansion $\hat{F}$ will help us in a crucial way. The expansion, when combined with the Cauchy–Riemann equations, will later provide the key theoretical result of this chapter, specifically that the path is monotonic in t. Before considering that, however, we must present the Cauchy–Riemann equations.

CAUCHY–RIEMANN

Complex functions have many fascinating properties, but perhaps the most extraordinary is the famed *Cauchy–Riemann* (C-R) equations: If F is continuously differentiable, then

$$\frac{\partial F_i^r}{\partial u_j} = \frac{\partial F_i^s}{\partial v_j}$$

and

$$\frac{\partial F_i^r}{\partial v_j} = \frac{-\partial F_i^s}{\partial u_j}.$$

The proof is left as Exercise 6, but let us illustrate this result for a simple case. Suppose that $F: \mathbb{C}^1 \rightarrow \mathbb{C}^1$ is as follows:

$$F(z) = z^2 + 2z + 3.$$

Thus

$$F(z) = (u + \hat{i}v)^2 + 2(u + \hat{i}v) + 3 = F^r(u, v) + \hat{i}F^s(u, v)$$

and an easy calculation shows that

$$F^r = u^2 - v^2 + 2u + 3$$
$$F^s = 2uv + 2v.$$

Taking derivatives, we obtain

$$\frac{\partial F^r}{\partial u} = 2u + 2$$

$$\frac{\partial F^r}{\partial v} = -2v$$

$$\frac{\partial F^s}{\partial u} = 2v$$

$$\frac{\partial F^s}{\partial v} = 2u + 2,$$

and we see the C-R hold, as, of course, they must.

DETERMINANT OF ONE SIGN

The C-R underlie many powerful results, and so too they underlie our all solutions proof. As we now show, they ensure that $\det \hat{F}' \geq 0$ and that fact will be used later to show that our homotopy path must always increase in t.

Suppose $\hat{F}$ is the expansion of an F that satisfies the C-R. Suppose also that $F: \mathbb{C}^1 \longrightarrow \mathbb{C}^1$. Then, using the C-R, (18.1.5), and defining a and b,

$$\frac{\partial \hat{F}_1}{\partial u} = \frac{\partial \hat{F}_2}{\partial v} = a$$

$$\frac{\partial \hat{F}_1}{\partial v} = \frac{-\partial \hat{F}_2}{\partial u} = b.$$

Then for $\hat{F}$, its Jacobian determinant becomes

$$\det \hat{F}' = \det \begin{bmatrix} \dfrac{\partial \hat{F}_1}{\partial u} & \dfrac{\partial \hat{F}_1}{\partial v} \\ \dfrac{\partial \hat{F}_2}{\partial u} & \dfrac{\partial \hat{F}_2}{\partial v} \end{bmatrix} = \det \begin{bmatrix} a & b \\ -b & a \end{bmatrix} = a^2 + b^2 \geq 0. \quad (18.1.8)$$

We see, at least in this case, that $\det \hat{F}' \geq 0$.

The same result applies in the more general instance also. Let $F: \mathbb{C}^m \longrightarrow \mathbb{C}^m$, so that $\hat{F}: R^n \longrightarrow R^n$ for $2m = n$. The C-R and (18.1.5) provide that, where we are defining a_{ij} and b_{ij},

$$\frac{\partial \hat{F}_{2i-1}}{\partial u_j} = \frac{\partial \hat{F}_{2i}}{\partial v_j} = a_{ij} \qquad \begin{aligned} i &= 1, \ldots, m \\ j &= 1, \ldots, m \end{aligned}$$

and

$$\frac{\partial \hat{F}_{2i-1}}{\partial v_j} = -\frac{\partial \hat{F}_{2i}}{\partial u_j} = b_{ij}.$$

An instant's reflection shows that det $\hat{F}'$ has the form

$$
\det
\begin{bmatrix}
a_{11} & b_{11} & a_{12} & b_{12} & \cdots & a_{1m} & b_{1m} \\
-b_{11} & a_{11} & -b_{12} & a_{12} & \cdots & -b_{1m} & a_{1m} \\
a_{21} & b_{21} & a_{22} & b_{22} & \cdots & a_{2m} & b_{2m} \\
-a_{21} & b_{21} & -b_{22} & a_{22} & \cdots & -b_{2m} & a_{2m} \\
\cdot & \cdot & \cdot & \cdot & & \cdot & \cdot \\
\cdot & \cdot & \cdot & \cdot & & \cdot & \cdot \\
\cdot & \cdot & \cdot & \cdot & & \cdot & \cdot \\
a_{m1} & b_{m1} & a_{m2} & b_{m2} & \cdots & a_{mm} & b_{mm} \\
-b_{m1} & a_{m1} & -b_{m2} & a_{m2} & \cdots & -b_{mm} & a_{mm}
\end{bmatrix}
\qquad (18.1.9)
$$

Carefully observe the special form of 2×2 blocks, as they are quite important. The calculation of this determinant is straightforward (see Exercise 1). Specifically, as one evaluates the determinant, it will be seen that the 2×2 block form ensures that

$$\det \hat{F}' \geq 0,$$

which means the general case is just like the special case of (18.1.8). Indeed, this is precisely the result we were seeking, and we have obtained the following theorem. (Recall from complex variables that continuous differentiability of a function on complex space ensures that the C-R hold.)

THEOREM 18.1.1   Let $F: \mathbb{C}^m \longrightarrow \mathbb{C}^m$ be continuously differentiable, and let $\hat{F}: R^n \longrightarrow R^n$ be its expansion. Then

$$\det \hat{F}' \geq 0.$$

An easy corollary we will need is:

COROLLARY 18.1.2   Let $F: \mathbb{C}^1 \longrightarrow \mathbb{C}^1$ be

$$F(z) = z^q - 1 \qquad (18.1.10)$$

with expansion $\hat{F}$. Then at any solution $z^* = u^* + \hat{i}v^*$ to $F(z) = 0$,

$$\det \hat{F}'(u^*, v^*) > 0.$$

Proof.   Since $F(z) = \hat{F}_1(u, v) + \hat{i}\hat{F}_2(u, v)$,

$$\frac{dF(z)}{dz} = \left(\frac{\partial \hat{F}_1}{\partial u}\frac{du}{dz} + \frac{\partial \hat{F}_1}{\partial v}\frac{dv}{dz}\right) + \hat{i}\left(\frac{\partial \hat{F}_2}{\partial u}\frac{du}{dz} + \frac{\partial \hat{F}_2}{\partial v}\frac{dv}{dz}\right).$$

Using the C-R, let

$$\frac{\partial \hat{F}_1}{\partial u} = \frac{\partial \hat{F}_2}{\partial v} = a, \quad \frac{\partial \hat{F}_1}{\partial v} = -\frac{\partial \hat{F}_2}{\partial u} = b,$$

so that

$$\frac{dF(z)}{dz} = a\frac{du}{dz} + b\frac{dv}{dz} - \hat{\imath}b\frac{du}{dz} + \hat{\imath}a\frac{dv}{dz}. \tag{18.1.11}$$

Also, $dF/dz = qz^{q-1}$. Since at any solution to (18.1.10), $z \neq 0$,

$$\frac{dF(z)}{dz} \neq 0 \qquad \text{if } z \text{ solves } F(z) = 0. \tag{18.1.12}$$

Comparing (18.1.11) and (18.1.12), we see that it is impossible for both $a = 0$ and $b = 0$, so $a^2 + b^2 > 0$. Then via (18.1.8), at any solution to $F(z) = 0$,

$$\det \hat{F}' = a^2 + b^2 > 0. \qquad \square$$

18.2 DEVELOPMENT OF THE HOMOTOPY

The results we need about functions of complex variables have been stated, and now the homotopy must be specified.

Suppose that we are to solve $F(z) = 0$, where $F: \mathbb{C}^m \longrightarrow \mathbb{C}^m$. For q_i a positive integer, let our homotopy be $H: \mathbb{C}^m \times T \longrightarrow \mathbb{C}^m$, where

$$H_i(z, t) = (1 - t)((z_i)^{q_i} - 1) + tF_i(z) = 0, \qquad i = 1, \ldots, m. \tag{18.2.1}$$

This homotopy has a number of useful properties. First note that at $t = 0$, the system becomes

$$(z_1)^{q_1} - 1 \qquad\qquad\qquad = 0$$
$$(z_2)^{q_2} - 1 \qquad\qquad\qquad = 0$$
$$\cdot$$
$$\cdot$$
$$(z_m)^{q_m} - 1 = 0$$

From (18.1.4) each equation i has q_i solutions. Thus if $z = (z_i) \in H^{-1}(0)$, each component z_i has q_i values. But this means that there are

$$Q = \prod_1^m q_i \text{ points } z \in H^{-1}(0).$$

As a quick example, suppose $m = 2$, $q_1 = 3$, $q_2 = 2$, so that at $t = 0$ the system becomes

$$(z_1)^3 - 1 = 0$$
$$(z_2)^2 - 1 = 0.$$

This has six solutions $z = (z_1, z_2)$, as follows:

$$(1, 1), \quad \left(-\frac{1}{2} + \hat{i}\frac{\sqrt{3}}{2}, 1\right), \quad \left(-\frac{1}{2} - \hat{i}\frac{\sqrt{3}}{2}, 1\right)$$

$$(1, -1), \quad \left(-\frac{1}{2} + \hat{i}\frac{\sqrt{3}}{2}, -1\right), \quad \left(-\frac{1}{2} - \hat{i}\frac{\sqrt{3}}{2}, -1\right).$$

These six points are in $H^{-1}(0)$, and Q, the number of points in $H^{-1}(0)$, is $Q = q_1 q_2 = 6$.

Several other properties of the homotopy follow easily from our previous discussion. For instance, the homotopy $H: \mathbb{C}^m \times T \rightarrow \mathbb{C}^m$ can be expanded into $\hat{H}: R^n \times T \rightarrow R^n$, where $n = 2m$. In detail, if $H_i = H_i^r + \hat{i}H_i^s, i = 1, \ldots, m$, then for $z_i = u_i + \hat{i}v_i$ and $w = (u_1, v_1, \ldots, u_m, v_m)$,

$$\hat{H}_1(w, t) = H_1^r(z, t)$$
$$\hat{H}_2(w, t) = H_1^s(z, t)$$
$$\vdots \qquad \qquad \vdots$$
$$\hat{H}_{n-1}(w, t) = H_m^r(z, t)$$
$$\hat{H}_n(w, t) = H_m^s(z, t).$$

As earlier mentioned, H and $\hat{H}$ are equivalent. For instance, $z \in H^{-1}(t)$ if and only if there is a corresponding $w \in \hat{H}^{-1}(t)$. In particular, let $w^0 \in \hat{H}^{-1}(0)$ correspond to $z^0 \in H^{-1}(0)$, where $w^0 = (u_1^0, v_1^0, \ldots, u_m^0, v_m^0)$, and $z^0 = u^0 + \hat{i}v^0$. Then Corollary 18.1.2 provides that, if $w^0 \in \hat{H}^{-1}(0)$,

$$\det \hat{H}_w'(w^0, 0) > 0.$$

As this holds for any point $w^0 \in \hat{H}^{-1}(0)$, automatically $\hat{H}$ is regular at $t = 0$.

Further, since H and $\hat{H}$ are equivalent, Theorem 18.1.1 immediately applies. Let $\hat{H}_w'$ be the $n \times n$ Jacobian of $\hat{H}$ with respect to w. Then for any $t = \bar{t}$ fixed,

$$\det \hat{H}_w'(w, t) \geq 0 \qquad \text{for any } t = \bar{t}. \tag{18.2.2}$$

Note especially that we have achieved the previously mentioned observation. Specifically, via the BDE theorem of Chapter 2, a path in $\hat{H}^{-1}$ must then be monotonic in t; that is, the path cannot reverse itself in t.

The all-solutions result will be almost immediate given our discussion so far. As usual the proof will be by path following, but it will also employ the idea of path finiteness. Given $H: \mathbb{C}^m \times T \to \mathbb{C}^m$ and a set $S \subset T$, we say that H is *path finite* on S if it cannot run off to infinity at any $\bar{t} \in S$. This means that if it runs off to infinity, it can do so only by approaching some $\bar{t}$, where $\bar{t}$ is not in S.

For instance, suppose that H is path finite for $0 \leq t < 1$. Then a path can run off to infinity only as t approaches 1. In Figure 18.2.1, path A then could not occur since it diverges to infinity as it approaches some $\bar{t} < 1$. However, all other paths could occur, as they either do not diverge to infinity at all or do so as $t \to 1$. As a different example, and recalling that $T = \{t \mid 0 \leq t \leq 1\}$, suppose that H were path finite for T. Then no path could run off to infinity. In Figure 18.2.1, A, B, and G would then be eliminated.

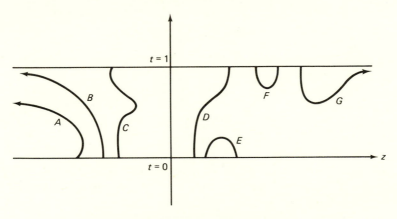

Path A diverges to infinity at $\bar{t} < 1$. Paths B and G diverge to infinity at $t = 1$.

Figure 18.2.1

Conditions that guarantee when H (or equivalently $\hat{H}$) is path finite will be given shortly. But first we present the main result of this chapter, the all-solutions theorem.

THEOREM 18.2.1 (All Solutions). Let $F: \mathbb{C}^m \to \mathbb{C}^m$, $F \in C^2$, and suppose that we wish to find all solutions to

$$F(z) = 0, \qquad\qquad (18.2.3)$$

where $F = 0$ has a finite number of solutions. Suppose that the homotopy

$$H_i(z, t) = (1 - t)((z_i)^{q_i} - 1) + tF_i(z)$$

is regular and path finite for $0 \le t < 1$. Then, by starting from the $Q = \prod_1^m q_i$ solutions to $\hat{H}^{-1}(0)$, and following the corresponding paths, we obtain all solutions of (18.2.3).

Proof. From (18.2.2) each path must be monotonic in t, so that paths which reverse themselves, such as E, F, and G of Figure 18.2.1, cannot occur. Under the path-finite condition, no path can run off to infinity for some $\bar{t} < 1$, so path A cannot occur. Indeed, only paths such as B, C, and D can occur.

All paths starting from $\hat{H}^{-1}(0)$ must either diverge to infinity, such as B, or go to a solution $w^1 \in \hat{H}^{-1}(1)$. However, any point in $\hat{H}^{-1}(1)$ must connect via a path to a point in $\hat{H}^{-1}(0)$. In other words, if we follow all Q paths from all points $w^0 \in \hat{H}^{-1}(0)$, some paths might diverge but the remainder will find all the points in $\hat{H}^{-1}(1)$. [No point $w^1 \in \hat{H}^{-1}(1)$ can be left out, as it must have a path going to some point $w^0 \in \hat{H}^{-1}(0)$.] Thus we can find all solutions to (18.2.3). $\square$

Under the hypothesis of Theorem 18.2.1 all solutions are found. Notice, however, that some of the paths starting from $t = 0$ may be extra. More precisely, of the paths from $t = 0$, some go to find all solutions of $F(z) = 0$, while others, the extra ones, diverge to infinity. There are instances such as this for which we may have to start with too many paths and have a few left over that diverge to infinity. Under certain other circumstances, however, we know beforehand exactly how many paths are needed and none are extra. The following corollary specifies that.

COROLLARY 18.2.2 Suppose that H is regular and path finite for T. Then there exist exactly $Q = \prod_1^m q_i$ solutions to

$$F(z) = 0$$

and each path from a start $w^0 \in \hat{H}^{-1}(0)$ leads to a solution (i.e., no paths run off to infinity).

Proof. Under the hypothesis of path finite for T, no path can diverge to infinity. Consequently, each $w^0 \in \hat{H}^{-1}(0)$ connects to a point $w^1 \in \hat{H}^{-1}(1)$, and conversely. As there are Q starts in $\hat{H}^{-1}(0)$, there must be Q solutions to $F(z) = 0$. $\square$

Overall, we see that Theorem 18.2.1 provides only an upper bound on the number of solutions. However, under the path-finite condition of the corollary, the exact number of solutions to $F(z) = 0$ is known beforehand, and there are no extra paths. Certainly, knowing there are no extra paths will reduce work, but knowing that depends upon the path-finite condition, which we now must explore.

18.3 CONDITIONS FOR PATH FINITENESS

Path finiteness was essential to our proofs. Let us now analyze it and the properties of the F and H that guarantee it. The key here is in the choice of q_i for each H_i in (18.2.1). So far, these q_i have not been specified and their proper choice will, in many instances, ensure that an appropriate path finiteness holds.

We begin with a path-finite condition for T. Suppose that $F = (F_i)$ has the form

$$F_i(z) = (z_i)^{q_i} + G_i(z), \qquad i = 1, \ldots m, \tag{18.3.1}$$

where, if $\| z_i \| \longrightarrow \infty$,

$$\frac{G_i(z)}{(z_i)^{q_i}} \longrightarrow 0. \tag{18.3.2}$$

Define H_i in (18.2.1) using the same q_i as is already given by the F_i in (18.3.1). Let us show H is then path finite for T. Since

$$H_i(z, t) = (1 - t)((z_i)^{q_i} - 1) + tF_i(z)$$
$$H_i(z, t) = (z_i)^{q_i} - (1 - t) + tG_i(z).$$

Dividing yields

$$\frac{H_i}{(z_i)^{q_i}} = 1 - \frac{(1 - t)}{(z_i)^{q_i}} + \frac{tG_i(z)}{(z_i)^{q_i}}.$$

Therefore, as $\| z_i \| \longrightarrow \infty$ by (18.3.2),

$$\frac{H_i}{(z_i)^{q_i}} \longrightarrow 1 \neq 0. \tag{18.3.3}$$

But this means that $H(z, t) = 0$ cannot have a solution for $\| z \|$ large, and therefore $z \in H^{-1}(t)$ cannot diverge to infinity. In short, suppose that F has the form of (18.3.1), where (18.3.2) holds for all i. Then, using the same q_i, H is path finite for T.

If the form of F is (18.3.1) where (18.3.2) holds, H will be path finite for T. Certain other forms of F may not yield H path finite for T, but will yield path finite for $0 \leq t < 1$. Here is an instance of that.

Let a function $F = (F_i)$ be given. Suppose that for each i we can select an integer q_i so that the ratio

$$\frac{F_i(z)}{(z_i)^{q_i} - 1} \not< 0 \qquad \text{if } \|z_i\| \longrightarrow \infty. \tag{18.3.4}$$

The notation means that the ratio cannot approach a real and negative number as $\|z_i\| \longrightarrow \infty$. The ratio limit can be real and nonnegative, pure imaginary, or mixed (even with the real part negative). However, it cannot be pure real and negative.

If for each i we can determine a q_i so that (18.3.4) holds, then that guarantees path finiteness for $0 \le t < 1$, as we now show. Using the same q_i in H_i, since

$$H_i(z, t) = (1 - t)((z_i)^{q_i} - 1) + tF_i(z),$$

we have

$$\frac{H_i(z, t)}{(z_i)^{q_i} - 1} = 1 - t + \frac{tF_i(z)}{(z_i)^{q_i} - 1}.$$

Yet here, because (18.3.4) holds,

$$\frac{H_i(z, t)}{(z_i)^{q_i} - 1}$$

does not approach zero if $\|z_i\| \longrightarrow \infty$, and t approaches $\bar{t}, 0 \le \bar{t} < 1$. Immediately, $H(z, t) = 0$ cannot hold if $\|z\| \longrightarrow \infty$ and t approaches $\bar{t}, 0 \le \bar{t} < 1$. We therefore have path finiteness for $0 \le t < 1$.

To apply the foregoing results in actual practice, suppose we are given an arbitrary $F = (F_i)$. If each F_i is of the form (18.3.1), where (18.3.2) holds, then using the same q_i, H is path finite for T. If (18.3.1) and (18.3.2) do not hold, then try to select the q_i for each i so that (18.3.4) holds. If that can be done, then H is path finite for $0 \le t < 1$.

CONDITIONS FOR POLYNOMIALS

It is often easy to verify that one of the foregoing path-finite conditions hold, especially if each F_i is polynomial. In particular, suppose that each F_i is a polynomial in m variables. That is, each F_i is the sum of terms of the form

$$a_{ik}(z_1)^{l_{1k}^i}(z_2)^{l_{2k}^i}, \cdots, (z_m)^{l_{mk}^i},$$

where each l_{ik}^h is a nonnegative integer and a_{ik} a complex number. Notationally,

$$F_i(z) = \sum_k a_{ik} \prod_{h=1}^{m} (z_h)^{l_{ik}^h}, \qquad i = 1, \ldots, m. \tag{18.3.5}$$

For example, let $z = (z_1, z_2, z_3)$ and

$$F_1(z) = 2z_1^2 z_2^3 + 14z_1^4 z_2^2 z_3^4 + z_3$$

$$F_2(z) = 3z_1^2 z_2 + z_1^2 z_3^2$$

$$F_3(z) = z_3^2.$$

Here $a_{11} = 2$, $a_{12} = 14$, $a_{13} = 1$, $l_{11}^1 = 2$, $l_{11}^2 = 3$, $l_{11}^3 = 0$, $l_{12}^1 = 4$, $l_{12}^2 = 2$, $l_{12}^3 = 4$, $l_{13}^1 = 0$, $l_{13}^2 = 0$, $l_{13}^3 = 1$, $a_{21} = 3$, $a_{22} = 1$, and so on.

Now assuming that each F_i is a polynomial in m variables, specify

$$\bar{l}_{ik} = \sum_{h=1}^{m} l_{ik}^h \tag{18.3.6}$$

as the power of a term, and define

$$q_i = 1 + \max_k \{\bar{l}_{ik}\}. \tag{18.3.7}$$

Note that q_i is one more than the power of the highest-powered term in F_i. In the example, $\bar{l}_{11} = 5$, $\bar{l}_{12} = 10$, $\bar{l}_{13} = 1$, so $q_1 = 11$. Similarly, $q_2 = 5$, $q_3 = 3$.

By definition of q_i in (18.3.7), it follows that as $\|z_i\| \to \infty$,

$$\frac{F_i(z)}{(z_i)^{q_i} - 1} \longrightarrow 0.$$

Yet, then, (18.3.4) holds.

Observe what all this means. If each F_i is a polynomial in m variables, to form each H_i, use the q_i of (18.3.7). Then H will be path finite for $0 \le t < 1$.

In certain other situations we can be sure (18.3.2) will hold so that H is path finite for T. More precisely, let us assume that each F_i has the form

$$F_i(z) = (z_i)^{q_i} + G_i(z), \tag{18.3.8}$$

where G_i is the polynomial

$$G_i(z) = \sum_k a_{ik} \prod_{h=1}^{m} (z_h)^{l_{ik}^h}.$$

Assume also that for each i

$$q_i \ge 1 + \max_k \{\bar{l}_{ik}\}, \tag{18.3.9}$$

where $\bar{l}_{ik}$ is defined by (18.3.6). Then, since q_i is at least one greater than the highest powered term in G_i, (18.3.2) holds. In short, suppose that each F_i has the form (18.3.8), where each q_i satisfies (18.3.9). Then, using these q_i, H is path finite for T.

We have hereby given several conditions and function forms that ensure path finiteness. These conditions are easily obtained, especially if F is a polynomial. The basic idea is to make the q_i so large that the paths are forced to be finite. Intuitively, when q_i is sufficiently large, the $(z_i)^{q_i}$ term will dominate and thereby force the finiteness of the solution paths. Of course, the q_i that guarantees path finiteness need not always exist. Also, other conditions for path finiteness can be developed, but the ones presented here are quite useful in practice.

EXAMPLES

Let us apply our all-solutions methodology to some simple examples.

EXAMPLE 18.3.1.

Suppose that $F(z) = z^2 + 2z + 5 = 0$. Let us use (18.3.8) and (18.3.9), so that

$$H(z, t) = (1 - t)(z^2 - 1) + tF(z) = 0$$

or

$$H(z, t) = (z^2 - 1) + t(2z + 6) = 0.$$

This has two solutions:

$$z = -t \pm \sqrt{t^2 - 6t + 1}, \qquad 0 \le t \le 1.$$

There are two starting points at $t = 0$, $z = +1$ and $z = -1$. Following the corresponding two paths leads to the two solutions $z = -1 + \hat{i}2$ and $z = -1 - \hat{i}2$ (see Figure 18.3.1). Thus, starting from the two solutions $z^0 \in$

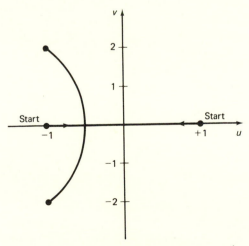

Starting at $z = \pm 1$, the paths lead to solutions $z = -1 \pm \hat{i}2$.

Figure 18.3.1

$H^{-1}(0)$, we obtain the two solutions to the problem. Note that for $0 \leq t \leq 3 - 2\sqrt{2}$, z is real, whereas for $3 - 2\sqrt{2} < t \leq 1$, z is mixed real and imaginary. Also observe that the paths are not regular everywhere, but the procedure still works. Regularity is a very minor assumption and, as discussed in Chapter 22, rarely affects us.

EXAMPLE 18.3.2.

This example is curious and shows that even though the roots to the function

$$F(z) = z^2 - 5z + 6 = 0$$

are real, the paths are not necessarily so. Formulating the homotopy,

$$H(z, t) = (1 - t)(z^2 - 1) + tF(z) = 0$$

or

$$H(z, t) = (z^2 - 1) + t(-5z + 7) = 0$$

the solution paths are

$$z = \frac{5t \pm \sqrt{25t^2 - 28t + 4}}{2}, \qquad 0 \leq t \leq 1.$$

Note that z is real for

$$0 \leq t \leq \frac{14 - 4\sqrt{6}}{25} \quad \text{and} \quad \frac{14 + 4\sqrt{6}}{25} \leq t \leq 1,$$

but is mixed for

$$\frac{14 - 4\sqrt{6}}{25} < t < \frac{14 + 4\sqrt{6}}{25}.$$

The two paths go from $+1$ to -1 to the solutions $+2$ and $+3$ (see Figure 18.3.2). Again, the loss of regularity does not affect us, as discussed in Chapter 22.

THE PROCEDURE

To summarize, suppose that for $F: \mathbb{C}^m \rightarrow \mathbb{C}^m$, we must solve

$$F(z) = 0.$$

We would then determine the q_i (assuming they exist) so that the homotopy

$$H_i(z, t) = (1 - t)((z_i)^{q_i} - 1) + tF(z)$$

is path finite for either $0 \leq t < 1$ or $0 \leq t \leq 1$. Conditions (18.3.2) and (18.3.4) can be helpful in determining the q_i.

The homotopy H would then be expanded into $\hat{H}$ by separating the real

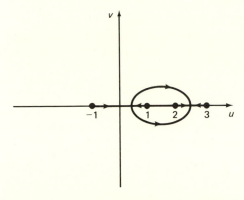

Although the solutions are real, the paths are not necessarily so.

Figure 18.3.2

and imaginary parts. After that the path following could be implemented. Specifically, there would be $Q = \prod_1^m q_i$ starting points in $\hat{H}^{-1}(0)$. The paths from each of these starting points could then be followed until $t = 1$. Although some paths may diverge to infinity, the rest will determine all the solutions to $F(z) = 0$.

18.4 FURTHER CONSIDERATIONS

REAL VARIABLE PROBLEMS

Recall that to find all solutions, our spaces must be complex. Yet suppose that we are given a problem that is real-valued, say $F: R^m \longrightarrow R^m$, and we must solve

$$F(x) = 0.$$

To handle this, first substitute z for x and then solve the revised problem, where

$$F(z) = 0.$$

For example, given

$$F(x) = x^2 + 2x - 1 = 0,$$

transform this to

$$F(z) = z^2 + 2z - 1 = 0,$$

where now $F: \mathbb{C}^m \longrightarrow \mathbb{C}^m$. The function will then be defined appropriately and we may apply the all solutions procedure to it.

The question arises: Must we substitute z for x; that is, why not just use $F(x) = 0$ where everything is real-valued? The answer is because the procedure will not work.

EXAMPLE 18.4.1

For instance, consider the situation of Example 18.3.2, but now require the variable to be real. Then

$$F(x) = x^2 - 5x + 6 = 0$$

and

$$
\begin{aligned}
H(x, t) &= (1 - t)(x^2 - 1) + tF(x) \\
&= (x^2 - 1) + t(-5x + 7).
\end{aligned}
$$

The solution paths are

$$x = \frac{5t \pm \sqrt{25t^2 - 28t + 4}}{2}, \qquad 0 \leq t \leq 1.$$

By following these paths where, recall, the paths must be real, we obtain the situation in Figure 18.4.1. Note the two starts $x^0 = +1$ and $x^0 = -1$ connect to each other. Thus the approach does not work. If we start a path from an $x^0 \in H^{-1}(0)$, instead of leading us to a point $x^1 \in H^{-1}(1)$, it leads us to another point in $H^{-1}(0)$.

By following the all-solutions procedure, however, it does work. In particular, substitute z for x and then solve

$$F(z) = z^2 - 5z + 6 = 0.$$

Now the procedure does work and has already been solved in Example 18.3.2.

The reason the procedure does not work in the reals is that

$$\det H'_x(x, t)$$

can then be positive, negative, or both. Thus the path can reverse itself and return to $t = 0$, as in Figure 18.4.1. However, by substituting z and then taking the expansion $\hat{H}(w, t)$, we obtain by (18.2.2) that

$$\det \hat{H}'_w(w, t) \geq 0.$$

The path is now monotonic in t and cannot reverse, so the procedure succeeds. Of course, after substituting z for x, we must ensure that $F(z)$ is analytic as it need not be even if $F(x)$ is continuously differentiable (see Exercises 7 and 8), and that the other hypotheses of our theorems hold. Nevertheless, the fact that the path is monotonic in t is, as mentioned, the key theoretical observation that underlies our all-solutions procedure.

In sum, suppose that we are given a problem in the reals to solve, so that $F: R^m \longrightarrow R^m$ and $F(x) = 0$. First substitute z for x so that $F: \mathbb{C}^m \longrightarrow \mathbb{C}^m$ and

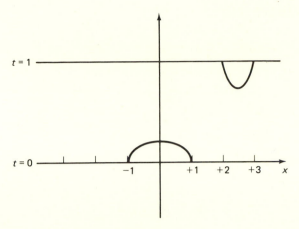

The solution paths to $H(x, t) = (1 - t) (x^2 - 1) + t(x^2 - 5x + 6) = 0, x \in R^1$

Figure 18.4.1

$F(z) = 0$. Then apply the procedure to this problem; that is, form H, take its expansion $\hat{H}$, and follow the various paths.

GLOBAL OPTIMUM

Consider the problem of maximizing a function $f: R^n \longrightarrow R^1$. The usual practice is to find a point that solves the gradient of f equals zero,

$$\nabla f(x) = 0.$$

However, many points may solve these equations, and we have the complication of relative maxima or other types of stationary points arising.

Traditionally, the problem of relative optima has been attacked by making a concavity assumption. However, that assumption is no longer necessary because the technique of this chapter finds all solutions. Specifically, determine all stationary points. Then evaluate which of the pure real solutions obtained yields the largest f. That solution produces the global optimum.

SUMMARY

The all-solutions procedure presented in this chapter is powerful both practically and theoretically, and no longer need we be content with finding one fixed point or one solution. Rather, under reasonable conditions we can calculate all of them, compare what happens at the different ones, and determine the best one. Furthermore, the methodology underlying our demonstration is, as usual, path following.

EXERCISES/CHAPTER 18

1. Show that the determinant of any matrix of the form (18.1.9) is nonnegative.

2. The fundamental theorem of algebra states the following: Let

$$f(z) = a_n z^n + a_{n-1} z^{n-1} + \ldots + a_0 = 0, \qquad a_n \neq 0 \qquad \text{(E.18.1)}$$

   where $a_n, \ldots, a_0$ are complex numbers and z a complex variable. Then there are n roots to $f(z) = 0$. Prove the fundamental theorem of algebra by a path-following argument.

3. Consider equation (E.18.1), where $n \geq 2$. Suppose that the map H is defined to be

$$H(z, t) = (1 - t)(z - 1) + t f(z) = 0.$$

   Assume that H is regular. What would H^{-1} be like?

4. In Example 18.3.1, note that H is not regular. Suppose that the homotopy equations are perturbed to

$$H(z, t) = \epsilon_1 + \hat{i}\epsilon_2.$$

   a. If $(\epsilon_1, \epsilon_2) = (0.01, 0.01)$, what are the starting points? Where will each path go?
   b. If $(\epsilon_1, \epsilon_2) = (0.01, -0.01)$, what are the starting points? Where will each path go?

5. Repeat Exercise 4 for H defined as in Example 18.3.2.

6. Let $F: \mathbb{C}^m \longrightarrow \mathbb{C}^m$ be continuously differentiable. Prove that the C-R hold.

7. Let $F: R^1 \longrightarrow R^1$ be $F(x) = x^2 + \max\{x, 2\}^2 - 4$. Here F is real. Show that we cannot use the procedure in the chapter to find all solutions. [*Hint:* Show that $F(z)$ is not defined.] Show that even if $F(x)$ is continuously differentiable, that $F(z)$ need not be (so the C-R need not hold). What conditions are required on $F(x)$ to ensure, after substituting z for x, that $F(z)$ is continuously differentiable?

8. Give an example of a well-defined map where the C-R does not hold.

9. Discuss what happens when the all-solutions procedure is used for solving

$$z^2 + 2z + e^z = 50.$$

10. a. Show that the map H of Example 18.4.1 is regular.
    b. Compare H of Example 18.4.1 to H of Example 18.3.2 and discuss why the former is regular whereas the latter is not.

11. (*Multiple solutions*) Consider a polynomial function $F: \mathbb{C}^1 \longrightarrow \mathbb{C}^1$. Suppose that a solution $z = \bar{z}$ of $F(z) = 0$ is a solution of multiplicity k, for some positive integer k. Prove that the all-solutions procedure will have exactly k paths approaching $z = \bar{z}$ as t approaches 1. (Assume that almost all perturbations of the homotopy equations yield a regular H.)

12. Explain how the all-solutions procedure can be extended to find:
    a. all fixed points.
    b. all equilibrium points.

13. What happens in Theorem 18.2.1 if $F(z) = 0$ does not have a finite number of solutions and we follow the paths H^{-1} starting from $t = 0$? (Assume that H is regular by perturbing the homotopy equations, if necessary.)

NOTES

18.1. Independently and almost simultaneously, Drexler [1977, 1978] and Garcia and Zangwill [1979a, 1980b] developed the all-solutions algorithm. Drexler's approach is based on the classical algebraic "Bezout's method of elimination," and is somewhat cumbersome. See van der Waerden [1931]. Garcia and Zangwill's approach, on the other hand, is topological and geometrical, and is the basis of the approach of this chapter.

18.2. The theorem is based on Garcia and Zangwill [1980b].

18.3. For polynomial functions of the form (18.3.5), Chow, Mallet-Paret, and Yorke [1978] showed that the extra $+1$ is not needed in (18.3.7) for the path finiteness condition to hold. Garcia and Li [1980] gave an exact count of the number of solutions to polynomial systems and showed a generalization of the fundamental theorem of algebra.

18.4. For determining all solutions of polynomial functions in one variable, see Kuhn [1976, 1977]. For computation of a unique solution, see Garcia and Zangwill [1979c].

19

The Linear
Complementarity Problem

This chapter presents the linear complementarity problem (LC), a problem that has now become classical. One reason for its stature is that LC solves many diverse problems, including quadratic programming and bimatrix games, for example. Another reason is that its key algorithm, Lemke's algorithm, introduced the ideas of complementarity and of path following which historically spawned much of the field and underpin this entire book. Without a doubt, LC and Lemke's algorithm must be studied carefully and fully, which we do in this chapter and in Chapter 20.

19.1 THE LINEAR COMPLEMENTARITY PROBLEM

Given a vector $q \in R^n$ and a square $n \times n$ matrix M, the *linear complementarity problem* (LC) is as follows: Find a vector $x \in R^n$ such that for

$$z \equiv q + Mx$$

we have

$$z \geq 0, \quad x \geq 0, \quad z^\tau x = 0. \qquad (19.1.1)$$

In other words, we must find an x that satisfies (19.1.1), where $z \equiv q + Mx$. (Recall that the triple equality is "equality by definition." Also, τ is the transpose.)

Many problems can be formulated via the LC and here we present two examples, quadratic programming and bimatrix games.

EXAMPLE 19.1.1 (Quadratic Programming)

A quadratic program (QP) is the problem of:

$$\min c^\tau x + \tfrac{1}{2} x^\tau Q x$$

subject to

$$Ax \leq b$$

$$x \geq 0 \tag{19.1.2}$$

where $c \in R^n$, Q is a symmetric $n \times n$ real matrix, A is $m \times n$, and $b \in R^m$ are all given. The Kuhn–Tucker necessary conditions for QP are that there must exist vectors $u \in R^n$, $v \in R^m$, and $\lambda \in R^m$ such that

$$c + Qx + A^\tau \lambda - u = 0$$

$$Ax + v = b \tag{19.1.3}$$

$$u \geq 0, \quad v \geq 0, \quad x \geq 0, \quad \lambda \geq 0, \quad u^\tau x = 0, \quad v^\tau \lambda = 0.$$

Clearly, this can be rewritten as

$$\begin{pmatrix} u \\ v \end{pmatrix} \equiv \begin{pmatrix} c \\ b \end{pmatrix} + \begin{pmatrix} Q & A^\tau \\ -A & 0 \end{pmatrix} \begin{pmatrix} x \\ \lambda \end{pmatrix} \tag{19.1.4}$$

$$u \geq 0, \quad v \geq 0, \quad x \geq 0, \quad \lambda \geq 0, \quad u^\tau x = 0, \quad v^\tau \lambda = 0.$$

Quadratic programming is therefore an example of an LC where

$$q = \begin{pmatrix} c \\ b \end{pmatrix} \quad \text{and} \quad M = \begin{pmatrix} Q & A^\tau \\ -A & 0 \end{pmatrix}.$$

EXAMPLE 19.1.2 (Bimatrix Games)

The bimatrix game, or two-person game, was introduced in Chapter 8. Recall that there are two players, where player 1 solves

$$\max (x^1)^\tau A^1 x^2$$

subject to

$$\sum_{j=1}^{n_1} x_j^1 = 1$$

$$x^1 \geq 0 \tag{19.1.5A}$$

given x^2;

and player 2 solves

$$\max (x^1)^\tau A^2 x^2$$

subject to

$$\sum_{j=1}^{n_2} x_j^2 = 1$$

$$x^2 \geq 0 \qquad\qquad (19.1.5B)$$

given x^1.

Recall also that a Nash equilibrium point is pair of vectors $(\bar{x}^1, \bar{x}^2)$ such that $\bar{x}^1$ solves (19.1.5A) given $\bar{x}^2$, and $\bar{x}^2$ and solves (19.1.5B) given $\bar{x}^1$. In addition, we presume that each element of A^1 and A^2 is negative. If not, simply subtract a scalar from each element and clearly the solution to (19.1.5) is unchanged. (Exercise 13). Explicitly, then, we suppose that

$$A^i < 0, \qquad i = 1, 2. \qquad\qquad (19.1.6)$$

FORMULATION AS AN LC

Let us now formulate the bimatrix game (19.1.5) as an LC. Denote e as the vector of all 1's. First suppose $(\bar{x}^1, \bar{x}^2)$ is a Nash equilibrium point. Then for all x^i, $i = 1, 2$, such that

$$e^\tau x^i = 1, \qquad x^i \geq 0, \quad i = 1, 2, \qquad\qquad (19.1.7)$$

$$(\bar{x}^1)^\tau A^1 \bar{x}^2 \geq (x^1)^\tau A^1 \bar{x}^2, \qquad (\bar{x}^1)^\tau A^2 \bar{x}^2 \geq (\bar{x}^1)^\tau A^2 x^2. \qquad (19.1.8)$$

That is $\bar{x} = (\bar{x}^1, \bar{x}^2)$ is a Nash equilibrium point if and only if (19.1.8) holds for all x satisfying (19.1.7).

Next, for any x satisfying (19.1.7), observe that (19.1.8) is equivalent to

$$[(\bar{x}^1)^\tau A^1 \bar{x}^2]e \geq A^1 \bar{x}^2, \qquad [(\bar{x}^1)^\tau A^2 \bar{x}^2]e \geq (A^2)^\tau \bar{x}^1. \qquad (19.1.9)$$

To see this, let $x^i = e^j$, $i = 1, 2$, where e^j is the vector with all zeros except for a 1 in the jth position. Then (19.1.8) implies that

$$(\bar{x}^1)^\tau A^1 \bar{x}^2 \geq (e^j)^\tau A^1 \bar{x}^2, \qquad \text{all } j = 1, \ldots, m$$

$$(\bar{x}^1)^\tau A^2 \bar{x}^2 \geq (\bar{x}^1)^\tau A^2 e^j, \qquad \text{all } j = 1, \ldots, n.$$

and hence (19.1.9) holds. Conversely, from (19.1.9),

$$[(\bar{x}^1)^\tau A^1 \bar{x}^2](x^1)^\tau e \geq (x^1)^\tau A^1 \bar{x}^2$$

$$[(\bar{x}^1)^\tau A^2 \bar{x}^2]e^\tau x^2 \geq (\bar{x}^1)^\tau A^2 x^2. \qquad (19.1.10)$$

But $(x^i)^\tau e = 1$ for $i = 1, 2$ by (19.1.7), so (19.1.8) holds. Consequently, $\bar{x} = (\bar{x}^1, \bar{x}^2)$ is a Nash equilibrium point if and only if (19.1.7) and (19.1.9) hold.

The LC equivalent to the bimatrix games can now be expressed as follows:

$$\begin{pmatrix} z^1 \\ z^2 \end{pmatrix} = \begin{pmatrix} -e \\ -e \end{pmatrix} + \begin{pmatrix} 0 & -A^1 \\ -(A^2)^\tau & 0 \end{pmatrix} \begin{pmatrix} x^1 \\ x^2 \end{pmatrix}. \tag{19.1.11}$$

$$z \ge 0, \qquad x \ge 0, \qquad z^\tau x = 0.$$

We must demonstrate that solving (19.1.11) solves the bimatrix game, and for that it is only necessary to verify a one-to-one correspondence between points $(\bar{x}^1, \bar{x}^2)$ satisfying (19.1.7) and (19.1.9) and the points (x^1, x^2) satisfying (19.1.11).

Suppose that (x^1, x^2) satisfies (19.1.11). Then clearly, $x^i \ne 0$, $x^i \ge 0$, $i = 1, 2$, so $e^\tau x^i > 0$, $i = 1, 2$. Specify

$$\bar{x}^i = \frac{x^i}{e^\tau x^i}, \qquad i = 1, 2. \tag{19.1.12}$$

Obviously, $\bar{x}^i$ satisfies (19.1.7). Let us also show that $\bar{x}$ satisfies (19.1.9).

From (19.1.11),

(A) $\qquad\qquad A^1 x^2 \le -e, \qquad (A^2)^\tau x^1 \le -e$

(B) $\qquad\qquad (x^1)^\tau A^1 x^2 = -e^\tau x^1, \qquad (x^1)^\tau A^2 x^2 = -e^\tau x^2. \tag{19.1.13}$

Plugging in (19.1.12), the system (19.1.13) becomes

(A) $\qquad\qquad A^1 \bar{x}^2 \le \dfrac{-e}{e^\tau x^2}, \qquad (A^2)^\tau \bar{x}^1 \le \dfrac{-e}{e^\tau x^1}$

(B) $\qquad\qquad (e^\tau x^1)(e^\tau x^2)(\bar{x}^1)^\tau A^i \bar{x}^2 = -(e^\tau x^i), \qquad i = 1, 2. \tag{19.1.14}$

Simplifying (19.1.14B) and since $e^\tau \bar{x}^i = 1$,

$$(\bar{x}^1)^\tau A^1 \bar{x}^2 = -\frac{1}{e^\tau x^2}, \qquad (\bar{x}^1)^\tau A^2 \bar{x}^2 = -\frac{1}{e^\tau x^1}. \tag{19.1.15}$$

But then (19.1.14A) is equivalent to

$$A^1 \bar{x}^2 \le [(\bar{x}^1)^\tau A^1 \bar{x}^2]e, \qquad (A^2)^\tau \bar{x}^1 \le [(\bar{x}^1)^\tau A^2 \bar{x}^2]e, \tag{19.1.16}$$

which is exactly (19.1.9).

Conversely, suppose that $(\bar{x}^1, \bar{x}^2)$ satisfies (19.1.7) and (19.1.9). Then let

$$x^1 = -\frac{\bar{x}^1}{(\bar{x}^1)^\tau A^2 \bar{x}^2}, \qquad x^2 = -\frac{\bar{x}^2}{(\bar{x}^1)^\tau A^1 \bar{x}^2}. \tag{19.1.17}$$

It is an easy exercise to show that (x^1, x^2) satisfies (19.1.11) (Exercise 1).

Overall, given (x^1, x^2) a solution to the LC, we obtain a solution to (19.1.7) and (19.1.9). Conversely, if $(\bar{x}^1, \bar{x}^2)$ solves (19.1.7) and (19.1.9), a solution to the LC is obtained. We conclude that solving LC (19.1.11) is equivalent to finding a Nash equilibrium point for the bimatrix game.

OTHER APPLICATIONS

The LC has also been used to analyze numerous problems, including problems involving reinforced concrete beams, plastic structures, free-boundary problems for journal bearings, finance models, portfolio selection, and actuarial graduation. References are given in the chapter notes.

19.2 SOLVING THE LC BY PATH FOLLOWING

With the applications for LC introduced, we now present how to solve the LC by path following. The idea is to first transform the LC into an equivalent system of equations. Then a homotopy is stated which, as usual, is a trivial system at $t = 0$, while at $t = 1$ yields the system whose solution solves the LC.

THE SYSTEM OF EQUATIONS

Begin by noticing that the LC (19.1.1) can be restated as a problem of finding a solution to a system of equations. Define $F: R^n \longrightarrow R^n$ by

$$F_i(x) = \min\{x_i, z_i\}, \qquad i = 1, 2, \ldots, n, \qquad (19.2.1)$$

where $z \equiv q + Mx$. Then see that for $x \in R^n$,

$$F(x) = 0 \quad \text{if and only if} \quad x \geq 0, \quad z \geq 0, \quad z^\tau x = 0.$$

Explicitly, finding a solution of (19.2.1) is equivalent to finding a solution of the LC.

This type of formulation, slightly revised, will form the basis of our homotopy.

THE HOMOTOPY SYSTEM

Given q and M, first choose a positive vector $d \in R^n$ such that $d + q > 0$. Then consider the LC:

$$z = d + q + Mx \qquad (19.2.2)$$
$$z \geq 0, \quad x \geq 0, \quad z^\tau x = 0.$$

Clearly, (19.2.2) has a *trivial* solution $x = 0$, where $z = d + q$.

Now define the homotopy:

$$H_i(x, t) = \min\{x_i, z_i\} = 0, \qquad i = 1, \ldots, n \qquad (19.2.3)$$

for $z \equiv (1 - t)d + q + Mx$. As in (19.2.1), we see that (19.2.3) is precisely equivalent to the LC problem

$$z^\tau x = 0, \qquad x \geq 0, \qquad z \geq 0$$

where

$$z \equiv (1 - t)d + q + Mx. \qquad (19.2.4)$$

At $t = 0$, notice that the trivial LC (19.2.2) with trivial solution $x = 0$ is obtained, whereas when $t = 1$, we have the original LC (19.1.1).

Equation (19.2.3), or equivalently (19.2.4), provides the homotopy system for our procedure. Lemke's algorithm, which we will discuss shortly, operates on it and generates a path in H^{-1} starting from $(x, t) = (0, 0)$. The path will either diverge to infinity or reach $t = 1$. If the path reaches $t = 1$, then the Lemke algorithm obtains a solution to the LC (19.1.1).

PATH EXISTENCE

Before presenting Lemke's algorithm, we must first verify that the solution set to the homotopy system H defined by (19.2.3) is indeed a collection of paths. The difficulty is that this particular definition of H is not differentiable, and differentiability is required for the path theorem, Theorem 1.4.2. Fortunately, H can be redefined so that it becomes differentiable. In fact, H can be defined so that it is C^k, for any k. (Recall that C^k means that kth partial derivatives exist and are continuous.)

We will employ the functions defined in Chapter 4, namely the functions

$$\alpha_i^+ = \max\{0, \alpha_i\}^{k+1}, \qquad \alpha_i^- = \max\{0, -\alpha_i\}^{k+1}, \qquad i = 1, \ldots, n. \qquad (19.2.5)$$

Explicitly, the transformation of variables utilized is

$$z_i = \alpha_i^+, \qquad i = 1, 2, \ldots, n$$
$$x_i = \alpha_i^-, \qquad i = 1, 2, \ldots, n, \qquad (19.2.6)$$

where $\alpha = (\alpha_1, \ldots, \alpha_n) \in R^n$. Recall that if $\alpha_i \geq 0$, we have $z_i \geq 0$ and $x_i = 0$, whereas if $\alpha_i < 0$, we have $z_i = 0$ and $x_i > 0$. In other words, for any α_i,

$$z_i \geq 0, x_i \geq 0 \quad \text{and} \quad z_i x_i = 0, \qquad \text{for all } i. \qquad (19.2.7)$$

Further, the transformation is one-to-one, that is, given any α_i, there is a unique pair (x_i, z_i) that satisfies (19.2.7), and conversely.

Via transformation (19.2.6), the homotopy system equivalent to (19.2.3) or (19.2.4) becomes

$$H(\alpha, t) = \alpha^+ - (1 - t)d - q - M\alpha^- = 0. \qquad (19.2.8)$$

Here $\alpha^+ = (\alpha_1^+, \ldots, \alpha_n^+)$, and similarly for α^-.

The solutions $(\alpha, t) \in H^{-1}$ of (19.2.8) have unique corresponding (x, t) which are solutions to (19.2.3), and conversely. Indeed, (19.2.3) and (19.2.8) are entirely equivalent systems. By (19.2.8), H is C^k. So, under regularity, the solutions $(\alpha, t) \in H^{-1}$ of (19.2.8) form differentiable paths. But as (19.2.3) is one-to-one with (19.2.8), the solutions to (19.2.3) must also be paths (although not necessarily differentiable paths, see Exercise 12).

In the remainder of this chapter, for specificity we have

$$\alpha_i^+ = \max\{0, \alpha_i\}^3, \qquad \alpha_i^- = \max\{0, -\alpha_i\}^3. \qquad (19.2.9)$$

EXAMPLE 19.2.1

Let $q = -2$, $M = (1)$, $d = 3$. Then (19.2.3) becomes

$$z = (1 - t)3 - 2 + x$$
$$z \geq 0, \qquad x \geq 0, \qquad zx = 0,$$

which has a solution of

$$(z, x) = \begin{cases} (1 - 3t, 0) & \text{for } t \leq \frac{1}{3} \\ (0, 3t - 1) & \text{for } t \geq \frac{1}{3}. \end{cases} \qquad (19.2.10)$$

This solution path is clearly piecewise linear (PL).

System (19.2.8) becomes

$$\max\{0, \alpha\}^3 = (1 - t)3 - 2 + \max\{0, -\alpha\}^3$$

and yields as a solution

$$(\alpha, t) = \left(\alpha, \frac{1 - \alpha^3}{3}\right), \qquad \alpha \in R^1, \qquad (19.2.11)$$

which is a differentiable path. Further, via the transformation (19.2.6) we see that (19.2.10) and (19.2.11) are equivalent.

STARTING

Since $d + q > 0$ from (19.2.8), we have that at $t = 0$,

$$\alpha_i = \sqrt[3]{d_i + q_i}, \qquad i = 1, \ldots, n$$

and the algorithm will commence there. Note that the start $t = 0$, $\alpha_i = \sqrt[3]{d_i + q_i}$ for (19.2.8) is precisely equivalent to the start $t = 0$, $x = 0$ for (19.2.3).

NO LOOPS

The correspondence between (19.2.3) and (19.2.8) lets us analyze the paths easily and we now show that there cannot be any loops. If $t \leq 0$ and using (19.2.8), the points on the path are

$$\alpha_i = \sqrt[3]{(1-t)d_i + q_i}$$

since for all i, $d_i > 0$ and $d_i + q_i > 0$. In particular the path can be thought of as "starting at infinity" for $t = -\infty$. That is, the path can be considered to "start" at $t = -\infty$. Then as t increases from $-\infty$, it satisfies (19.2.11) until t is at least zero. As a main observation, perceive that this particular path cannot have a loop. (A loop could not go to $t = -\infty$.) The algorithm itself operates on a portion of this path (the portion for $t \geq 0$). Consequently, the algorithm cannot loop either, and the following theorem has been established.

THEOREM 19.2.1   Given q, M, and d where $d > 0$ and $q + d > 0$, let H defined by (19.2.8) be regular. Then H^{-1} is a collection of differentiable paths. Moreover, the path in H^{-1} through the point

$$t = 0, \qquad \alpha_i = \sqrt[3]{d_i + q_i}, \qquad i = 1, \ldots, n \qquad (19.2.12)$$

cannot be a loop.

COMMENTS

The regularity condition on H is quite weak and can be eliminated in a straightforward manner (Exercise 2). We know because (19.2.3) and (19.2.8) are equivalent that the H^{-1} of (19.2.3) also consists of paths. Further, the (19.2.3) path through $t = 0$, $x = 0$, cannot be a loop since that path corresponds to the path through (19.2.12).

At this juncture a procedure for solving the LC has been obtained: namely, solve (19.2.8) via one of our previously discussed algorithms. Although that approach is valid, it might be inefficient, since (19.2.8) is nonlinear. Although (19.2.8) permitted use of Theorem 1.4.2 and helped us analyze the paths, (19.2.3) is far better for computation. System (19.2.3) has a great deal of linearity and Lemke's algorithm, as we see next, exploits that linearity ingeniously.

19.3 LEMKE'S METHOD

Lemke's method traces the path in H^{-1} starting from $(x, t) = (0, 0)$ of the homotopy

$$H_i(x, t) = \min \{x_i, z_i\} = 0, \qquad i = 1, \ldots, n, \qquad (19.3.1)$$

where

$$z \equiv (1 - t)d + q + Mx. \tag{19.3.2}$$

As noted, this system is equivalent to (19.2.4).

To avoid degeneracies a regularity condition is required:

> For each $(x, t) \in H^{-1}$, at least $n - 1$ of
> the variables x, z are greater than zero. $\tag{19.3.3}$

This condition is extremely weak and is conceptually similar to other regularity assumptions on H since it assists in providing a nice path (see Exercises 3 to 5). In particular, (19.3.3) implies that (x, z) can have at most $n + 1$ variables equal to zero. Since (19.3.1) requires that (x, z) must have at least n variables zero, the regularity condition ensures either that $n - 1$ variables are positive and $n + 1$ are zero, or that n variables are positive and n are zero.

To develop Lemke's algorithm, the system (19.2.4) can be restated as

$$Iz - Mx + td = d + q$$
$$x \geq 0, \quad z \geq 0$$
$$z^{\tau}x = 0, \tag{19.3.4}$$

where I is the identity matrix. This system will turn out to be very convenient, and to understand it let us first analyze the requirement that for i, $z_i x_i = 0$.

ZERO SET

We term x_i the *complement* of z_i, and vice versa. Clearly, there are n pairs of complementary variables, and by (19.3.1) either $x_i = 0$ or $z_i = 0$, or possibly both. That is, at least one of the two complementary variables x_i or z_i must be equal to zero, and this holds for all n pairs.

At any step Lemke's algorithm for each i designates either x_i or z_i to be zero throughout that step. Explicitly, there is a *zero set* at step k

$$B^k = \{u_1, u_2, \ldots, u_n\},$$

where u_i is set to either x_i or z_i. Throughout the entire step k Lemke's algorithm requires that all $u_i = 0$. To reiterate, on step k, u_i is set equal to x_i or z_i, and $u_i = 0$ throughout that step.

The zero set is the key to Lemke's algorithm. To understand it, call w_i the complement of u_i. If $u_i \equiv x_i$, then $w_i \equiv z_i$, and conversely. Because u_i is in the zero set, it must be zero throughout the step; however, the w_i can be zero or positive. In particular, given B^k, notice that (19.3.4) can be expressed as the $n \times (n + 1)$ system

$$A^k w + td = d + q$$
$$w \geq 0, \qquad t \in R^1 \tag{19.3.5}$$

where for $i = 1, \ldots, n$ if

$$u_i \equiv x_i = 0,$$

then

$$w_i \equiv z_i \quad \text{and} \quad A_i^k \equiv e^i$$

and if

$$u_i \equiv z_i = 0,$$

then

$$w_i \equiv x_i \quad \text{and} \quad A_i^k \equiv -M_i.$$

Here A_i^k is the ith column of A^k. The column e^i (a 1 in the ith position and zeros elsewhere) is associated with $w_i \equiv z_i$, and $-M_i$ is associated with $w_i \equiv x_i$ and we choose the appropriate one. Essentially, since the variables u_i, $i = 1, \ldots, n$, are zero throughout step k, we drop them from (19.3.4) and what is left is (19.3.5).

EXAMPLE 19.3.1

As an example, suppose that $n = 2$ and $M = (m_{ij})$, $d = (d_i)$, and $q = (q_i)$. System (19.3.4) then becomes

$$z_1 \quad - m_{11}x_1 - m_{12}x_2 + td_1 = d_1 + q_1$$
$$z_2 - m_{21}x_1 - m_{22}x_2 + td_2 = d_2 + q_2$$
$$z_1 x_1 = 0, \qquad z_2 x_2 = 0, \qquad \text{all variables nonnegative.}$$

If $B^k = \{x_1, x_2\}$, then since $u_1 \equiv x_1$ and $u_2 \equiv x_2$, the corresponding system (19.3.5) becomes

$$z_1 \quad + td_1 = d_1 + q_1$$
$$z_2 + td_2 = d_2 + q_2$$
$$z_1 \geq 0, \qquad z_2 \geq 0$$

where $w_1 \equiv z_1$ and $w_2 \equiv z_2$.

If $B^k = \{x_1, z_2\}$, then the corresponding (19.3.5) is

$$z_1 - m_{12}x_2 + td_1 = d_1 + q_1$$
$$-m_{22}x_2 + td_2 = d_2 + q_2$$
$$z_1 \geq 0, \qquad x_2 \geq 0,$$

where $w_1 \equiv z_1$ and $w_2 \equiv x_2$.

The main point is that given the zero set B^k, it immediately identifies a corresponding (19.3.5). At a step, because B^k is given and the variables in it are held at zero, we need only work with the corresponding (19.3.5).

Notice that by regularity at least $n - 1$ of the w_i are positive at any one time. Also, assuming that (19.3.5) has full row rank, then since it is $n \times (n + 1)$, its solutions will generate a straight-line segment.

STEP k

We now present the algorithmic step. On step k we are at a point (x^k, t^k) and the zero set B^k is given. Also, a *distinguished* variable w_l is specified such that $w_l = 0$. Thus at (x^k, t^k) the variables $u_i = 0$, $i = 1, \ldots, n$, because these variables are in the zero set and also $w_l = 0$.

From B^k we obtain the corresponding (19.3.5). Moreover, at (x^k, t^k) there are already $n + 1$ variables at zero, so by (19.3.3),

$$w_i > 0, \qquad i \neq l.$$

This means that the distinguished variable w_l can be increased in (19.3.5) (at least a little) by adjusting the other w_i, $i \neq l$ and t.

Now in (19.3.5) increase w_l and suppose that for some positive value of w_l a variable w_j becomes zero. The point where that occurs is (x^{k+1}, t^{k+1}) and we start the next step, $k + 1$. Thus increase w_l in (19.3.5) until some variable w_j hits zero, and that starts the next step, $k + 1$.

For step $k + 1$ the new zero set B^{k+1} is formed as follows. The variable w_j that just hit zero goes into the zero set and its complement u_j comes out. All other variables that were in B^k remain in B^{k+1}. Thus B^{k+1} is the same as B^k except the variable that hit zero goes in and its complement comes out. Moreover, the complement of w_j becomes the new distinguished variable that is to be increased.

In brief, the variable that hits zero enters the zero set. Its complement both comes out and is the new distinguished variable to be increased.

From the new B^{k+1} we can form the corresponding (19.3.5). Then increase the new distinguished variable just as before. The process continues in this manner.

FIRST STEPS

To commence the algorithm, let $(x^0, t^0) = (0, 0)$, and $B^0 = \{x_1, x_2, \ldots, x_n\}$ is the zero set initially. Then (19.3.4) is satisfied and

$$z^0 = d + q > 0.$$

By increasing t we now gradually move from (x^0, t^0) but remaining in H^{-1}. Throughout the step, we keep $x = x^0 = 0$ (i.e., all variables in B^0 are kept at zero) and the corresponding (19.3.5) is kept satisfied. In detail, for the initial step, (19.3.5) reduces to

$$z + td = d + q$$
$$z \geq 0, \qquad t \in R^1.$$

Moreover, because $z^0 > 0$, t can be increased at least a little from $t = t^0 = 0$, keeping

$$z = (1 - t)d + q \geq 0,$$

so (19.3.5) is still satisfied.

As we increase t in (19.3.5) from $(x^0, t^0) = (0, 0)$, two possibilities arise. First, if t can increase to 1, $z = q \geq 0$ and the point $x = 0$, $z = q$ is a solution to the LC. Otherwise, some z_i becomes zero in (19.3.5) when t hits a value $t = t^1 < 1$. That is, t is increased until at $t = t^1$ some z_i becomes zero. This brings us to the new point $(x^1, t^1) = (0, t^1)$.

At $t = t^1$, suppose that z_l became zero. Since x_l, the complement of z_l, has been at zero, now both $x_l = z_l = 0$. To form the new zero set B^1, x_l is replaced by z_l. Thus

$$B^1 = \{x_1, x_2, \ldots, x_{l-1}, z_l, x_{l+1}, \ldots, x_n\}.$$

Also, x_l is the distinguished variable to be increased.

EXAMPLE 19.3.2

Let

$$q = \begin{pmatrix} -2 \\ -1 \end{pmatrix}, \qquad M = \begin{bmatrix} 2 & 1 \\ -1 & 3 \end{bmatrix}, \qquad d = \begin{pmatrix} 3 \\ 2 \end{pmatrix}.$$

Then to start the algorithm, (19.3.5) is the system

$$z_1 + 3t = 1$$
$$z_2 + 2t = 1$$
$$z_1 \geq 0, \qquad z_2 \geq 0.$$

The initial zero set is $B^0 = \{x_1, x_2\}$, and these variables are kept at zero. We can increase t only up to $t = \frac{1}{3}$. Thus $(x^1, t^1) = (0, 0, \frac{1}{3})$.

The new zero set $B^1 = \{z_1, x_2\}$. Further letting $B^1 = \{u_1, u_2\}, u_1 \equiv z_1 = 0$, $u_2 \equiv x_2 = 0$. Also the distinguished variable is $w_1 \equiv x_1$, which is the complement of $u_1 \equiv z_1$. Notice that w_1 had been at zero for the entire initial step.

INCREASING w_l

During the general step of the algorithm we must increase w_l from $w_l = 0$. Notice that three possibilities can occur. First, as w_l is increased, if t becomes 1, the corresponding point solves the LC. Second, if w_l can be increased to infinity in (19.3.5) with t less than 1, then again we terminate. This means that the method has generated a path that diverges to infinity.

Finally, we have the standard case: that is, w_l can only be increased to some finite positive value $\bar{w}_l$ because some $w_j, j \neq l$, becomes zero in (19.3.5). Increasing w_l further is impossible because then w_j becomes negative. [Note that t can be negative in (19.3.5).] Then the point (x^{k+1}, t^{k+1}) corresponding to $w_l = \bar{w}_l$ in (19.3.5) will be the new point. The complement of w_j is the new distinguished variable. Also, $B^{k+1} = \{u_1, \dots, u_{j-1}, w_j, u_{j+1}, \dots, u_n\}$ is the new zero set.

In essence, the standard step of the algorithm is extremely simple. Whichever variable hits zero, that enters the zero set. Its complement, call it w_l, leaves the zero set and is increased. Any other variable in the zero set remains at zero.

SUMMARY OF LEMKE'S ALGORITHM

Step 0

Initially, $(x^0, t^0) = (0, 0)$, $B^0 = \{x_1, \dots, x_n\}$. Increase t from zero in the system

$$z + td = d + q$$
$$z \geq 0, \qquad t \in R^1 \qquad\qquad (19.3.6)$$

1. If t can be increased to 1, then $x = 0$, $z = q \geq 0$, is an LC solution.

2. Otherwise, some z_l becomes zero in (19.3.6) for $t = t^1$. Let $(x^1, t^1) = (0, t^1)$, $w_l = x_l$ be the distinguished variable, and

$$B^1 = \{x_1, \dots, x_{l-1}, z_l, x_{l+1}, \dots, x_n\}.$$

Go to step 1.

Step k, $k \geq 1$.

Let (x^k, t^k) be the current point, w_l the distinguished variable, and $B^k = \{u_1, \dots, u_n\}$ the zero set. Set $u_1 = \dots = u_n = 0$. Then (19.3.4) becomes

$$A^k w + td = d + q$$
$$w \geq 0, \qquad t \in R^1 \qquad\qquad (19.3.7)$$

where A^k and w are defined by (19.3.5). Increase w_l from zero in (19.3.7).

1. If t becomes equal to 1, terminate. We have an LC solution at hand.

2. If w_l increases to infinity in (19.3.7), terminate. We have a path diverging to infinity.

3. Otherwise, some w_j becomes zero in (19.3.7) when $w_l = \bar{w}_l > 0$.

Let (x^{k+1}, t^{k+1}) be the new point corresponding to $w_l = \bar{w}_l$, the complement of w_j be the distinguished variable, and $B^{k+1} = \{u_1, \ldots, u_{j-1}, w_j, u_{j+1}, \ldots, u_n\}$. Go to step $k + 1$.

Repeat this process until $t = 1$ or the path diverges to infinity. (Recall that, as was mentioned earlier, the path cannot be a loop.)

Observe that as w_l is increased from (x^k, t^k), only one variable can hit zero. By regularity (19.3.3), at least $n - 1$ variables must be positive. All n variables in the zero set remain zero, so $u_i = 0$, $i = 1, \ldots, n$. If two or more w_j hit zero at once, there would be $n - 2$ or fewer variables positive. Consequently, as we increase w_l, only one variable can hit zero.

EXAMPLE 19.3.3

Let

$$q = \begin{pmatrix} -2 \\ -1 \end{pmatrix}, \qquad M = \begin{bmatrix} 2 & 1 \\ -1 & 3 \end{bmatrix}, \qquad d = \begin{pmatrix} 3 \\ 2 \end{pmatrix}.$$

Then the system $z = (1 - t)d + q + Mx$ is

$$z_1 - 2x_1 - x_2 + 3t = 1$$
$$z_2 + x_1 - 3x_2 + 2t = 1.$$

Let us compute the sequence $\{(x^k, t^k)\}$ by Lemke's method.

STEP 0. $(x^0, t^0) = (0, 0, 0)$. $B^0 = \{x_1, x_2\}$. The system $A^0w + td = d + q$

is

$$z_1 = 1 - 3t$$
$$z_2 = 1 - 2t.$$

We increase t to $t^1 = \frac{1}{3}$. Then $(x^1, t^1) = (0, 0, \frac{1}{3})$, the distinguished variable is x_1, and $B^1 = \{z_1, x_2\}$.

STEP 1. The system $A^1w + td = d + q$ is

$$-2x_1 + 3t = 1 \qquad\qquad t = \frac{1}{3} + \frac{2}{3}x_1$$
$$\qquad\qquad\qquad\qquad \text{or}$$
$$z_2 + x_1 + 2t = 1 \qquad\qquad z_2 = \frac{1}{3} - \frac{7}{3}x_1.$$

Increase x_1 to $x_1^2 = \frac{1}{7}$. Then $(x^2, t^2) = (\frac{1}{7}, 0, \frac{3}{7})$, the distinguished variable is x_2 and $B^2 = \{z_1, z_2\}$.

STEP 2. The system $A^2w + td = d + q$ is

$$t = \frac{3}{7} + x_2$$
$$x_1 = \frac{1}{7} + x_2.$$

Increase x_2 to $x_2^3 = \frac{4}{7}$ so that $(x^3, t^3) = (\frac{5}{7}, \frac{4}{7}, 1)$. The point $(\frac{5}{7}, \frac{4}{7})$ is an LC solution.

Theorem 19.2.1 provides that the algorithm generates a path. However, using the algorithm itself, we can verify not only that it indeed generates a path but also show that the path is a PL path. First consider the $n \times n$ matrix A^k of (19.3.5). Under (19.3.3), at least $n - 1$ variables w_i must be positive; thus A^k must have rank at least $n - 1$ (Exercise 4).

Next, for (19.3.5) examine the $n \times (n + 1)$ matrix (A^k, d). This has rank at least $n - 1$ since A^k does. Further, because d must be selected only so that $d > 0$ and $q + d > 0$, it is easy to also choose d so that (A^k, d) has rank n. Then (19.3.5) has a straight-line segment as a solution.

We call the LC *regular* if (19.3.3) holds and d is selected so that $d > 0$, $d + q > 0$, and the matrix (A^k, d) is rank n for all k. If the LC is regular, we are assured, during a step, that the algorithm is generating a straight-line segment.

The question remains: What happens as we change the distinguished variable at a step? When we change variables we come in on one line segment and go out on another. More precisely, suppose that we increase w_i until w_j becomes zero at (x^{k+1}, t^{k+1}). Then both $u_j = 0$ and $w_j = 0$. Also, by (19.3.3) for all other i, if $u_i = 0$, then $w_i > 0$, $i \neq j$. Thus to maintain (19.3.4), only one variable can be increased, namely the complement of w_j. That means that there is only one line segment out, namely the line segment formed by increasing the complement of w_j. In short, we come in on a line segment by increasing w_i and go out on the only line segment available by increasing the complement of w_j. The path is thus piecewise linear.

Finally, recall that by Theorem 19.2.1 we could think of the path as starting from $t = -\infty$, so it can have no loops. Let us confirm this also. Starting from $(x^0, t^0) = (0, 0)$, suppose that we decrease t and make it negative. Because $d > 0$, we have $z = d + q - td \geq d + q > 0$ for all $t \leq 0$. Thus the point $x^0 = 0$, $z = d + q - td$ solves (19.3.4) for all $t \leq 0$. Since this holds for all $t \leq 0$, we can think of the path as "starting" at $t = -\infty$ and then, as t increases, passing through $t = 0$, $x = 0$. The path consequently cannot be a loop.

To review our discussion above, notice that:

1. During a step the algorithm generates a straight-line segment.

2. As it switches steps at any (x^k, t^k), there is only one line segment in and one line segment out.

3. There can be no loop.

The following theorem has been verified.

THEOREM 19.3.1 Suppose that the LC is regular. Then the Lemke algorithm generates a PL path with no loops.

SUMMARY

This chapter presented the LC and Lemke's algorithm, which historically form the basis for path-following procedures. The LC homotopy is not differentiable but does have a special structure which by use of Lemke's algorithm produces a piecewise-linear path.

LC is highly useful and in this chapter we showed, for example, how it subsumes quadratic programming and the bimatrix game. Chapter 20 presents the details on how it solves these problems.

EXERCISES/CHAPTER 19

1. Prove that if $(\bar{x}^1, \bar{x}^2)$ is a Nash equilibrium point, then (x^1, x^2) defined by (19.1.17) satisfies (19.1.11).

2. Let H be defined by (19.2.8). Show that for almost every $q \in R^n$, H is regular.

3. Let A^k defined by (19.3.5) have rank at least $n - 1$ for all A^k. Show that for almost every $q \in R^n$, $d \in R^n$, the regularity condition (19.3.3) holds.

4. Show that regularity condition (19.3.3) ensures that A^k of (19.3.5) has rank at least $n - 1$.

5. Suppose that the regularity condition (19.3.3) does not hold. Discuss what difficulties could arise when applying Lemke's method. Give concrete examples.

6. Suppose that H defined by (19.2.8) is regular. Will the regularity condition (19.3.3) hold? Explain your answer.

7. Let
$$q = \begin{pmatrix} -2 \\ -1 \end{pmatrix}, \qquad M = \begin{bmatrix} 2 & -1 \\ -1 & 3 \end{bmatrix}, \qquad d = \begin{pmatrix} 3 \\ 2 \end{pmatrix}.$$
Use Lemke's method on this problem.

8. Suppose that in Lemke's algorithm as w_l is increased at step k, t remains fixed at $t = t^k$. Show that A^k of (19.3.5) must have rank less than n.

9. Let
$$q = \begin{pmatrix} -1 \\ 2 \\ -3 \end{pmatrix}, \qquad M = \begin{pmatrix} 2 & 0 & 2 \\ -1 & 5 & -1 \\ 2 & -1 & 0 \end{pmatrix}, \qquad d = \begin{pmatrix} 4 \\ 4 \\ 4 \end{pmatrix}.$$
Use Lemke's algorithm on this LC problem.

10. Show that H as defined by (19.2.3) is a piecewise-linear function.

11. Given an LC (19.1.1), suppose that Lemke's algorithm terminates at an LC solution at $t = 1$. Prove that A^k of (19.3.5) corresponding to the terminal step has positive determinant.

12. Prove that if the solution set to either (19.2.3) or (19.2.8) consists of paths, the

solution set to the other must also consist of paths. Discuss why the solution path to (19.2.3) need not be differentiable even if the path to (19.2.8) is.

13. Prove that if a scalar is subtracted from each element of A^1 and A^2, the solution to (19.1.5) is unchanged.

NOTES

The linear complementarity problem was formally identified by Cottle and Dantzig [1968], who called it "the fundamental problem." Although there are other algorithms, such as the method of Cottle and Dantzig [1968] and Graves' algorithm [1967], the most interesting known in the literature is Lemke's algorithm [1968]. Lemke's algorithm, is preceded by the Lemke–Howson algorithm [1964], which is designed especially for bimatrix games (discussed in Chapter 20). Since Lemke's algorithm was given, numerous variants and related approaches were developed. For example, see Chandrasekaran [1970], McCammon [1970], van de Panne [1974]. Mangasarian [1976] characterizes certain LCs as LPs. For applications, see Dantzig, Eaves, and Gale [1979], Maier [1970], and Pang, Kaneko, and Hallman [1979].

20

Linear Complementarity in Action

Chapter 19 introduced the linear complementarity (LC) problem, Lemke's algorithm, and certain key applications, such as quadratic programming and bimatrix games. This chapter relates the specifics as to how to actually solve these applications. Section 20.1 articulates some general theorems which are used in Section 20.2 to validate that the Lemke algorithm solves the quadratic programming problem. Then in Section 20.3 we analyze the bimatrix game and for that create a new algorithm, the Lemke–Howson.

20.1 EXISTENCE OF AN LC SOLUTION

Now that Lemke's method has been presented (Chapter 19), we can detail how it finds solutions to the linear complementarity (LC) problem. To review, recall that, given q and M, we choose a positive vector $d \in R^n$ such that $d + q > 0$, and then consider the homotopy

$$H_i(x, t) = \min \{x_i, z_i\} = 0, \qquad i = 1, \ldots, n, \qquad (20.1.1)$$

where $z \equiv (1 - t)d + q + Mx$.

At $t = 0$, homotopy (20.1.1) becomes the trivial system

$$z = d + q + Mx \qquad (20.1.2)$$

$$z \geq 0, \qquad x \geq 0, \qquad z^{\tau}x = 0,$$

while at $t = 1$, the LC problem

$$z = q + Mx$$
$$z \geq 0, \qquad x \geq 0, \qquad z^\tau x = 0 \tag{20.1.3}$$

is obtained.

Lemke's method is initiated at $(x, t) = (0, 0)$ and generates a piecewise-linear path of solutions to (20.1.1). The path either diverges to infinity or reaches an LC solution at $t = 1$. We now provide conditions that ensure the latter case. Such conditions imply not only that an LC solution exists, but that Lemke's method is *guaranteed* to find an LC solution. These conditions are quite practical, and in the next section are used to prove that the Lemke method will solve the quadratic programming problem with convex objective function.

CONDITIONS

As our first set of conditions for the Lemke algorithm, suppose that:

    (A) M has nonzero principal minors.

    (B)   The trivial LC (20.1.2) has only the solution $x = 0$.   (20.1.4)

Recall that a principal minor of an $n \times n$ matrix M is constructed as follows. Let J be a nonempty subset of the first n integers. That is, J is a nonempty subset of $\{1, 2, \ldots, n\}$. Then consider the submatrix formed by deleting the rows and columns of M for $j \notin J$. The determinant of such a submatrix is called a *principal minor*. To form the submatrix note that both rows and columns must be deleted for $j \notin J$. The principal minor is then the determinant of the submatrix.

For example, let $n = 3$ and if $J = \{1\}$, the submatrix is

$$M_{JJ} = (m_{11})$$

since rows and columns 2 and 3 are deleted. If $J = \{1, 3\}$, the submatrix is

$$M_{JJ} = \begin{pmatrix} m_{11} & m_{13} \\ m_{31} & m_{33} \end{pmatrix}.$$

Also, if $J = \{1, 2, \ldots, n\}$, since nothing is deleted,

$$M_{JJ} = M.$$

To repeat, the submatrices are formed by deleting both rows and columns for $j \notin J$. The determinant of such a submatrix is a principal minor. (Notationally, M_{KL} indicates the submatrix formed by deleting the rows not in K and columns not in L, where K and L are nonempty subsets of $\{1, 2, \ldots, n\}$.)

Returning to our conditions, condition (20.1.4A) states that for all J,

$$\det M_{JJ} \neq 0.$$

Then for $K = \{1, 2, \ldots, n\} - J$ and letting

$$A^k = \begin{pmatrix} I & -M_{KJ} \\ 0 & -M_{JJ} \end{pmatrix}, \tag{20.1.5}$$

clearly, $\det A^k \neq 0$. But A^k is simply a matrix (19.3.5), that is, a matrix generated by the algorithm. The valuable conclusion obtained is that under (20.1.4A), all matrices A^k generated by the algorithm have their determinant nonzero.

Since $\det A^k \neq 0$, on any linear part of the PL path, by the BDE, $\dot{t} = (-1)^{n+1} \det A^k \neq 0$. Thus as we move along any linear part of the path, t can never be constant but will either continuously increase or continuously decrease. Further, with only a finite number of linear segments on the path, should the path diverge to infinity, it must do so on a ray. Yet because t must keep increasing or keep decreasing along the ray, no horizontal ray can exist. Consequently, if the path diverges to infinity, either $t \to +\infty$ or $t \to -\infty$. In sum, under (20.1.4A) if the path diverges to infinity, $t \to +\infty$ or $t \to -\infty$.

Since the start is $t = 0$, if $t \to +\infty$ that is fine, because when $t = 1$ we terminate. However, what happens should Lemke's algorithm diverge to infinity for $t \to -\infty$? Notice that the path starting from $(x^0, t^0) = (0, 0)$ has t increasing at least a little from $t = 0$ (Exercise 1). Thus if $t \to -\infty$, the path must cross $t = 0$ again at some other point $\bar{x} \neq x^0$. Yet condition (20.1.4B) prevents that, as then there would be two solutions to the trivial LC. We have arrived at the following.

THEOREM 20.1.1   Assuming that the LC is regular, let M have nonzero principal minors and suppose that the LC (20.1.2) has only the trivial LC solution $x = 0$. Then the LC (20.1.3) has at least one solution.

Proof.   The path cannot diverge to infinity for $0 < t < 1$, as that would require a horizontal ray. Hence if it does diverge to infinity, it must do so for $t \to +\infty$ or $t \to -\infty$. Yet as mentioned, the path cannot cross $t = 0$ again, so it must go to $t = 1$. $\square$

COPOSITIVE

As an example of when $x = 0$ is the only solution of (20.1.2), suppose that M satisfies

$$x^\tau M x \geq 0, \quad \text{all } x \geq 0. \tag{20.1.6}$$

Then M is said to be a *copositive* matrix. For example, positive-semidefinite matrices are copositive.

Suppose that M is copositive. Let us show that $x = 0$ is the only solution to (20.1.2). From (20.1.2), we have

$$x \geq 0, \quad d + q + Mx \geq 0, \quad x^\tau(d + q + Mx) = 0.$$

Yet since M is copositive, given any $x \geq 0$,

$$0 = x^\tau(d + q) + x^\tau Mx \geq x^\tau(d + q).$$

But since $d + q > 0$, this clearly requires that $x = 0$. We therefore see that if M is copositive, (20.1.2) has $x = 0$ as its unique solution. This result and Theorem 20.1.1 immediately provide:

COROLLARY 20.1.2   Let the LC be regular. Let M have nonzero principal minors and let M be copositive. Then the LC (20.1.3) has at least one solution.

20.2   QUADRATIC PROGRAMS

The requirement that M have nonzero principal minors is interesting but needs to be weakened somewhat. For example, consider the QP (Example 19.1.1). If the objective function is convex, it is known from nonlinear programming that the matrix Q is positive semidefinite. Moreover, then the matrix $M \equiv \begin{pmatrix} Q & A^\tau \\ -A & 0 \end{pmatrix}$ is also positive semidefinite (Exercise 2). Notice, however, that the lower right-hand block matrix of M is a zero matrix, and hence M has a zero principal minor. Theorem 20.1.1 therefore does not apply to the convex QP.

Yet the convex quadratic programming problem is an extremely nice problem and generally has a solution, so we now show all is well and that Lemke's algorithm solves the problem.

COPOSITIVE-PLUS MATRICES

The idea is to replace the nonzero principal minor requirement of Theorem 20.1.1 with a weaker condition:

$$\text{If } x \geq 0 \quad \text{and} \quad x^\tau Mx = 0, \quad \text{then } (M + M^\tau)x = 0. \quad (20.2.1)$$

A copositive matrix M with property (20.2.1) is said to be *copositive-*

plus. Thus a copositive-plus matrix is a square matrix that satisfies both (20.1.6) and (20.2.1). A positive-semidefinite matrix M is copositive; moreover, it satisfies (20.2.1) (Exercise 4). Consequently, the matrices M for convex quadratic programming are copositive-plus matrices.

Suppose that we apply the Lemke algorithm on a copositive-plus M. Starting from $(x^0, t^0) = (0, 0)$, Lemke's algorithm will initiate a PL path. Since M is copositive, as discussed, we know that the PL path cannot return to $t = 0$. However, since M can have zero principal minors, it is possible that we diverge to infinity horizontally. This can occur when for the current system,

$$A^k w + td = d + q$$
$$w \geq 0, \qquad t \in R^1$$
$$B^k = \{u_1, u_2, \ldots, u_n\} \qquad \text{the zero set,} \tag{20.2.2}$$

the matrix A^k has rank $n - 1$. Then for a fixed $t = \bar{t}$, w might diverge to infinity. Geometrically, we would then be diverging to infinity on a horizontal ray where $t = \bar{t}$ is fixed, $0 < \bar{t} < 1$.

What is truly remarkable, however, is that if Lemke's algorithm diverges to infinity, the LC has *no* solution. To prove this, a lemma is required.

LEMMA 20.2.1   Let the LC (20.1.3) be given. Suppose that there exists an $\tilde{x}$ such that

(A) $$\tilde{x} \geq 0, \qquad M\tilde{x} \geq 0$$
(B) $$(M + M^\tau)\tilde{x} = 0$$
(C) $$\tilde{x}^\tau q < 0. \tag{20.2.3}$$

Then the LC (20.1.3) has no solution.

Proof.   Take any $x \geq 0$, $z \equiv q + Mx$. Clearly,

$$\tilde{x}^\tau z = \tilde{x}^\tau q + \tilde{x}^\tau Mx = \tilde{x}^\tau q + x^\tau M^\tau \tilde{x}.$$

By (20.2.3A and B),

$$\tilde{x}^\tau z = \tilde{x}^\tau q - x^\tau M\tilde{x} \leq \tilde{x}^\tau q < 0,$$

the last inequality due to (20.2.3C).

Yet, since $\tilde{x} \geq 0$, we have $z \not\geq 0$. In other words, for any $x \geq 0$, $z \equiv q + Mx$, we have $z \not\geq 0$. But then our LC has no solution, as it requires $x \geq 0$ and $z \geq 0$. Consequently, if (20.2.3) holds, we know that the LC has no solution.   □

Lemma 20.2.1 states that under (20.2.3), the LC (20.1.3) has no solution. The lemma will be used in the following way. Suppose that Lemke's method is applied on a copositive-plus M. Also suppose that if Lemke's method diverges to infinity, an $\tilde{x}$ exists such that (20.2.3) holds. Then the LC cannot have a solution. This fact will help us prove that Lemke's method solves the QP, but first, an algebraic lemma will describe what occurs if the algorithm diverges to infinity on a horizontal ray.

LEMMA 20.2.2 Suppose that for $\bar{t}$ fixed, $0 < \bar{t} < 1$, Lemke's method diverges to infinity. Then points $\bar{x}$ and $\tilde{x}$ are generated such that, if we define

(A) $\qquad\qquad \bar{z} \equiv (1 - \bar{t})d + q + M\bar{x}$

(B) $\qquad\qquad \tilde{z} \equiv M\tilde{x},$ $\qquad\qquad\qquad\qquad$ (20.2.4)

then

(A) $\qquad\qquad \bar{x} \geq 0,\ \tilde{x} \geq 0,\ \bar{z} \geq 0,\ \tilde{z} \geq 0$

(B) $\qquad\qquad \bar{z}^\tau \bar{x} = \bar{z}^\tau \tilde{x} = \tilde{z}^\tau \bar{x} = \tilde{z}^\tau \tilde{x} = 0$

(C) $\qquad\qquad \tilde{x} \neq 0.$ $\qquad\qquad\qquad\qquad$ (20.2.5)

Proof. Suppose that for $\bar{t}$ fixed, $0 < \bar{t} < 1$, Lemke's method diverges to infinity. Then for $t = \bar{t}$, (20.2.2) has a solution $w = \bar{w} + \lambda\tilde{w}$, $\lambda \geq 0$, where $\bar{w} \geq 0$, $\tilde{w} \geq 0$, and $\tilde{w} \neq 0$. This solution is a ray of solutions to (20.2.2), since $\lambda \longrightarrow \infty$ implies that $\|w\| \longrightarrow \infty$.

Since $\lambda \geq 0$ is arbitrary, we must then have

$$A^k \bar{w} = (1 - \bar{t})d + q$$
$$A^k \tilde{w} = 0. \qquad\qquad (20.2.6)$$

Now, let us translate $\bar{w}$ and $\tilde{w}$ back into the original variables z and x. That is, if $w_i \equiv z_i$, then $\bar{z}_i = \bar{w}_i$, $\tilde{z}_i = \tilde{w}_i$, whereas if $w_i \equiv x_i$, then $\bar{x}_i = \bar{w}_i$, $\tilde{x}_i = \tilde{w}_i$. Also for the variables in the zero set B^k, $\bar{x}_i$, $\bar{z}_i$, $\tilde{x}_i$, $\tilde{z}_i$ are all zero.

Then (20.2.6) is the same as

(A) $\qquad\qquad (I, -M)\begin{pmatrix} \bar{z} \\ \bar{x} \end{pmatrix} = (1 - \bar{t})d + q$

$\qquad\qquad\qquad\qquad\qquad\qquad\qquad\qquad$ (20.2.7)

(B) $\qquad\qquad (I, -M)\begin{pmatrix} \tilde{z} \\ \tilde{x} \end{pmatrix} = 0,$

where

$$\bar{z}_i\bar{x}_i = \bar{z}_i\tilde{x}_i = \tilde{z}_i\bar{x}_i = \tilde{z}_i\tilde{x}_i = 0 \qquad \text{for } i = 1, \ldots, n. \qquad (20.2.8)$$

Finally, since $\tilde{w} \neq 0$, we have $\tilde{x} \neq 0$; otherwise, (20.2.7B) implies that $\tilde{z} = \tilde{x} = 0$ or $\tilde{w} = 0$. This proves the lemma. $\qquad \square$

We are now ready to establish:

THEOREM 20.2.3 Let M be copositive-plus and assume that the LC is regular. Then either the Lemke algorithm finds an LC solution, or if the algorithm diverges to infinity, it demonstrates that the LC has no solution.

Proof. Because M is copositive, the trivial system (20.1.2) has only one solution at $t = 0$. Thus either Lemke's algorithm terminates at $t = 1$ with an LC solution, or it diverges to infinity for a fixed $t = \bar{t}$, $0 < \bar{t} < 1$. Suppose that it diverges to infinity. Then for $\bar{t}$ fixed, $0 < \bar{t} < 1$, there exist $\bar{x}$ and $\tilde{x}$ such that (20.2.4) and (20.2.5) hold by Lemma 20.2.2. From (20.2.4) and (20.2.5), $M\tilde{x} \geq 0$. Also, $\tilde{x} \geq 0$ and $\tilde{x}^\tau M\tilde{x} = \tilde{x}^\tau\tilde{z} = 0$, which implies, via (20.2.1), that

$$(M + M^\tau)\tilde{x} = 0.$$

We have exhibited an $\tilde{x}$ such that (20.2.3A and B) hold. It remains only to demonstrate that (20.2.3C) holds and we are done.

Premultiply (20.2.4A) by $\tilde{x}$,

$$\tilde{x}^\tau q = (\bar{t} - 1)\tilde{x}^\tau d + \tilde{x}^\tau \bar{z} - \tilde{x}^\tau M\bar{x}. \qquad (20.2.9)$$

Via (20.2.5B), $\tilde{x}^\tau\bar{z} = 0$. Also, from (20.2.3B), (20.2.4B), and (20.2.5B),

$$-\tilde{x}^\tau M\bar{x} = +\bar{x}^\tau M\tilde{x} = \bar{x}^\tau\tilde{z} = 0.$$

Hence (20.2.9) reduces to

$$\tilde{x}^\tau q = (\bar{t} - 1)\tilde{x}^\tau d < 0,$$

since $\bar{t} < 1$, $\tilde{x} \geq 0$, $\tilde{x} \neq 0$, and $d > 0$. Therefore, Lemma 20.2.1 ensures that if Lemke's algorithm diverges to infinity, no solution exists.

We conclude that Lemke's algorithm either finds a solution or verifies that no solution exists by diverging to infinity. $\qquad \square$

THE QP EXAMPLE

Theorem 20.2.3 immediately applies to the convex QP example because, as mentioned, the matrix Q is then positive semidefinite and so the matrix

$M \equiv \begin{pmatrix} Q & A^\tau \\ -A & 0 \end{pmatrix}$ is also positive semidefinite. Specifically, suppose that Lemke's algorithm finds an LC solution for (20.1.3). This then corresponds to the minimum point of the QP, since (20.1.3) are simply the Kuhn–Tucker conditions. Otherwise, if Lemke's algorithm diverges to infinity, the LC has no solution, and hence the QP has no minimum point. The Lemke algorithm thus tells us precisely whether the convex quadratic programming problem has a solution or not.

20.3   BIMATRIX GAMES

The LC problem corresponding to the bimatrix game problem was formulated in Example 19.1.2. This LC, recall, had the form

$$\begin{pmatrix} z^1 \\ z^2 \end{pmatrix} = \begin{pmatrix} -e \\ -e \end{pmatrix} + \begin{pmatrix} 0 & -A^1 \\ -(A^2)^\tau & 0 \end{pmatrix} \begin{pmatrix} x^1 \\ x^2 \end{pmatrix} \tag{20.3.1}$$

$$z^1 \geq 0, \qquad z^2 \geq 0, \qquad x^1 \geq 0, \qquad x^2 \geq 0$$

$$(z^1)^\tau x^1 = (z^2)^\tau x^2 = 0,$$

where $A^i < 0$ is $n_1 \times n_2$, $i = 1, 2$.

We have already shown in Chapter 8, that a Nash equilibrium point exists, and hence the equivalent LC formulation (20.3.1) must have a solution. Unfortunately, Lemke's method cannot be employed to find that solution. Although the matrix $M \equiv \begin{pmatrix} 0 & -A^1 \\ -(A^2)^\tau & 0 \end{pmatrix}$ is a copositive matrix (Exercise 5), it has zero principal minors, so that Corollary 20.1.2 is inapplicable. Also, the matrix M is not copositive-plus; hence the theory of Section 20.2 does not apply either.

Fortunately, there is a different method for finding the LC solution of (20.3.1), the *Lemke–Howson* method. It not only predates the Lemke approach historically but is designed especially to solve (20.3.1).

THE SYSTEM OF EQUATIONS

Given the LC (20.3.1), define

$$F_j(x^1, x^2) = \min \{x_j^1, z_j^1\} = 0, \qquad j = 1, \ldots, n_1$$

$$F_{n_1 + j}(x^1, x^2) = \min \{x_j^2, z_j^2\} = 0, \qquad j = 1, \ldots, n_2. \tag{20.3.2}$$

System (20.3.2) consists of n equations and n variables (x^1, x^2), where $n = n_1 + n_2$. As we have seen earlier, a solution to (20.3.2) will be in one-to-one correspondence with an LC solution to (20.3.1).

THE LEMKE–HOWSON APPROACH

Because $A^i < 0$, $i = 1, 2$, it is not necessary to set up a homotopy system by introducing a vector $d > 0$ and a parameter t. Rather, the Lemke–Howson approach operates on the system (20.3.2) directly.

The method is initiated at an $\mathring{x}$ satisfying

$$F_1(\mathring{x}^1, \mathring{x}^2) \geq 0$$
$$F_j(\mathring{x}^1, \mathring{x}^2) = 0, \qquad j = 2, \ldots, n.$$

Starting from $\mathring{x}$, a PL path is generated that satisfies

$$F_1(x) \geq 0$$
$$F_j(x) = 0, \qquad j = 2, \ldots, n. \tag{20.3.3}$$

for all points x along the path. The method is terminated as soon as $F_1(x)$ becomes zero, or when x diverges to infinity. Since $F_j(x) = 0$, $j = 2, \ldots, n$, is maintained throughout, when F_1 becomes zero, an LC solution to (20.3.1) is achieved.

Observe that the PL path is formed by relaxing the first component F_1 to $F_1 \geq 0$. The remaining equations

$$F_2 = 0, \ldots, F_n = 0$$

become a system of $n - 1$ equations and n variables. Since there is one more variable than equation, the solution to this smaller system will yield the PL path.

REGULARITY

Just as in the Lemke method, to avoid degeneracy we assume a regularity condition; specifically, suppose that

> For each x satisfying (20.3.3),
> at least n of the variables x, z are positive. $\qquad$ (20.3.4)

Since $F_j(x) = 0$, $j = 2, \ldots, n$, must hold, at least $n - 1$ of the variables x, z are zero. Hence, by (20.3.4), $n - 1$ or n variables are zero, and $n + 1$ or n variables are positive.

THE INITIAL POINT

As for the initial point $\mathring{x}$, let

$$A^2_{1s} = \max_j \{A^2_{1j}\}, \tag{20.3.5}$$

where A^2 is the $n_1 \times n_2$ matrix of (20.3.1), and specify

(A)
$$\dot{x}_1^1 = -\frac{1}{A_{1s}^2} > 0$$

(B)
$$\dot{x}_j^1 = 0, \quad j = 2, \ldots, n_1. \tag{20.3.6}$$

Note that for $\dot{x}^1$ we have, from (20.3.1),

(A)
$$\dot{z}_s^2 = 0$$

(B)
$$\dot{z}_j^2 \geq 0, \quad j = 1, \ldots, s-1, s+1, \ldots, n_2. \tag{20.3.7}$$

Next call

$$A_{rs}^1 = \max_i \{A_{is}^1\},$$

where A^1 is the $n_1 \times n_2$ matrix of (20.3.1), and let

(A)
$$\dot{x}_s^2 = -\frac{1}{A_{rs}^1} > 0$$

(B)
$$\dot{x}_j^2 = 0, \quad j = 1, \ldots, s-1, s+1, \ldots, n_2. \tag{20.3.8}$$

For $\dot{x}^2$ satisfying (20.3.8), it is easily seen that

(A)
$$\dot{z}_r^1 = 0$$

(B)
$$\dot{z}_j^1 \geq 0, \quad j = 1, \ldots, r-1, r+1, \ldots, n_1. \tag{20.3.9}$$

The initial point for the Lemke–Howson method is now provided. Summarized,

$$\dot{x}_1^1 = -\frac{1}{A_{1s}^2} > 0, \quad \dot{x}_j^1 = 0, \quad j = 2, \ldots, n_1,$$

$$\dot{x}_s^2 = -\frac{1}{A_{rs}^1} > 0, \quad \dot{x}_j^2 = 0, \quad j = 1, \ldots, s-1, s+1, \ldots, n_2. \tag{20.3.10}$$

Also,

$$\dot{z}_r^1 = 0, \quad \dot{z}_j^1 \geq 0, \quad j = 1, \ldots, r-1, r+1, \ldots, n_1$$
$$\dot{z}_s^2 = 0, \quad \dot{z}_j^2 \geq 0, \quad j = 1, \ldots, s-1, s+1, \ldots, n_2.$$

From (20.3.2), certainly $F_1(\dot{x}) \geq 0$. Moreover, $F_j(\dot{x}) = 0, j = 2, \ldots, n$. In fact, if $r = 1$, then $F_1(\dot{x}) = 0$, so that $\dot{x}$ is an LC solution of (20.3.1), and we are done.

THE ZERO SET
AND THE DISTINGUISHED VARIABLE

From the initial point $\dot{x}$ we will generate a PL path that satisfies (20.3.3) for all points x along the path. Moreover, the Lemke–Howson procedure will be seen to be very much like the Lemke procedure.

Of course, if $r = 1$, then $\dot{x}$ is an LC solution and we are done. Hence suppose that $r \geq 2$.

Initially, at $\dot{x}$ the *zero set* is defined to be

$$B^0 = \{x_2^1, \ldots, x_{r-1}^1, z_r^1, x_{r+1}^1, \ldots, x_{n_1}^1, x_1^2, \ldots, x_{s-1}^2, z_s^2, x_{s+1}^2, \ldots, x_{n_2}^2\} \quad (20.3.11)$$

and the *distinguished variable* is

$$x_r^1.$$

Note that B^0 has $n - 1$ elements. At the point $\dot{x}$ all variables in the zero set are zero, and the distinguished variable is also zero.

More generally, consider the typical step k. A zero set

$$B^k \doteq \{u_2, \ldots, u_n\}, \quad (20.3.12)$$

where u_j is either x_j or z_j, will be given. Also, there will be a distinguished variable w_l, $l \geq 2$. At the current point, all variables in B^k will be zero, with the distinguished variable w_l also currently zero. All other variables are strictly positive by regularity (20.3.4).

Just as in the Lemke step, increase the distinguished variable w_l from zero. All $n - 1$ variables in B^k are kept at zero. Then either w_l goes to infinity, in which case the algorithm is terminated, or some variable not in B^k becomes zero.

Suppose that some variable, say w_j, not in B^k becomes zero. By regularity (20.3.4), w_j is the unique variable that becomes zero. If w_j is x_1^1 or z_1^1, then an LC solution to (20.3.1) is obtained and we terminate. If w_j is neither x_1^1 nor z_1^1, let

$$B^{k+1} = \{u_2, \ldots, u_{j-1}, w_j, u_{j+1}, \ldots, u_n\} \quad (20.3.13)$$

and let u_j be the distinguished variable, where u_j is the complement of w_j. Thus the variable that hits zero goes into the zero set, and its complement becomes the distinguished variable. We then proceed to step $k + 1$.

Observe that, throughout, the points x generated satisfy (20.3.3). Also, the new distinguished variable u_j at step $k + 1$ was zero throughout step k. At step $k + 1$, u_j will then be increased from zero, while all variables in B^{k+1} will be kept at zero.

In summary, the Lemke–Howson method is as follows.

Initialization

Construct $\dot{x}$ as defined by (20.3.10). If $r = 1$, then $\dot{x}$ is an LC solution. If not, let B^0 be defined by (20.3.11) and let the distinguished variable be x_r^1. Go to step 0.

Step k, k ≥ 0

Let the zero set be

$$B^k = \{u_2, \ldots, u_n\}$$

and the distinguished variable be w_l, $l \geq 2$.
Increase w_l from zero.

(A)   If w_l diverges to infinity, terminate.

(B)   Otherwise, some unique w_j becomes zero.
     (B.1)   If w_j is either x_1^1 or z_1^1, terminate with an LC solution.
     (B.2)   If not, let

$$B^{k+1} = \{u_2, \ldots, u_{j-1}, w_j, u_{j+1}, \ldots, u_n\}$$

     and the distinguished variable be u_j. Go to step $k + 1$.

The PL Path Cannot Be a Loop

The PL path generated from $\dot{x}$ cannot be a loop. This is because there exists a ray starting from $\dot{x}$ that satisfies (20.3.3) for all points in the ray. To see this, with (20.3.10) satisfied, suppose that $\dot{x}_s^2$ is increased from $\dot{x}_s^2 = -1/A_{rs}^1$ to $x_s^2 \rightarrow \infty$. Then z_j^1, for all $j = 1, 2, \ldots, n_1$, will increase to $z_j^1 = \infty$. Hence, for x satisfying (20.3.10) since

$$x_s^2 \geq -\frac{1}{A_{rs}^1} \qquad \text{and} \qquad x_s^2 \longrightarrow \infty, \qquad (20.3.14)$$

system (20.3.3) is satisfied, and in fact $z_j^1 \rightarrow \infty$, for all j.

We therefore see that the PL path can be thought of as starting from "$x_s^2 = \infty$," and hence cannot be a loop.

EXISTENCE OF A UNIQUE RAY

We have just shown that there is a ray (20.3.14) that satisfies (20.3.3) for all points in the ray. Let us now prove that, in fact, (20.3.14) is the only ray that satisfies (20.3.3).

LEMMA 20.3.1   There is a unique ray satisfying (20.3.3), namely the ray (20.3.14).

Proof.   Consider any ray satisfying (20.3.3). On this ray some x_j^i goes to infinity. [Otherwise, if all x_j^i are bounded, then all z_j^i are bounded by (20.3.1), so there is no ray.]

If x_k^1 for some k goes to infinity on a ray satisfying (20.3.3), then $z_j^2 \longrightarrow \infty$ for all $j = 1, \ldots, n_2$. Hence since (20.3.3) holds, we have $x^2 = 0$. This implies that $z^1 = -e$, a contradiction of (20.3.3).

Thus x_k^1 must be bounded for all $k = 1, \ldots, n_1$, and x_k^2 goes to infinity along the ray, for some $k = 1, \ldots, n_2$. Since $x_k^2 \longrightarrow \infty$ for some k, $z_j^1 \longrightarrow \infty$ for all $j = 1, \ldots, n_1$. By (20.3.3),

$$x_j^1 = 0, \qquad j = 2, \ldots, n_1. \tag{20.3.15}$$

Further, from (20.3.15) and (20.3.3),

$$x_1^1 = -\frac{1}{A_{1s}^2} \tag{20.3.16}$$

since if $x_1^1 < -1/A_{1s}^2$, we have $z_s^2 < 0$, and if $x_1^1 > -1/A_{1s}^2$, we have $z^2 > 0$, implying $x^2 = 0$, a contradiction.

Equations (20.3.15) and (20.3.16) establish that $x^1 = \mathring{x}^1$ along the ray. Moreover, since $x^1 = \mathring{x}^1$, recall from (20.3.7) that $z^2 = \mathring{z}^2$, where by regularity (20.3.4),

$$\mathring{z}_s^2 = 0, \qquad \mathring{z}_j^2 > 0, \qquad j = 1, \ldots, s-1, s+1, \ldots, n_2. \tag{20.3.17}$$

Yet (20.3.17) and (20.3.3) imply that

$$x_j^2 = 0, \qquad j = 1, \ldots, s-1, s+1, \ldots, n_2. \tag{20.3.18}$$

Hence along the ray, (20.3.15) to (20.3.18) hold and therefore it must be that $x_s^2 \longrightarrow \infty$. Indeed, note that this is the ray (20.3.14). Thus (20.3.14) is the unique ray satisfying (20.3.3). $\square$

Using the lemma, we can now show that the Lemke–Howson method will obtain an LC solution without fail.

THE LEMKE–HOWSON METHOD
OBTAINS AN LC SOLUTION

If the Lemke–Howson method is applied on the LC (20.3.1), we know that the method either diverges to infinity or an LC solution is obtained. In fact, however, the Lemke–Howson method *cannot diverge to infinity*. If it did, a ray different from the initial ray (20.3.14) would have to be generated. The new ray would satisfy (20.3.3) since all points on the PL path satisfy (20.3.3). But this is impossible since Lemma 20.3.1 states that the ray (20.3.14) is the unique ray satisfying (20.3.3).

We therefore have obtained the result.

THEOREM 20.3.2 Let the LC (20.3.1) be given and assume that the regularity condition (20.3.4) holds. Then the Lemke–Howson method will generate an LC solution to (20.3.1).

SUMMARY

This chapter showed how the Lemke method is employed for finding an LC solution. In the specific case of a convex quadratic program, we proved that the method finds the minimum point or shows that it does not exist. We then described the Lemke–Howson method for finding the Nash equilibrium point of the bimatrix game. Thus we see that the Lemke and Lemke–Howson methods are powerful procedures; moreover, historically they form the basis of path-following theory.

EXERCISES/CHAPTER 20

1. Suppose that the LC is regular. Prove that the path from $x = 0$, $t = 0$ must increase at least a little from $t = 0$.

2. Consider an objective function

$$c^\tau x + \tfrac{1}{2} x^\tau Q x$$

   where Q is $n \times n$.
   a. If Q is not symmetric, show how the objective function can be transformed to an equivalent objective $c^\tau x + \tfrac{1}{2} x^\tau \bar{Q} x$, where $\bar{Q}$ is symmetric.
   b. Suppose that Q is symmetric. Show that if the objective function is a convex function, then Q is positive semidefinite.
   c. If Q is positive semidefinite, show that

$$M \equiv \begin{pmatrix} Q & A^\tau \\ -A & 0 \end{pmatrix}$$

   is positive semidefinite.

3. Prove that a matrix M has positive principal minors if and only if there is a unique LC solution to (20.1.3) for all $q \in R^n$.

4. Prove that a positive-semidefinite matrix is copositive-plus.

5. Prove that the matrix M of (20.3.1) is a copositive matrix but need not be copositive-plus.

6. Consider the bimatrix game where the payoff matrices are

$$A^1 = \begin{pmatrix} 0 & 2 & 1 \\ 1 & 0 & 2 \end{pmatrix}, \qquad A^2 = \begin{pmatrix} -3 & -3 & -4 \\ -4 & -3 & -3 \end{pmatrix}.$$

   a. Set up the LC corresponding to this game.
   b. Solve the LC by the Lemke–Howson method.
   c. Determine the Nash equilibrium point and the payoff to each player for the bimatrix game.

7. Consider the linear program

$$\min c^\tau x$$
$$Ax \le b$$
$$x \ge 0. \qquad \text{(E.20.1)}$$

   Suppose that the constraint set of (E.20.1) is bounded and x^0 is a point satisfying

$$Ax^0 < b, \qquad x^0 > 0.$$

   Suppose that we set up the quadratic program

$$\min \frac{1-t}{2} \|x - x^0\|^2 + tc^\tau x$$
$$Ax \le b$$
$$x \ge 0. \qquad \text{(E.20.2)}$$

   a. Set up the LC corresponding to (E.20.2), for t given.
   b. Describe a procedure for solving the LC, for $0 \le t \le 1$.
   c. Prove that the LC solutions generated will be monotone in t.

8. Let

$$q = \begin{pmatrix} 5 \\ -9 \\ 3 \\ -5 \end{pmatrix}, \qquad M = \begin{pmatrix} 1 & -1 & -1 & -1 \\ 1 & 0 & 1 & 2 \\ -1 & -1 & 1 & -1 \\ 1 & 2 & 1 & 0 \end{pmatrix}.$$

   Solve the LC by Lemke's method.

9. Given M, suppose that for some $d > 0$ and for $k = 0, 1$, the LC

$$z = kd + Mx$$
$$z \ge 0, \qquad x \ge 0, \qquad z^\tau x = 0$$

   has only the trivial solution $x = 0$. Prove that the LC (20.1.3) has a solution for any q.

10. Construct an LC of dimension $n = 3$ that has an even number of solutions.

11. Write the LC corresponding to

$$\max c^\tau x + \tfrac{1}{2} x^\tau Q x$$
$$x \ge 0.$$

12. Assuming the hypothesis of Theorem 20.1.1, prove that the number of LC solutions is odd.

NOTES

The original Lemke–Howson algorithm of Section 20.3 can be found in Lemke and Howson [1964]. Lemke [1968] proved that if M is copositive-plus, then Lemke's method finds an LC solution or shows that no solution exists. Murty [1971] demonstrated that the LC has a unique solution for every q if and only if M is a P-matrix. Eaves [1971c] and Garcia [1973] showed classes of matrices processed by Lemke's method. Orientation in LC problems was introduced by Shapley [1974]. Generalizations were made by Lemke and Grotzinger [1976], Eaves and Scarf [1976], and Todd [1976b]. Cottle and Dantzig [1968] have shown that the principal pivoting method finds an LC solution if M is a P-matrix and q is arbitrary; also see Cottle [1968] and Ingleton [1966]. Kelly and Watson [1979] use spherical geometry to certain classes of matrices M. More recent results include those of Aganacic and Cottle [1979], Cottle [1980], Cottle and von Randow [1979], Garcia and Gould [1980b], van der Heyden [1980], Howe [1980], Kaneko [1979], Kojima and Saigal [1979], and Pang [1979].

The LC problem has also been generalized to nonlinear and other forms. See Cottle and Dantzig [1970], Cottle, Habetler, and Lemke [1967], Habetler and Kostreva [1980], Habetler and Price [1971], Karamardian [1969, 1972] Lemke [1970], Megiddo and Kojima [1977], Moré [1974], and Saigal and Simon [1973].

Algorithms for the nonlinear complementarity problem are given by Fisher and Gould [1974], Fisher and Tolle [1977], Habetler and Kostreva [1978], Kojima [1974], Kostreva [1976], and Lüthi [1976], among others.

21

The Kakutani Theorem and the Economic Equilibrium Revisited

Many operations are well-modeled by a function, which means that the process takes a point and yields a new point. Frequently, however, the operation is not a function. Instead of a point, the operation yields a set and we have what is called a point-to-set map. This chapter delves into point-to-set maps and their properties. First, by introducing the idea of upper hemicontinuity, their continuity is established. Since point-to-set maps are not functions, that very essential fixed-point theorem, the Brouwer, no longer holds. Luckily, the upper-hemicontinuity idea lets us establish a new fixed-point theorem, the Kakutani theorem.

Finally, by orchestrating all these concepts simultaneously, we reprove the existence of the economic equilibrium point. The proof is of the classical variety usually used to establish equilibrium and is somewhat more technical than our previous demonstration in Chapter 7. It nevertheless yields important insights, and equilibrium is so fundamental that viewing it from two different perspectives is itself beneficial.

21.1 POINT-TO-SET MAPS

Up to now in our discussion we have dealt with functions, that is, operations which given a point x, yield another point, $F(x)$. By definition of a function, for any x the function value $F(x)$ must be unique. For instance, let $F: R^1 \longrightarrow R^1$ be defined by $F(x) = x^2$. Then if $x = 3$, there must be a unique value $F(x) =$

9. If $x = 5$, then $F(x) = 25$ uniquely. For any x there can be only one single value for $F(x)$. That is demanded by definition of a function.

In many circumstances, however, requiring F to be a function is too stringent because $F(x)$ is not unique but is an entire set. Examine:

1. $F(x) = \{u \mid x \le u \le x + 3\}$
2. $F(x) = \{\sqrt{x}\} \cup \{\sqrt{x + 1}\}$
3. $F(x) = \{x\}$
4. $F(x_1, x_2) = \{(u_1, u_2) \mid x_1 \le u_1, x_2 \le u_2\}$

For 1 to 3, $x \in R^1$ and $F(x) \subset R^1$; for 4, $x \in R^2$ and $F(x) \subset R^2$. Notice in 3 that $F(x)$ is a single point, whereas in 2, the set $F(x)$ may be empty.

These examples introduce the concept of a point-to-set map. For $D \subset R^n$, $F: D \longrightarrow R^l$ is a *point-to-set* map if it takes points $x \in D$ and yields a set $F(x) \subset R^l$.

As in example 3, $F(x)$ could be a set consisting of a single point. That is permissible because it is required only that $F(x)$ be some set in R^l, and a set consisting of one point suffices. Also, we denote the points u in the set $F(x)$ by $u \in F(x)$.

With many applications the operation $F(x)$ is not unique and we must let F be a point-to-set map. For example, in game theory, given a strategy x, the preferences of a person often are not unique, but instead form an entire set. Concave functions may not be differentiable, but only subdifferentiable. Then at any x the subdifferentials may form a set. Also consider maximizing a function $f(x)$. The optimal point may not be unique, but instead there might be an entire set of optimal points.

In short, many operations on x produce an entire set $F(x)$. Then F does not satisfy the definition of a function and we must define it as a point-to-set map.

UPPER HEMICONTINUITY

Given a point-to-set map F, to study it requires that we know how $F(x)$ varies as x varies. As x changes, $F(x)$ usually changes also, but the difficulty is that $F(x)$ could jump all over the place. Such excessive wildness of $F(x)$ can be limited by an appropriate smoothness condition, and let us now examine that. Essentially, we will develop a property for a point-to-set map that is analogous to the continuity property for a function.

Given a sequence $\{x^k\}$, consider the corresponding sequence $\{u^k\}$ formed by taking

$$u^k \in F(x^k) \qquad \text{all } k. \tag{21.1.1}$$

[Here u^k can be any point in the set $F(x^k)$. All that is required is that for each x^k we select u^k as some point in $F(x^k)$.]

Now suppose that both sequences converge; that is,

$$x^k \longrightarrow \bar{x} \qquad \text{as } k \longrightarrow \infty$$

and

$$u^k \longrightarrow \bar{u} \qquad \text{as } k \longrightarrow \infty.$$

The question arises: Is it true that

$$\bar{u} \in F(\bar{x})? \tag{21.1.2}$$

In general, it need not be. For example, suppose that

$$F(x) = \begin{cases} \{u \mid 1 \leq u \leq 2\}, & x \neq 1 \\ \{4\}, & x = 1. \end{cases} \tag{21.1.3}$$

Let

$$x^k = 1 - \frac{1}{k}.$$

Then for $u^k \in F(x^k)$ we may select $u^k = \frac{3}{2}$. Here $x^k \to 1 = \bar{x}$ and $u^k \to \frac{3}{2} = \bar{u}$. But

$$\frac{3}{2} \notin F(1) = \{4\}.$$

On the other hand, suppose that

$$F(x) = \begin{cases} \{u \mid 1 \leq u \leq 2\}, & x \neq 1 \\ \{u \mid 0 \leq u \leq 3\}, & x = 1. \end{cases} \tag{21.1.4}$$

Then, again, let $x^k = 1 - 1/k$ and $u^k = \frac{3}{2}$, so $\bar{x} = 1$ and $\bar{u} = \frac{3}{2}$. But now

$$\bar{u} \in F(\bar{x}).$$

In short, suppose that $x^k \to \bar{x}$ and $u^k \to \bar{u}$, where $u^k \in F(x^k)$. Then depending on F, condition (21.1.2) may or may not hold. (Figure 21.1.1 provides a graphical interpretation.)

Condition (21.1.2) is quite basic and if it holds, F is called *upper hemicontinuous* (uhc). Specifically, for $D \subset R^n$, a point-to-set map $F: D \to R^l$ is upper hemicontinuous at $\bar{x} \in D$ if wherever

$$x^k \longrightarrow \bar{x} \qquad \text{and} \qquad u^k \longrightarrow \bar{u},$$

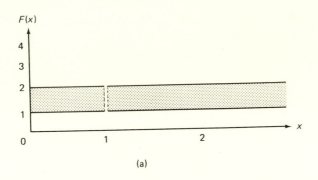

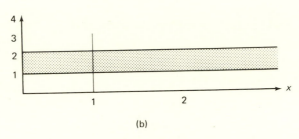

The set $F(x)$ is the vertical area indicated above x. Part (a) depicts (21.1.3), and (b) depicts (21.1.4).

Figure 21.1.1

where

$$u^k \in F(x^k),$$

we have

$$\bar{u} \in F(\bar{x}).$$

In words, for $x^k \longrightarrow \bar{x}$, select $u^k \in F(x^k)$ such that $u^k \longrightarrow \bar{u}$. Check if $\bar{u} \in F(\bar{x})$. If this holds, then F is uhc at $\bar{x}$. Note that the result must hold for all such sequences. This means that for any sequence $x^k \longrightarrow \bar{x}$ and any sequence $u^k \longrightarrow \bar{u}$, where $u^k \in F(x^k)$, we must have $\bar{u} \in F(\bar{x})$.

The map F is called uhc on D if it is uhc at all $x \in D$. The upper-hemi-continuity condition is sometimes referred to as upper semicontinuity. Since the latter term is also employed for totally unrelated properties and may be confusing, it will not be utilized here. Intuitively, F is uhc at $\bar{x}$ if the set $F(\bar{x})$ is big enough to "catch" all limit points $\bar{u}$, where $u^k \longrightarrow \bar{u}$ (see Figure 21.1.2).

Let us now examine some common instances where uhc arises.

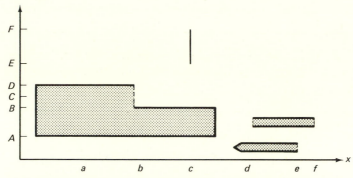

The point-to-set map has $F(x)$ as sets in R^1. All indicated points above any particular x constitute the set $F(x)$. For example, $F(a) = \{u \mid A \leq u \leq D\}$, $F(b) = \{u \mid A \leq u \leq B\}$. The dashed line indicates that the boundary line is not included. Also, $F(c) = \{u \mid A \leq u \leq B\} \cup \{u \mid E \leq u \leq F\}$.

The map is uhc at all points except b, d, e, and f. For example, at b consider a sequence of points $b^k \to b$, where the b^k are to the left of b. Let $u^k = C \in F(b^k)$ for all k. Then $u^k \to \bar{u} = C \notin F(b)$.

Observe that F is uhc at c even though part of $F(c)$ is the single line segment $\{u \mid E \leq u \leq F\}$. Letting $c^k \to c$, where $c^k \neq c$, $F(c^k)$, does not include $\{u \mid E \leq u \leq F\}$. However, for any $u^k \in F(c^k)$ where $u^k \to \bar{u}, \bar{u} \in \{u \mid A \leq u \leq B\}$, so $\bar{u} \in F(c)$. The uhc at point c illustrates that under uhc the set $F(x)$ can suddenly become bigger.

Figure 21.1.2

MAXIMIZATION AND UHC

A frequent application of uhc is to optimization problems. Letting f: $R^{n_1} \times R^{n_2} \longrightarrow R^1$ and $V \subset R^{n_2}$, consider the problem

$$\max_y f(x, y)$$

$$y \in V. \tag{21.1.5}$$

Here V is the feasible region and we are optimizing over $y \in V$ for x fixed. The variable x may be considered a parameter. We are thus solving (21.1.5) for the optimal point y^* as a "function" of the parameter x.

Our interest is in how the optimal point y^* changes as we vary x. However, as noted, the optimizing point may not always be unique and thus might not be a function of x. Consequently, define a point-to-set map $R^{n_1} \longrightarrow R^{n_2}$ as follows:

$$F(x) = \{y^* \mid y^* \text{ optimizes } (21.1.5) \text{ given } x\}. \tag{21.1.6}$$

At any x, $F(x)$ is the set of points y^* that optimize the problem for that x. Clearly, F is a point-to-set map and generally will change as x changes.

It turns out that under extremely weak assumptions the map F is uhc.

THEOREM 21.1.1   In the optimization problem (21.1.5), suppose that V is closed and f is continuous. Then $F(x)$, the set of optimizing points, is a uhc map.

Proof.   Let $x^k \longrightarrow \bar{x}$ where

$$y^k \in F(x^k), \qquad \text{all } k$$

and suppose that

$$y^k \longrightarrow \bar{y}.$$

Here y^k optimizes (21.1.5) given x^k. We must prove that

$$\bar{y} \in F(\bar{x}).$$

Verbally, we must prove that $\bar{y}$ optimizes the problem given $\bar{x}$.

Quickly notice that since V is closed, $\bar{y} \in V$, so we know that $\bar{y}$ is a feasible point.

Now take an arbitrary point $y \in V$. Since y is feasible and y^k is optimal given x^k,

$$f(x^k, y^k) \geq f(x^k, y).$$

Taking limits and via continuity,

$$f(\bar{x}, \bar{y}) \geq f(\bar{x}, y).$$

But this means that $\bar{y}$ yields a value of the objective at least as high as any feasible point y. Since $\bar{y}$ is feasible, it must be optimal. Thus

$$\bar{y} \in F(\bar{x}). \qquad \square$$

In brief, as long as the objective function is continuous and the feasible set is closed, the set of optimizing points $F(x)$ varies "nicely" as x changes.

Concavity Added

Often in practice the set V is convex. Also, for x fixed, $f(x, y)$ is concave in y. It then turns out that the set of optimizing points $F(x)$ is itself convex. That is, for x fixed, $F(x)$ is a convex set.

LEMMA 21.1.2   For the optimization problem (21.1.5), suppose that V is convex and f is concave in y for x fixed. Then the set of optimizing points $F(x)$ is a convex set for any x.

Proof.   Let $y^1 \in F(x)$ and $y^2 \in F(x)$, so that both y^1 and y^2 are optimal for the problem given x. Obviously,

$$f(x, y^1) = f(x, y^2). \tag{21.1.7}$$

We must prove, defining

$$\tilde{y} = (1 - \lambda)y^1 + \lambda y^2 \qquad \text{for } 0 \leq \lambda \leq 1,$$

that

$$\tilde{y} \in F(x).$$

Because V is convex, $\tilde{y} \in V$, so $\tilde{y}$ is feasible. By concavity of $f(x, y)$ in y for x fixed,

$$f(x, \tilde{y}) \geq (1 - \lambda)f(x, y^1) + \lambda f(x, y^2)$$

and, using (21.1.7),

$$f(x, \tilde{y}) \geq f(x, y^1) = f(x, y^2).$$

Thus $\tilde{y}$ must also optimize and

$$\tilde{y} \in F(x). \qquad \square$$

A map F that is a convex set $F(x)$ for all x is termed *convex-valued*.

With some key properties of point-to-set maps and uhc introduced, we will establish, in the next section, a very general fixed-point theorem.

21.2   THE KAKUTANI THEOREM

Point-to-set maps occur quite frequently, and it is therefore important to understand their fixed points and when they arise. For continuous functions, of course, the Brouwer theorem provides fixed points. Yet point-to-set maps are not functions and Brouwer's theorem cannot be applied. The Kakutani theorem, fortunately, will come to our rescue and provide the fixed point.

The Kakutani theorem is widely utilized, and in fact later in this chapter it will be used to establish economic equilibria. Also, its proof relies heavily upon the triangulation algorithms of Chapter 13, which the reader is strongly urged to reread.

FIXED POINT DEFINED

First, let us define a fixed point for point-to-set maps. Given $D \subset R^n$, consider a point-to-set map $F \colon D \longrightarrow D$. (Here F takes points in D into sets in D, that is $F(x) \subset D$ for all $x \in D$.) The map F has a fixed point at $\bar{x}$ if

$$\bar{x} \in F(\bar{x}). \tag{21.2.1}$$

A point $\bar{x}$ is a fixed point if $\bar{x}$ is in the set $F(\bar{x})$.

EXAMPLE 21.2.1

Suppose that $D = \{u \,|\, 0 \le x \le 10\}$ and

$$F(x) = \begin{cases} \{y \,|\, 10 - x \le u \le 10\}, & 0 \le x \le 6 \\ \{u \,|\, \quad 0 \le u \le 10 - x\}, & 6 < x \le 10. \end{cases}$$

Then any $\bar{x}$, $5 \le \bar{x} \le 6$, is a fixed point (see Figure 21.2.1).

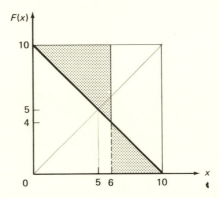

The point to set map $F(x)$ for Example 21.2.1. At any x the set $F(x)$ is the shaded portion above x. Here any $\bar{x}$, $5 \le \bar{x} \le 6$, satisfies $\bar{x} \in F(\bar{x})$. Thus these $\bar{x}$ are the fixed points for this point-to-set map.

Figure 21.2.1

THE APPROACH

To proceed with the Kakutani theorem, suppose that $D \subset R^n$ is a non-empty, compact, convex set. We assume that D has an interior point; otherwise, because D is convex, we can easily reduce the dimension of the problem so that it has an interior (Exercise 6). Further suppose that the set $F(x)$ is nonnull for all $x \in D$.

The proof of the Kakutani theorem will be shown by following a path in a triangulation, as was done with triangulation algorithms. From Chapter 13, recall there will be a labeling function G and a piecewise-linear path will connect n-simplices [facets of $(n + 1)$-simplices]. Each n-simplex $\{v^i\}_0^n$ on the path will have the property that

$$\sum_{i=0}^{n} \lambda_i G(v^i) = 0, \qquad \sum_{0}^{n} \lambda_i = 1, \qquad \lambda_i \ge 0.$$

To construct the triangulation, first select a point x^0 interior to D. Then triangulate $R^n \times [0, 1]$, but ensure that $(x^0, 0)$ is interior to an n-simplex of $R^n \times \{0\}$. This particular simplex will turn out to be the starting simplex. Require the starting simplex to be interior to D and let the simplices get smaller and smaller as t increases until as $t \longrightarrow 1$ they get extremely small (see Figure 21.2.2).

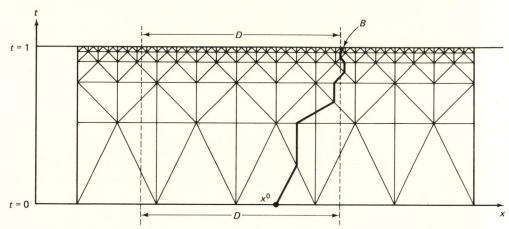

A triangulation for the Kakutani theorem. The set D is indicated. Notice that the triangulation extends beyond $D \times [0, 1]$. The point $(x^0, 0)$ is interior to the starting n simplex. [The triangulation of $R^n \times [0, 1]$ is by $(n + 1)$-simplices whose facets are n-simplices. Here $n = 1$.] The PL path starting from $(x^0, 0)$ is indicated. The path reaches $t = 1$ at a terminal n-simplex B.

Figure 21.2.2

LABELING

The label function of the triangulation is defined in a special way (see Figure 21.2.3). For each vertex $v = (x, t)$ of the triangulation, specify

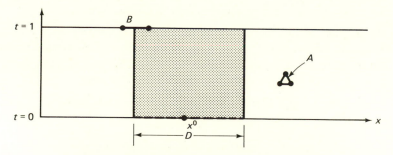

The labeling function G of (21.2.2) has $G(v) = x - x^0$ at $t = 0$ and if $x \notin D$ for $t > 0$. In the shaded area (which excludes $t = 0$) the function $G(v) = x - u$, where $u \in F(x)$.

Simplex A is entirely outside $D \times [0, 1]$, so cannot contain x^0 as a convex combination of the x^i. Thus the path can never have all its vertices entirely outside $D \times [0, 1]$. For $t = 0$, there is a unique starting n-simplex which contains x^0. Also, the terminal n-simplex B is entirely within $D \times \{1\}$ or straddles the boundary of $D \times \{1\}$.

Figure 21.2.3

$$G(v) = \begin{cases} x - u & \text{for some } u \in F(x) \quad \text{if } x \in D \quad \text{and} \quad t > 0 \\ x - x^0 & \text{otherwise.} \end{cases} \tag{21.2.2}$$

In words, $G(v) = x - x^0$ if x is outside D or if $t = 0$. $G(v) = x - u$ only if both $x \in D$ and $t > 0$. Here $u \in F(x)$ can be any point whatsoever in $F(x)$. Also, G is well-defined since $F(x)$ is nonempty for $x \in D$.

Perceive carefully that

$$G(v) = x - x^0$$

at $t = 0$ and any x outside D. This property of G is quite ingenious and yields two key results:

1. There is a unique starting simplex of the triangulation and the path can never return to $t = 0$. Consider any n-simplex $\{v^i = (x^i, 0)\}$ at $t = 0$. If it is on the path

    $$\sum \lambda_i G(v^i) = 0, \qquad \sum \lambda_i = 1, \qquad \lambda_i \geq 0.$$

    But by (21.2.2), since $t = 0$,

    $$\sum \lambda_i G(v^i) = \sum \lambda_i (x^i - x^0) = \sum \lambda_i x^i - x^0 = 0$$

    or

    $$\sum \lambda_i x^i = x^0. \tag{21.2.3}$$

    In words, at $t = 0$ any simplex on the path must satisfy (21.2.3), which states that x^0 is a convex combination of the vertices x^i. By construction at $t = 0$, x^0 is interior to a single n-simplex; hence x^0 can be a convex combination of those vertices only. At $t = 0$, then, there is a unique simplex on the path; the starting simplex. The path, consequently, starts from this unique starting simplex and can never return to $t = 0$.

2. No simplex on the path can have all its vertices outside $D \times [0, 1]$. That is, at least one vertex $v^i = (x^i, t^i)$ must have $x^i \in D$. Suppose for contradiction that there is an n-simplex $\{v^i = (x^i, t^i)\}$ on the path such that $x^i \notin D$ for all i. Because the n-simplex is on the path,

    $$\sum \lambda_i G(v^i) = 0, \qquad \sum \lambda_i = 1, \qquad \lambda_i \geq 0.$$

    Since $x^i \notin D$ for all i, by (21.2.2), we get

    $$\sum \lambda_i G(v^i) = \sum \lambda_i (x^i - x^0).$$

    Again

    $$\sum \lambda_i x^i = x^0$$

    and x^0 is a convex combination of the x^i. But x^0 is interior to D and if all x^i are outside D this cannot occur. (Any simplex outside of D will be "far" from x^0.) We conclude that any n-simplex on the path has at least one vertex in $D \times [0, 1]$.

Combining these two observations with our knowledge of t. algorithms, perceive the following. The path cannot return to $t = 0$. Moreov. because D is compact and no simplex on the path can be entirely outside $D \times [0, 1]$, the path cannot run off to infinity. Only one other possibility remains, and the crucial result is obtained that the path must reach $t = 1$. Conclude, therefore, that the terminal n-simplex of the path must be at $t = 1$ and also that that simplex cannot be entirely outside D.

THE TERMINAL SIMPLEX

Let us now examine the terminal n-simplex of the path, denoted B, which since it occurs at $t = 1$, $B = \{v^i = (x^i, 1)\}$. As mentioned, B must have at least one $x^i \in D$.

Since the terminal simplex is on the path

$$\sum \lambda_i G(v^i) = 0, \qquad \sum \lambda_i = 1, \qquad \lambda_i \geq 0.$$

Denote I as the set of indices for vertices $x^i \in D$, and J as the set of indices for all other vertices, that is, $x^i \notin D$. Then (21.2.2) yields

$$\sum_{i \in I} \lambda_i(x^i - u^i) + \sum_{i \in J} \lambda_i(x^i - x^0) = 0. \tag{21.2.4}$$

Also, since $F: D \longrightarrow D$,

$$u^i \in D. \tag{21.2.5}$$

Now we are ready for the Kakutani theorem. It observes what happens as the terminal simplex is made arbitrarily small in the limit.

THEOREM 21.2.1 (KAKUTANI)   Let $D \subset R^n$ be a nonempty compact convex set. Suppose that for an uhc point-to-set map $F: D \longrightarrow D$, $F(x)$ is nonempty and convex for all $x \in D$. Then F has a fixed point

$$\bar{x} \in F(\bar{x}).$$

Proof.   As presented above, we can obtain a terminal simplex $B = \{v^i\}$ such that (21.2.4) holds where I is the set of indices for $x^i \in D$ and J is the set of indices for $x^i \notin D$.

Consider now a sequence of such terminal simplices $B^k = \{v^{i,k}\}$ that get arbitrarily small as $k \longrightarrow \infty$. Select an infinite subsequence k such that the I and J are the same for each simplex. For those k

$$\sum_{i \in I} \lambda_{i,k}(x^{i,k} - u^{i,k}) + \sum_{i \in J} \lambda_{i,k}(x^{i,k} - x^0) = 0, \tag{21.2.6}$$

where $u^{i,k} \in F(x^{i,k})$ and $v^{i,k} = (x^{i,k}, 1)$.

Since D is compact and $0 \le \lambda_{i,k} \le 1$, the $\lambda_{i,k}$, $x^{i,k}$ and $u^{i,k}$ are in compact sets. This implies (on some subsequence if necessary) that

(A) $\qquad\qquad\quad \lambda_{i,k} \longrightarrow \bar\lambda_i, \qquad \sum \bar\lambda_i = 1, \qquad \bar\lambda_i \ge 0.$

(B) $\qquad\qquad\quad u^{i,k} \longrightarrow \bar u^i \qquad i \in I,$ $\qquad\qquad\qquad\qquad$ (21.2.7)

where by (21.2.5), since $u^{i,k} \in D$ all k,

$$\bar u^i \in D.$$

(C) $\qquad\qquad\quad x^{i,k} \longrightarrow \bar x \qquad$ for all i, where $\bar x \in D$ $\qquad\qquad$ (21.2.8)

because the simplices converge to a single point $\bar x$ and D closed ensures that $\bar x \in D$.

Hence we obtain

$$\sum_{i \in I} \bar\lambda_i(\bar x - \bar u^i) + \sum_{i \in J} \bar\lambda_i(\bar x - x^0) = 0$$

or

$$\bar x = \sum_{i \in I} \bar\lambda_i \bar u^i + \sum_{i \in J} \bar\lambda_i x^0. \qquad\qquad (21.2.9)$$

We now prove that $\sum_{i \in J} \bar\lambda_i = 0$. Suppose not. All $\bar u^i \in D$, and x^0 is interior to D, where D is convex. So if $\sum_{i \in J} \bar\lambda_i \ne 0$, (21.2.9) would make $\bar x$ interior to D. But from (21.2.8), for k large enough the $x^{i,k}$ would then be interior to D. In other words, for k large, all vertices of B^k would have to be interior to $D \times [0, 1]$. Yet then with no vertex $v^{i,k}$ outside $D \times [0, 1]$, obviously $J = \varnothing$. We conclude, consequently, that $\sum_{i \in J} \bar\lambda_i = 0$ and therefore

$$\bar x = \sum_{i \in I} \bar\lambda_i \bar u^i, \qquad \sum_{i \in I} \bar\lambda_i = 1, \qquad \bar\lambda_i \ge 0. \qquad (21.2.10)$$

Finally, since

$$u^{i,k} \in F(x^{i,k}), \qquad \text{all } k$$

via (21.2.7) and (21.2.8), and F uhc

$$\bar u^i \in F(\bar x), \qquad \text{all } i \in I.$$

Convexity of $F(\bar x)$ and (21.2.10) thereby ensure that

$$\bar x \in F(\bar x). \qquad \square$$

The demonstration does more than just prove the existence of a fixed point. If we operate a simplicial algorithm using labeling (21.2.2), the terminal simplex gives an approximate fixed point. Specifically, given the terminal simplex $B = \{v^i\}$, let

$$\bar x = \sum \lambda_i x^i.$$

The point $\bar x$ is the approximate fixed point to F, and our proof is thus constructive.

The Kakutani theorem is clearly the direct extension of the Brouwer. Both operate on convex, compact sets. Instead of continuity of a function, however, Kakutani uses uhc and convexity of a point-to-set map.

As mentioned, the Kakutani theorem is very useful, and in the next section we apply it to examine economic equilibria.

21.3 ECONOMIC EQUILIBRIUM EXISTENCE REVISITED

In Chapter 7 we demonstrated the existence of a competitive equilibrium. The proof did not require a fixed-point theorem, only path-following arguments. This section articulates the classical proof of equilibrium existence, which invokes the Kakutani fixed-point theorem.

ECONOMIC EQUILIBRIUM

Recall for an EE there are M agents, where $x^i \in R^n$ is a bundle of n-goods for agent, i, $f^i(x^i)$ is his utility, and $w^i \in R^n$ is his endowment. Also, $p \in R^n$ are the prices of the n goods. Here $(x^1, \ldots, x^M, p) = (x, p) \in R^{(M+1)n}$.

The goal in EE is to determine a competitive equilibrium $(\bar{x}, \bar{p})$ for the problem

$$\max_{x^i} f^i(x^i)$$

$$px^i \leq pw^i, \qquad i = 1, \ldots, M$$

$$0 \leq x^i \leq W,$$

where

$$W = \sum_{i=1}^{M} w^i \qquad\qquad (21.3.1)$$

$$\sum_{i=1}^{M} x^i \leq W$$

and

$$p \geq 0, \qquad \sum_{j=1}^{n} p_j = 1.$$

We transform (21.3.1) to an EP by introducing a market player $M + 1$, thus yielding

(A)
$$\max_{x^i} f^i(x^i)$$

$$px^i \leq pw^i, \qquad i = 1, \ldots, M$$

$$0 \leq x^i \leq W$$

for p given; and   (21.3.2)

(B)
$$\max_{p} f^{M+1}(x, p) = p\left(\sum_{i=1}^{M} x^i - W\right)$$

$$\sum_{j=1}^{n} p_j = 1$$

$$p \geq 0$$

for x given.

Recall these optimizations were termed subproblem i, $i = 1, \ldots, M + 1$.

MAPPINGS

To validate that (21.3.2) has an equilibrium point, we construct mappings F^i to which the Kakutani theorem can be applied. These mappings will be very similar to those in Theorem 21.1.1.

Given $x = (x^1, \ldots, x^M)$ and $p \geq 0$, $\sum_{j=1}^{n} p_j = 1$, define for $i = 1, \ldots, M$

$$F^i(x, p) = \{\overset{*}{x}{}^i \,|\, \overset{*}{x}{}^i \text{ is optimal for subproblem } i \text{ given } p\}$$

and

$$F^{M+1}(x, p) = \{\overset{*}{p} \,|\, \overset{*}{p} \text{ is optimal for subproblem } M + 1 \text{ given } x\}.$$

Thus F^i is the set of all optimal points for subproblem i given the appropriate variable x or p. [Note that we write F^i as depending on (x, p), but during the actual subproblem calculation $F^i, i = 1, \ldots, M$, depends only upon p and F^{M+1} depends only on x.]

INTERPRETATION AS DEMANDS

Let us now examine each F^i and explain that it has meaning in terms of our original EE. Consider agent i, who must solve subproblem i given prices p. The optimal point $\overset{*}{x}{}^i$ for that subproblem would maximize his or her utility at those prices. Those goods $\overset{*}{x}{}^i$ are the goods that he or she would desire or demand at those prices.

The set $F^i, i = 1, \ldots, M$, is termed agent i's demand set because it is the set of all demands $\overset{*}{x}{}^i$ at prices p. At any p, F^i expresses the goods agent i wants or demands in order to maximize his or her utility.

Analogously, F^{M+1} are the prices $\overset{*}{p}$ the market player demands given that the M agents want x. That is, given that the agents will buy x, the market demands prices $\overset{*}{p}$.

The F^i therefore reflects the actions each agent or the market player would take depending upon the actions of the others. Economically, they express the demands of the players.

CARTESIAN PRODUCT MAP

With the economic meaning of the F^i better understood, let us define the Cartesian product of these maps

$$F = F^1 \times F^2 \times \cdots \times F^{M+1}. \tag{21.3.3}$$

In detail, given a point (x, p), calculate the sets F^i, $i = 1, \ldots, M + 1$. Their Cartesian product is the set $F(x, p)$.

Observe that a point-to-set map $F: R^{(M+1)n} \longrightarrow R^{(M+1)n}$ has been constructed. Each point $(x, p) \in R^{(M+1)n}$. Also, $F^i \subset R^n$, $i = 1, \ldots, M + 1$, so being a Cartesian product, $F \subset R^{(M+1)n}$. In short, F takes points in $R^{(M+1)n}$ into sets in $R^{(M+1)n}$.

But let us be even more precise. Specify

$$D = \{(x, p) \,|\, 0 \le x^i \le W, \quad i = 1, \ldots, M, \quad \sum_{j=1}^{n} p_j = 1, \quad p \ge 0\}. \tag{21.3.4}$$

If $(x, p) \in D$, then each subproblem i is feasible, and all $F^i(x, p)$, $i = 1, \ldots, M + 1$, are defined and not null. This means $F(x, p)$ is not null. Further, by examining the optimal points of the subproblem, note that $F(x, p) \subset D$. We conclude therefore that F is a nonnull point-to-set map $F: D \longrightarrow D$, where D is convex and compact.

PROPERTIES OF F

The map F is very important, and soon the Kakutani theorem will be applied to it. But to do that, first we must explore a few of its properties. For instance, at any (x, p), the set $F(x, p)$ is convex; or rephrased, F is convex-valued.

LEMMA 21.3.1   Suppose that for each subproblem i, $i = 1, \ldots, M$, f^i is a concave function. Then for any (x, p), the set $F(x, p)$ is convex.

Proof.   By applying Lemma 21.1.2 to subproblem i, $i = 1, \ldots, M$, see that F^i is itself a convex set. Similarly, for x fixed, F^{M+1} is a convex set. But the Cartesian product of convex sets is, as is easily shown, convex. Thus $F(x, p)$ is a convex set.   □

Perceive that thus far we have uncovered, for D a compact convex set, that $F: D \longrightarrow D$ is a point-to-set map which is nonnull and convex-valued. To utilize the Kakutani theorem, it remains only to verify that F is uhc.

Given a sequence $\{(x^k, p^k)\}$ approaching (x^∞, p^∞), and a corresponding sequence

$$(\tilde{x}^k, \tilde{p}^k) \in F(x^k, p^k), \quad \text{where } (\tilde{x}^k, \tilde{p}^k) \longrightarrow (\tilde{x}^\infty, \tilde{p}^\infty), \quad (21.3.5)$$

to prove uhc we must establish that

$$(\tilde{x}^\infty, \tilde{p}^\infty) \in F(x^\infty, p^\infty).$$

Let us interpret (21.3.5), and to do so some terminology will be helpful.

Given (x, p), a vector $(\tilde{x}, \tilde{p})$ is called feasible for (x, p) if each of its components $\tilde{x}^i$ or $\tilde{p}$ is feasible for its corresponding subproblem, that is, if $\tilde{x}^i$ is feasible for subproblem i given p and $\tilde{p}$ is feasible for subproblem $M + 1$ given x. Analogously, we say that $(\tilde{x}, \tilde{p})$ is optimal for (x, p) if each component of $(\tilde{x}, \tilde{p})$ is optimal for its corresponding subproblem.

Returning now to (21.3.5), $(\tilde{x}^k, \tilde{p}^k) \in F(x^k, p^k)$ means that $(\tilde{x}^k, \tilde{p}^k)$ is optimal for (x^k, p^k). [Explicitly, for $\tilde{x}^k = (\tilde{x}^{1,k}, \ldots, \tilde{x}^{M,k})$, $\tilde{x}^{i,k}$ is optimal for subproblem i given p^k, and $\tilde{p}^k$ is optimal for $M + 1$ given x^k.] To prove

$$(\tilde{x}^\infty, \tilde{p}^\infty) \in F(x^\infty, p^\infty)$$

we must establish that $(\tilde{x}^\infty, \tilde{p}^\infty)$ is optimal for (x^∞, p^∞) [that is, for $\tilde{x}^\infty = (\tilde{x}^{1,\infty}, \ldots, \tilde{x}^{M,\infty})$ that $\tilde{x}^{i,\infty}$ is optimal for subproblem i given p^∞, and $\tilde{p}^\infty$ is optimal for $M + 1$ given x^∞].

Two lemmas will help. The first shows that $(\tilde{x}^\infty, \tilde{p}^\infty)$ is feasible for (x^∞, p^∞). [Conceptually, because $(\tilde{x}^k, \tilde{p}^k) \in F(x^k, p^k)$ for all k, $(\tilde{x}^k, \tilde{p}^k)$ is feasible for (x^k, p^k). We must validate that this holds in the limit, that is, that $(\tilde{x}^\infty, \tilde{p}^\infty)$ is feasible for (x^∞, p^∞).]

LEMMA 21.3.2 Let $(x^k, p^k) \in D$, $(x^k, p^k) \longrightarrow (x^\infty, p^\infty)$, $(\tilde{x}^k, \tilde{p}^k) \in F(x^k, p^k)$, and $(\tilde{x}^k, \tilde{p}^k) \longrightarrow (\tilde{x}^\infty, \tilde{p}^\infty)$. Then $(\tilde{x}^\infty, \tilde{p}^\infty)$ is feasible for (x^∞, p^∞).

Proof. Specify that $\tilde{x}^k = (\tilde{x}^{1,k}, \ldots, \tilde{x}^{M,k})$. Since $\tilde{x}^{i,k}$ is feasible for subproblem i given p^k, examining the constraints of subproblem i, $i = 1, \ldots, M$,

$$p^k \tilde{x}^{i,k} \leq p^k w^i$$

$$0 \leq \tilde{x}^{i,k} \leq W.$$

Taking limits yields

$$p^\infty \tilde{x}^{i,\infty} \le p^\infty w^i$$
$$0 \le \tilde{x}^{i,\infty} \le W.$$

This means that $\tilde{x}^{i,\infty}$ is feasible for subproblem i, given p^∞.

Finally, since $\tilde{p}^k$ is feasible for subproblem $M + 1$,

$$\sum_{j=1}^{n} \tilde{p}_j^k = 1, \qquad \tilde{p}^k \ge 0.$$

Hence

$$\sum_{j=1}^{n} \tilde{p}_j^\infty = 1, \qquad \tilde{p}^\infty \ge 0.$$

Thus $\tilde{p}^\infty$ is also feasible. $\square$

The lemma showed that $\tilde{x}^{i,\infty}$ is feasible for subproblem i given p^∞, and $\tilde{p}^\infty$ is feasible for subproblem $M + 1$ given x^∞. But we do not yet know if these points are optimal. To demonstrate that requires examination of the optimal point $\overset{*}{x}{}^i$. The next lemma proves there is a sequence $\overset{*}{x}{}^{i,k}$ where each point $\overset{*}{x}{}^{i,k}$ is feasible for subproblem i and such that $\overset{*}{x}{}^{i,k} \to \overset{*}{x}{}^i$.

LEMMA 21.3.3   Suppose that $w_j^i > 0$ all i and j and that $p^k \to p^\infty$. Also let $\overset{*}{x}{}^i$ be feasible for subproblem i given p^∞. Then there is a sequence

$$\overset{*}{x}{}^{i,k} \longrightarrow \overset{*}{x}{}^i,$$

where each $\overset{*}{x}{}^{i,k}$ is feasible for subproblem i given p^k.

Proof.   Examining the subproblems, we see that the only concern is the constraint

$$p^\infty \overset{*}{x}{}^i \le p^\infty w^i \tag{21.3.6}$$

Also, the case

$$p^\infty \overset{*}{x}{}^i < p^\infty w^i \tag{21.3.7}$$

is straightforward and left as Exercise 4. Consequently, suppose that

$$p^\infty \overset{*}{x}{}^i = p^\infty w^i. \tag{21.3.8}$$

Writing (21.3.8) out yields

$$\sum_{j=1}^{n} p_j^\infty \overset{*}{x}{}_j^i = \sum_{j=1}^{n} p_j^\infty w_j^i. \tag{21.3.9}$$

As $\sum_{j=1}^{n} p_j^\infty = 1$, $p^\infty \ge 0$, at least one $p_j^\infty > 0$. Thus since by assumption $w^i > 0$,

$$p^\infty w^i > 0.$$

Examining (21.3.9), for some index, say 1, $p_1^\infty \overset{*}{x}_1^i > 0$, so $p_1^\infty > 0$ and equation (21.3.9) then provides

$$\overset{*}{x}_1^i = \frac{p^\infty w^i - \sum_{j=2}^{n} p_j^\infty \overset{*}{x}_j^i}{p_1^\infty} > 0. \tag{21.3.10}$$

Define

$$y_1^{i,k} = \frac{p^k w^i - \sum_{j=2}^{n} p_j^k \overset{*}{x}_j^i}{p_1^k}. \tag{21.3.11}$$

For k sufficiently large, $p_1^k > 0$, since $p_1^\infty > 0$, and thus (21.3.11) is well-defined. Furthermore, taking limits of (21.3.11), we obtain

$$y_1^{i,k} \longrightarrow \overset{*}{x}_1^i. \tag{21.3.12}$$

Now define

$$\overset{*}{x}_1^{i,k} = \min \{y_1^{i,k}, W_1\}. \tag{21.3.13}$$

Since $\overset{*}{x}^i$ is feasible, $\overset{*}{x}_1^i \le W_1$. Then from (21.3.12) and (21.3.13),

$$\overset{*}{x}_1^{i,k} \longrightarrow \overset{*}{x}_1^i, \tag{21.3.14}$$

where, by (21.3.10),

$$0 \le \overset{*}{x}_1^{i,k} \le W_1 \tag{21.3.15}$$

for k sufficiently large.

Finally, specify that for $j = 2, \ldots, n$,

$$\overset{*}{x}_j^{i,k} = \overset{*}{x}_j^i. \tag{21.3.16}$$

Clearly, as $0 \le \overset{*}{x}^i \le W$,

$$0 \le \overset{*}{x}_j^{i,k} \le W_j, \qquad j = 2, \ldots, n. \tag{21.3.17}$$

Moreover, applying (21.3.11), (21.3.13) and (21.3.16), for k large yields

$$p_1^k \overset{*}{x}_1^{i,k} + \sum_{j=2}^{n} p_j^k \overset{*}{x}_j^{i,k} \le p^k w^i. \tag{21.3.18}$$

We have constructed a sequence $\overset{*}{x}^{i,k}$ such that taking an infinite subsequence if necessary, $\overset{*}{x}^{i,k}$ is feasible for subproblem i given p^k. Moreover,

$$\overset{*}{x}^{i,k} \longrightarrow \overset{*}{x}^i. \qquad \square$$

Everything needed is now at hand, so let us prove that the mapping F defined by (21.3.3) is uhc.

THEOREM 21.3.4   Let $(x^k, p^k) \in D$, $(x^k, p^k) \longrightarrow (x^\infty, p^\infty)$, $(\tilde{x}^k, \tilde{p}^k) \in F(x^k, p^k)$, and $(\tilde{x}^k, \tilde{p}^k) \longrightarrow (\tilde{x}^\infty, \tilde{p}^\infty)$. Then $(\tilde{x}^\infty, \tilde{p}^\infty) \in F(x^\infty, p^\infty)$.

Proof. From Lemma 21.3.2, we know that $\tilde{x}^{i,\infty}$ is feasible for i given $\bar{p}^{\infty}$, and $\bar{p}^{\infty}$ is feasible for $M+1$ given x^{∞}. We must prove that these points are optimal for their respective subproblems.

Let $\overset{*}{\tilde{x}}{}^{i}$ be optimal for subproblem i given p^{∞}. We have, since $\tilde{x}^{i,\infty}$ is feasible,

$$f^{i}(\tilde{x}^{i,\infty}) \le f^{i}(\overset{*}{\tilde{x}}{}^{i}), \qquad i = 1, \ldots, M. \tag{21.3.19}$$

Similarly, if $\overset{*}{\tilde{p}}$ is optimal for $M+1$ given x^{∞},

$$f^{M+1}(x^{\infty}, \bar{p}^{\infty}) \le f^{M+1}(x^{\infty}, \overset{*}{\tilde{p}}). \tag{21.3.20}$$

Consider subproblem $M+1$ first. Since $\bar{p}^{k}$ is optimal given x^{k}, and $\overset{*}{\tilde{p}}$ is feasible,

$$f^{M+1}(x^{k}, \overset{*}{\tilde{p}}) \le f^{M+1}(x^{k}, \bar{p}^{k}).$$

Taking limits the preceding inequality yields

$$f^{M+1}(x^{\infty}, \overset{*}{\tilde{p}}) \le f^{M+1}(x^{\infty}, \bar{p}^{\infty}). \tag{21.3.21}$$

Using (21.3.20) and (21.3.21), $\bar{p}^{\infty}$ is optimal for $M+1$ given x^{∞}; that is, $\bar{p}^{\infty} \in F^{M+1}(x^{\infty}, p^{\infty})$.

Next, examine subproblem i, $i = 1, \ldots, M$. By Lemma 21.3.3, there is a sequence $\overset{*}{\tilde{x}}{}^{i,k}$ feasible for subproblem i given p^{k} such that

$$\overset{*}{\tilde{x}}{}^{i,k} \longrightarrow \overset{*}{\tilde{x}}{}^{i}. \tag{21.3.22}$$

Hence since $\tilde{x}^{i,k}$ is optimal for subproblem i given p^{k},

$$f^{i}(\overset{*}{\tilde{x}}{}^{i,k}) \le f^{i}(\tilde{x}^{i,k}).$$

Thus

$$f^{i}(\overset{*}{\tilde{x}}{}^{i}) \le f^{i}(\tilde{x}^{i,\infty}), \qquad i = 1, \ldots, M. \tag{21.3.23}$$

Via (21.3.19) and (21.3.23), $\tilde{x}^{i,\infty}$ is optimal, which means that $\tilde{x}^{i,\infty} \in F^{i}(x^{\infty}, p^{\infty})$.

Since $\tilde{x}^{i,\infty} \in F^{i}(x^{\infty}, p^{\infty})$, $i = 1, \ldots, M$, and $\bar{p}^{\infty} \in F^{M+1}(x^{\infty}, p^{\infty})$,

$$(\tilde{x}^{\infty}, \bar{p}^{\infty}) \in F(x^{\infty}, p^{\infty}). \qquad \square$$

We have now demonstrated that the mapping F satisfies all the hypotheses of Kakutani's theorem. Thus a fixed point of the map F exists. Precisely, there is an $(\bar{x}, \bar{p}) \in D$ such that

$$(\bar{x}, \bar{p}) \in F(\bar{x}, \bar{p}). \tag{21.3.24}$$

EQUILIBRIUM POINT

The fixed point is exactly what we were seeking, and as we now show, is indeed the equilibrium point. Since $(\bar{x}, \bar{p}) \in D$ and by definition of F as a Cartesian product, (21.3.24) states that

$$\bar{x}^{i} \in F^{i}(\bar{x}, \bar{p}), \qquad i = 1, \ldots, M.$$

This implies, writing it out in detail, that at $\bar{x}$,

$$f^i(\bar{x}^i) = \max f^i(x^i)$$
$$\bar{p}x^i \le \bar{p}w^i, \qquad i = 1, \ldots, M$$
$$0 \le x^i \le W.$$

Furthermore, $\bar{p} \in F^{M+1}(\bar{x}, \bar{p})$ or, equivalently,

$$f^{M+1}(\bar{x}, \bar{p}) = \max p\left(\sum_{i=1}^{M} \bar{x}^i - W\right)$$
$$\sum_1^n p_j = 1, \qquad p \ge 0.$$

Verbally, given $\bar{p}$, the point $\bar{x}^i$ maximizes subproblem i, and given $\bar{x}$, the point $\bar{p}$ maximizes subproblem $M + 1$. Yet this means that $(\bar{x}, \bar{p})$ is an equilibrium point of (21.3.2).

THEOREM 21.3.5   Consider the economic equilibrium problem (21.3.1). Suppose that each f^i is continuous and concave and that all $w_j^i > 0$. Then a competitive equilibrium exists.

COMMENT

The traditional proof for the economic equilibrium problem given in this chapter is more technical and less intuitive than the path-following proof in Chapter 7. Further, the F^i functions in the Kakutani map are often difficult to calculate, as they require the solution of optimization problems.

Overall, compared to the traditional fixed-point proofs, path following is generally simpler in terms of both proof and computation.

SUMMARY

This chapter presented point-to-set maps and some of their properties. Uhc was useful to establish smoothness, and a new fixed-point theorem due to Kakutani was provided. All these results are valuable in numerous applications, and indeed the final section illustrated how they are employed to obtain a competitive equilibrium for the economic equilibrium problem.

EXERCISES/CHAPTER 21

1. Define the *graph* of a point-to-set map $F: D \subset R^n \longrightarrow R^l$ to be the set

    $$\{(x, u) \,|\, x \in D, u \in F(x)\}.$$

    Prove that F is uhc if for each closed subset $A \subset D$ the graph of $F: A \longrightarrow R^l$ is closed.

2. Prove that the point-to-set map

$$F(x) = \{u \mid -x \leq u \leq x\}$$

   is uhc. Graph the map.

3. Prove or give a counterexample to the following: The point-to-set map

$$F(x) = \{y \mid y^{\tau}x \leq b, y \geq 0\}$$

   is uhc, where b is a given real number.

4. Prove Lemma 21.3.3 in the case where $p^{\infty}\overset{*}{x}{}^{i} < p^{\infty}w^{i}$.

5. Consider the problem

$$\min f(x, y) = (y_1 - 10)^2 + (y_2 - x)^2$$

$$\text{subject to}$$

$$0 \leq y_1 \leq 1$$

$$0 \leq y_2 \leq 1 \qquad\qquad\qquad \text{(E.21.1)}$$

   given $x \in R^1$. Show by actual calculation that the map

$$F(x) = \{y \mid y \text{ optimizes (E.21.1) given } x\}$$

   is uhc.

6. Suppose that $F: D \longrightarrow D$, where $D \subset R^n$, is a nonempty convex set. Also suppose that D has an empty interior and D is not a single point. Show by construction that F can be transformed to an equivalent mapping

$$\tilde{F}: \tilde{D} \longrightarrow \tilde{D}, \qquad \text{where } \tilde{D} \subset R^k, \quad \text{some } k \leq n - 1,$$

   where $\tilde{D}$ is a convex set with a nonempty interior.

7. Construct a uhc mapping $F: D \longrightarrow D$, where D is nonempty compact and convex and $F(x)$ is nonempty but not necessarily convex for all $x \in D$, such that F has no fixed point.

8. Consider the NLP

$$\min f(x) = c^{\tau}x, \qquad x \in R^n$$

$$g_i(x) \geq 0, \qquad i = 1, \ldots, m. \qquad\qquad \text{(E.21.2)}$$

   Let $D = \{x \mid g_i(x) \geq 0, i = 1, \ldots, m\}$ be compact. Suppose that a point-to-set map F defined on D satisfies

   a. If $x \in D$ is not a Kuhn–Tucker point of (E.21.2), then $u \in F(x)$ implies that $u \in D$ and $f(u) < f(x)$. If $x \in D$ is a Kuhn–Tucker point of (E.21.2), then $u \in F(x)$ implies that $u \in D$ and $f(u) = f(x)$.

   b. F is uhc for all $x \in D$. Then prove that any sequence x^k with $x^0 \in D$, $x^{k+1} \in F(x^k)$, will contain a subsequence that converges to a Kuhn–Tucker point of (E.21.2).

9. Prove that a (single-valued) continuous map is uhc.

NOTES

21.1. An excellent exposition on point-to-set mappings may be found in Berge [1963]. Applications of point-to-set mappings to nonlinear programming algorithms are given in Zangwill [1969], and to subdifferentials in Rockafellar [1970].

21.2. Our proof of the Kakutani theorem is markedly different from that of Kakutani [1941] and is more along the lines of Eaves [1971a]. Cohen [1980] provides a different proof.

21.3. The Kakutani theorem is an important tool in competitive equilibrium theory. For example, see Arrow and Hahn [1971], Debreu [1975], and Scarf and Hansen [1973].

22

Relaxation of Regularity and Differentiability

Up to this point in the book the theorems have needed assumptions about regularity and differentiability. This chapter eliminates these assumptions. All the theorems proved will thereby hold for continuous functions. This includes fixed-point theorems, homotopy invariance theorems, solutions results, equilibrium points, and so on.

The process required to eliminate the assumptions is technical and in a sense not needed. Regularity holds almost all the time. Further, any continuous function can be approximated to any desired degree by a twice-continuously differentiable one. The regularity and differentiability assumptions are thus extremely weak, and in practical application are generally innocuous. In short, there is little need theoretically or practically to eliminate them. Still this chapter rids us of them and verifies that our previous theorems hold for functions that are merely continuous.

Two powerful theorems will come to our aid. Sard's theorem is the first and it states that regularity holds almost always. The Weierstrass theorem, the second, permits us to approximate a continuous function by a polynomial. Using these two theorems, we can approximate any continuous function very accurately. It is then easy to establish the theorems for continuous functions.

As mentioned, this chapter is technical and, because our previous assumptions hold most of the time anyway, possibly not even necessary. The reader primarily interested in applications can omit it with no loss of practical understanding.

22.1 THE TWO MAIN THEOREMS

This chapter will demonstrate that our previous results hold for continuous functions so that regularity and differentiability are not required. Two potent theorems will be utilized to achieve this, Sard's theorem and the Weierstrass theorem. As their verifications are somewhat far afield, they will not be reproven here; Sard's theorem can be found in Guillemin and Pollack [1974], Milnor [1965], and Sternberg [1964] and the Weierstrass approximation theorem in Ortega and Rheinboldt [1970]. Nevertheless, we must analyze and understand their implications well. Let us begin with Sard's theorem.

SARD'S THEOREM

Sard's theorem presents a surprisingly deep result that will permit us to eliminate regularity. So far, three concepts of regularity have been utilized, which we review here. Let $D \subset R^n$ be the closure of an open bounded set and consider the functions $H: D \times T \longrightarrow R^n$ and $F: D \longrightarrow R^n$. Then

H is regular if its Jacobian $H'(y)$ is of rank n for all $y \in H^{-1}$.

H is regular at $\bar{t}$ if the partial Jacobian $H'_x(x, \bar{t})$ is of rank n,

$$\text{for all } x \in H^{-1}(\bar{t}).$$

F is regular if its Jacobian F' is of rank n for all $x \in F^{-1}$.

[Recall that $F^{-1} = \{x \mid F(x) = 0\}$.]   (22.1.1)

All three of these definitions concern the rank of the Jacobian when evaluated at the solutions to systems of equations. Sard produced a deep and general theorem relative to that. To understand it, consider a continuously differentiable function $L: B \longrightarrow R^n$, where $B \subset R^q$, and denote its $n \times q$ Jacobian by L'. Also examine that system of n equations in q unknowns:

$$L(w) = e,$$

where e is an n-vector indicating the right-hand side (rhs). For a given rhs e, call the solutions to this system L_e^{-1}. Thus

$$L_e^{-1} = \{w \in B \mid L(w) = e\}.$$

The given right-hand side e is important in this system, and using it Sard developed a powerful concept of regularity. The function L is called *regular for rhs e* if L' is of rank n for all $w \in L_e^{-1}$. Note that the Jacobian must be of rank n at each and every point $w \in L_e^{-1}$. Also, if for e, L is not regular, it is called *degenerate*. Specifically, L is degenerate for e if the Jacobian L' is not of rank n at even one point $w \in L_e^{-1}$.

Some useful considerations follow easily. If L is regular for e, then the function

$$\bar{L} = L - e$$

is regular for zero. This is obvious since $L_e^{-1} = \bar{L}_0^{-1}$. To maintain consistency with our previous definition, we term a function regular if it is regular for rhs zero.

If L is regular for $\bar{e}$, then L is also regular for e sufficiently near $\bar{e}$. In other words, suppose that a function is regular at a given rhs $\bar{e}$. Then it is regular for all rhs e in a neighborhood of $\bar{e}$. This follows because if the Jacobian L' is of rank n at a point, by continuous differentiability it will be of rank n in a neighborhood of that point (Exercise 1).

PREVIOUS RESULTS

The extension of our previous usage of regularity (22.1.1) to the new definition is immediate. Let

$$H(y) = e \qquad \text{and} \qquad F(x) = e$$

and define

$$H_e^{-1} = \{y \in D \times T \,|\, H(y) = e\}$$
$$H_e^{-1}(\bar{t}) = \{x \in D \,|\, H(x, \bar{t}) = e\}$$
$$F_e^{-1} = \{x \in D \,|\, F(x) = e\}.$$

Clearly, H will be regular for the rhs e if H' is of rank n for all $y \in H_e^{-1}$. Similarly, given rhs e, H will be regular at $\bar{t}$ if $H'(x, \bar{t})$ has rank n for all $x \in H_e^{-1}(\bar{t})$. Also, F will be regular for rhs e if F' has rank n for all $x \in F_e^{-1}$. Obviously our previous usage is totally consistent with the new definition of regularity. All we are doing now is employing an arbitrary rhs e, which is exactly what is needed for Sard's theorem.

MEASURE ZERO

Sard's theorem proclaims that L is regular for all e except on a set of measure zero. But wait a minute. Before explaining Sard's theorem, let us explain what "measure zero" means. We say that a set has *measure zero* in R^n if it has no volume in R^n. For example, a line segment does have positive measure in R^1 but not in R^n for $n \geq 2$, so a line segment has measure zero in R^n, $n \geq 2$. A three-dimensional cube will have volume in R^3 but will be of measure zero in R^n, $n \geq 4$. A point has measure zero in R^n, for $n \geq 1$. The finite union of sets of measure zero also has measure zero. For instance, a finite number of paths in H^{-1}, each of finite length, will have measure zero in R^{n+1}. This is because each path has dimension one and $n \geq 1$.

Certain properties hold at all points in a space except on a set of measure zero. Such properties are said to hold almost everywhere (or almost always). This means that at any point in the space, except for points in a given set of measure zero, the property holds. To reiterate, suppose that a property holds almost always. Then it holds everywhere except for a set of measure zero. Intuitively, it holds at virtually any point in the space (i.e., everywhere in the space except for a set of measure zero).

Sard's theorem states that L is regular for e except for e on a set of measure zero. Thus Sard's theorem states that L will be regular for almost all e in the space. More precisely

---

SARD'S THEOREM 22.1.1   Let $L: D \subset R^q \to R^n$, where D is the closure of an open set and L be C^k, where $k \geq 1 + \max \{0, q - n\}$. Then L is regular for almost all e.

For a convenient rephrasing, recall that if L is not regular, it must be degenerate. Thus, Sard's theorem proclaims that the set of degenerate values of L has measure zero.

Sard's theorem immediately applies to our situation (22.1.1).

COROLLARY 22.1.2   Let $D \subset R^n$ be the closure of an open set, $H: D \times T \to R^n$ be C^2, and $F: D \to R^n$ be C^1. Then

1. H will be regular for almost all e.
2. H will be regular at $\bar{t}$ for almost all e.
3. F will be regular for almost all e.

These results are almost too good to be true. If we pick an arbitrary right-hand side e, even $e = 0$, almost assuredly we will have regularity. That is because a set of measure zero is so minute in size. If by some chance the function is not regular at some $\bar{e}$, then there are other e arbitrarily close to $\bar{e}$ at which the function is regular. This is an extremely important conclusion and worth reemphasizing. If at some $\bar{e}$ the function is not regular, we can perturb the right-hand side an arbitrarily small amount and find an e for which the function is regular.

However, one must exercise caution as Sard's theorem speaks about regularity as a function of the right-hand side. It says nothing about the left-hand side. In terms of the left-hand side, that is, in terms of x, y, or t, the Jacobian can have rank less than n on sets of positive measure. Sard tells us about the selection of e, the right-hand side. (See Exercise 3. See also Appendix A to this chapter, which discusses certain left-hand-side perturbations.)

GEOMETRY

Because Sard's theorem is quite important, let us interpret it with some examples.

EXAMPLE 22.1.1

Suppose that $F: R^1 \longrightarrow R^1$. Then F is regular as long as its derivative $F' \neq 0$. In terms of the system

$$F(x) = e,$$

examine Figure 22.1.1. Observe that F is regular for all e except for two points e^1 and e^2. Thus F is regular for all e except on a set of measure zero. (The two points e^1 and e^2 have measure zero in R^1.) Also observe that the set of x on which F' is not regular has positive measure. As mentioned, Sard's theorem does not hold for the left-hand side.

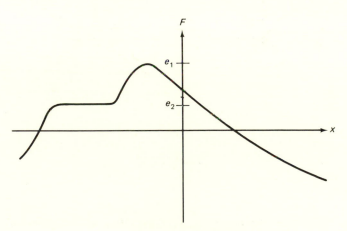

F is degenerate at e_1 and e_2.

Figure 22.1.1

EXAMPLE 22.1.2

Suppose that $H(x, t): R^1 \times T \longrightarrow R^1$ so that $x \in R^1$. In particular, let

$$H(x, t) = (x - t)(x + t) = x^2 - t^2.$$

Clearly,

$$H' = (2x, -2t).$$

Now examine

$$H(x, t) = x^2 - t^2 = 0.$$

The solution set H^{-1} consists of two straight lines $x = t$, $x = -t$, which intersect at the origin as in Figure 22.1.2(a). Whenever two paths cross we know that H cannot be regular. Indeed, the right-hand-side zero is a degenerate value for H because the point $(0, 0) \in H^{-1}$, and H' does not have rank 1 there.

Sard's theorem requires there must be e arbitrarily close to zero at which H is regular. Take an arbitrary $e > 0$ and solve

$$H(x, t) = (x - t)(x + t) = e.$$

This yields two separate paths as solutions:

$$x = \pm\sqrt{e + t^2}.$$

Figure 22.1.2(b) depicts these two paths. Note that they do not intersect. In fact, H is now regular. By perturbing the right-hand side we obtain regularity and separate the paths in H^{-1}, just as Sard's theorem implies. Other examples are given in Exercises 10 and 12.

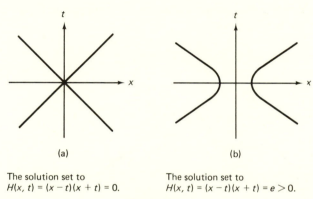

(a)

(b)

The solution set to
$H(x, t) = (x - t)(x + t) = 0.$

The solution set to
$H(x, t) = (x - t)(x + t) = e > 0.$

Figure 22.1.2

STABILITY

If a function is regular for $\bar{e}$, we know that it is regular for all e in a neighborhood of $\bar{e}$. However, if a function is degenerate for $\bar{e}$, then it is regular for virtually all points near $\bar{e}$. This is a crucial implication of Sard's theorem. Conceptually, regularity is a stable property. Minor changes do not disrupt it. However, degeneracy is an unstable situation. A slight change and we are back to regularity.

Regularity was assumed in our previous theorems because under regularity things are stable and we can then achieve very elegant proofs. Treating the unstable case, degenerate points, is more of a technical exercise. This is not to say that degenerate points do not arise; they do and we must consider them.

Nevertheless, regularity, because of its stability, is far more likely to arise naturally. As Sard's theorem states, regularity will occur almost all the time.

We now understand how Sard's theorem assists us with regularity. In the first place, regularity almost always occurs. Even if it does not, an arbitrarily small jiggle of the right-hand side can bring it about.

With Sard's theorem introduced, let us now turn to the other major theorem required.

22.2 THE WEIERSTRASS THEOREM

The Weierstrass theorem will help us eliminate the differentiability requirements in the theorems. Given any continuous function L, it states that L can be approximated to any desired degree by a polynomial.

First, we must consider how two functions can be close. Given two continuous functions $L: D \subset R^q \to R^n$ and $K: D \subset R^q \to R^n$, where D is compact, define the norm of the two functions by

$$\|L - K\|_\infty = \max_{x \in D} \|L(x) - K(x)\|.$$

At each x evaluate $\|L(x) - K(x)\|$. Then take the maximum over all x in D. The result is the norm of two functions or function norm.

Conceptually, at each x we determine how far apart $L(x)$ and $K(x)$ are by calculating $\|L(x) - K(x)\|$. The largest distance they are apart gives the function norm. To illustrate, suppose that

$$\|L - K\|_\infty < \epsilon.$$

The two functions are then never more than a distance ϵ apart. Mathematically this means that at any point $x \in D$

$$\|L(x) - K(x)\| < \epsilon.$$

The function norm expresses how close two functions are. To emphasize, $\|L - K\|_\infty < \epsilon$ if and only if at any $x \in D$, $L(x)$ and $K(x)$ are less than a distance ϵ apart. If two functions are close, they are close throughout their entire domain. They cannot be close at one point and far apart at another point.

CONVERGENCE

We can now define convergence of functions. Given $L: D \subset R^q \to R^n$, suppose that a sequence of functions $L^k: D \subset R^q \to R^n$ is also given. We say that $L^k \to L$ if

$$\lim_{k \to \infty} \|L^k - L\|_\infty \longrightarrow 0.$$

Verbally, the sequence L^k converges to L if as k gets large, L^k and L get arbitrarily close. This closeness is throughout their entire domain.

Now that the concept of two functions close is clear, we can formally state the following:

WEIERSTRASS THEOREM 22.2.1   Let $D \subset R^q$ be compact and $L\colon D \to R^n$ be continuous. Then, given any $\epsilon > 0$, there is a polynomial $P\colon D \to R^n$ such that

$$\| P - L \|_\infty < \epsilon.$$

The theorem says that a continuous function can be approximated as closely as we like by a polynomial. Of course, the closer we wish P to approximate L, the more terms, in general, P must have.

Note here that $L\colon D \subset R^q \to R^n$, where $L = (L_i)$ and $P\colon D \subset R^q \to R^n$, where $P = (P_i)$. Thus L_1 will be approximated by P_1, L_2 by P_2, L_3 by P_3, and so on. Moreover, each P_i is a polynomial of q variables. In other words, each P_i will be the sum of terms of the form

$$a_{i,k}(x_1)^{l^1_{i,k}}(x_2)^{l^2_{i,k}}, \ldots, (x_q)^{l^q_{i,k}},$$

where $l^j_{i,k}$ are nonnegative integers (possibly zero), a_{ik} a real number, and $x = (x_1, \ldots, x_q)$.

APPLICATION

In this chapter the Weierstrass theorem will be utilized to approximate homotopies and other functions. Explicitly, suppose that a continuous homotopy H is given. Then from Weierstrass's theorem a sequence of polynomials H^k exists such that

$$H^k \longrightarrow H.$$

Similarly, if a function F is continuous, a sequence of polynomials F^k exists such that

$$F^k \longrightarrow F.$$

This result that a continuous homotopy can be approximated by a sequence of polynomials (and similarly for a continuous F) will soon be quite helpful.

22.3 APPLYING THE SARD
AND WEIERSTRASS THEOREMS

With the Sard theorem and the Weierstrass theorem now developed, we can utilize them simultaneously. Given a continuous function, a function approximating it will be obtained where the approximating function is both a polynomial and regular. The approximating function, being a polynomial, will be infinitely differentiable. The continuous function will therefore be approximated by a function that is both regular and infinitely differentiable. The approximating function will later be used to reprove our theorems without the extra assumptions.

Let us show in detail how to obtain the approximation. For convenience it is done for a homotopy, although the process is identical for other functions.

Suppose that a homotopy $H: D \times T \longrightarrow R^n$ is given, where D is the closure of a bounded open set and H is continuous. We want to create another homotopy $\bar{H}: D \times T \longrightarrow R^n$ that is arbitrarily close to H but such that $\bar{H}$ is both regular and a polynomial.

First use Theorem 22.2.1 to approximate H by a polynomial $P: D \times T \longrightarrow R^n$. If P is regular at the right-hand-side zero, then let $\bar{H} = P$.

Suppose that P is not regular there. Since a polynomial is infinitely differentiable, Theorem 22.1.1 and Corollary 22.1.2 apply. Then for an e arbitrarily close to zero, the system

$$P(x, t) = e$$

or, equivalently,

$$P(x, t) - e = 0$$

will be regular. Define $\bar{H} = P - e$. The function $\bar{H}$ approximates H but is infinitely differentiable and regular at the right-hand-side zero. Moreover, since we can select P arbitrarily close to H and e arbitrarily small, $\bar{H}$ can be as close as we like to H.

The process is immediate from Sard and Weierstrass. Via Weierstrass we can approximate a continuous function by a polynomial. By Sard we can wiggle e if necessary to ensure that the polynomial is regular. Specifically, given any continuous function, there is an arbtrarily close approximation of it that is both polynomial and regular.

CONCLUSION

Taking this process one step further, consider a homotopy H and a function F both continuous. The approximations can be made arbitrarily close. There are thus sequences H^k and F^k of functions where all H^k and F^k are both polynomial and regular such that

$$H^k \longrightarrow H \quad \text{and} \quad F^k \longrightarrow F.$$

This result is precisely what we sought, and with it we can now proceed to revise our theorems.

22.4   ASSUMPTION RELAXATION FOR THE HOMOTOPY INVARIANCE THEOREM AND THE FIXED-POINT THEOREM

That our theorems hold for continuous functions with no regularity assumptions will now be demonstrated. The general procedure is the following. First approximate the continuous function by one that is polynomial and regular. The theorem already applies to the approximating function, because it is infinitely differentiable and regular. Then take limits, letting the approximation shrink to zero. But in the limit the approximating function becomes the initial continuous one. The theorem will then be shown to hold in the limit also, that is, for the initial continuous function.

To show the idea, this process will be demonstrated for some of the theorems in Chapter 3 that deal with degree, homotopy invariance, and fixed points. That will illustrate the general procedure; the proofs for our other theorems are left as exercises.

CHAPTER 3 REVISITED

At this point the reader is urged to review Chapter 3. One of the underlying themes of Chapter 3 is the notion of boundary-freeness. The following lemma sharpens our perception of that idea and of the limit process we will employ. The lemma states that if H is boundary-free and H^k is sufficiently close to H, then H^k is boundary-free also. (Recall that ∂D denotes the boundary of the set D.)

LEMMA 22.4.1   Given $D \subset R^n$, where D is the closure of an open bounded set, suppose that all H^k and H are continuous functions from $D \times T$ to R^n. Also suppose that $H^k \longrightarrow H$. If H is boundary-free, then for k sufficiently large, so is H^k.

Proof.   Because H is boundary-free, if x is in the boundary of D and $t \in T$,

$$H(x, t) \neq 0,$$

so that

$$\| H(x, t) \| > 0$$

for all $x \in \partial D$ and all $t \in T$. But ∂D is compact because D is the closure of an open bounded set. Also, T is compact and H is continuous. Thus for some $a > 0$,

$$\| H(x, t) \| > a$$

for all $x \in \partial D$ and $t \in T$.

Since $H^k \longrightarrow H$, if k is sufficiently large,

$$\| H^k(x, t) \| > \tfrac{1}{2}a$$

for all $x \in \partial D$ and $t \in T$. Therefore, $H^k(x, t) = 0$ cannot have a solution for any $x \in \partial D$ and $t \in T$. This means that for k sufficiently large, H^k is boundary-free. $\square$

With the boundary-free notion extended to sequences, we can begin our attack on degree.

DEGREE AND HOMOTOPY INVARIANCE

The first thing needed is a new definition of degree for continuous functions as our previous definition in Chapter 3 referred to regular and continuously differentiable functions. That is easily accomplished by exploiting the idea of approximations in the limit.

Given $D \subset R^n$, the closure of an open bounded set, let $F \colon D \longrightarrow R^n$ be continuous and assume that F^{-1} is in D^0 (the interior of D). Let $F^k \colon D \longrightarrow R^n$ be a sequence of continuously differentiable regular functions that converge to F. Then define

$$\deg (F) = \lim_{k \to \infty} \deg (F^k). \qquad (22.4.1)$$

(If necessary, take a subsequence so that the limit exists.)

Each F^k is continuously differentiable and regular, so by Chapter 3 its degree can be calculated. The limit of those degrees is the degree of F. More explicitly, the degree for a continuous function F is defined as the limit of the degree of the F^k, where the F^k are regular continuously differentiable functions that approximate F. (Note by definition (22.4.1) $F^{-1} \cap \partial D$ is required to be empty.)

At this point it is unresolved whether our degree definition is well-defined and unique. Specifically, suppose that two different sequences converge to F, say $F^k \longrightarrow F$ and $\bar{F}^k \longrightarrow F$. We must show that

$$\lim_{k \to \infty} \deg (F^k) = \lim_{k \to \infty} \deg (\bar{F}^k).$$

In other words, we must show that the limits given by the two different sequences are the same. Although that could be shown now, it is conceptually easier to assume it and first extend the homotopy invariance theorem to continuous functions. It will then easily follow that degree is well-defined. We therefore proceed immediately to the homotopy invariance theorem.

HOMOTOPY INVARIANCE

The homotopy invariance theorem makes an elegant usage of Sard's theorem by applying it to three functions at once.

Suppose that there are three continuously differentiable functions L^1: $A^1 \longrightarrow R^n$, $L^2: A^2 \longrightarrow R^n$, and $L^3: A^3 \longrightarrow R^n$. Also suppose that the set of degenerate values for L^i is the set C^i, $i = 1, 2, 3$, where each set C^i has measure zero. Notice that for rhs e, L^i is degenerate if $e \in C^i$ but is regular if $e \notin C^i$.

Now specify

$$C = C^1 \cup C^2 \cup C^3.$$

Let us examine C. If $e \in C$, then at least one L^i is degenerate for that rhs. However, if $e \notin C$, then all L^i, $i = 1, 2, 3$, are regular for e. To reiterate, if $e \notin C$, then

$$L^1, \quad L^2, \quad \text{and} \quad L^3$$

are all regular for e. Observe that the same e works for all three functions simultaneously. Understand then, give $e \notin C$, that

$$\bar{L}^1 = L^1 - e, \qquad \bar{L}^2 = L^2 - e, \qquad \text{and} \qquad \bar{L}^3 = L^3 - e$$

are all regular.

The set C has measure zero because it is the finite union of measure-zero sets. Also, all three functions will be regular for the same rhs e, except for $e \in C$. Restated, all three functions are regular for e, except for e on a set of measure zero. Further, e can be selected arbitrarily close to zero.

With these observations the following theorem is a quick consequence of the earlier homotopy invariance theorem of Chapter 3.

HOMOTOPY INVARIANCE THEOREM 22.4.2 Let $D \subset R^n$ be the closure of an open bounded set with $H: D \times T \longrightarrow R^n$ continuous. Suppose that H is boundary-free and

$$H(x, 0) = E(x) \qquad \text{and} \qquad H(x, 1) = F(x)$$

Then

$$\deg(E) = \deg(F).$$

Proof. Since H is boundary-free, setting $t = 0$ and $t = 1$, neither E^{-1} nor F^{-1} has points in ∂D. Thus the definition of degree would apply to E and to F.

Via Weierstrass's theorem there is a sequence of polynomials H^k such that

$$H^k \longrightarrow H. \tag{22.4.2}$$

Define

$$E^k(x) = H^k(x, 0) \qquad \text{and} \qquad F^k(x) = H^k(x, 1)$$

and notice that by (22.4.2),

$$E^k(x) \longrightarrow E(x) \qquad \text{and} \qquad F^k(x) \longrightarrow F(x). \tag{22.4.3}$$

For any k, by Sard's theorem, the set on which H^k is degenerate (i.e., does not have rhs that is regular) has measure zero, the set on which E^k is degenerate has measure zero, and the set on which F^k is degenerate has measure zero. But then our previous reasoning applies.

Specifically, we can select an e^k such that for e^k, all three are regular. The same e^k works for all three simultaneously. In particular, we can select e^k arbitrarily small for which

$$\bar{H}^k = H^k - e^k, \qquad \bar{E}^k = E^k - e^k, \qquad \text{and} \qquad \bar{F}^k = F^k - e^k$$

are all regular for rhs zero.

Recall that e^k can be selected close to zero; that is, we can select e^k so that

$$e^k \longrightarrow 0 \qquad \text{as } k \longrightarrow \infty.$$

But then

$$\bar{H}^k \longrightarrow H, \qquad \bar{E}^k \longrightarrow E \qquad \bar{F}^k \longrightarrow F. \tag{22.4.4}$$

Notice we have

$$\bar{H}^k(x, 0) = \bar{E}^k(x), \qquad \bar{H}^k(x, 1) = \bar{F}^k(x),$$

where all $\bar{H}^k$, $\bar{E}^k$, and $\bar{F}^k$ are polynomials and regular. Further, by Lemma 22.4.1, $\bar{H}^k$ is boundary-free for k sufficiently large. We therefore possess all the ingredients of Theorem 3.4.3, so that

$$\deg \bar{E}^k = \deg \bar{F}^k.$$

But using (22.4.4) and by definition of degree for a continuous function,

$$\deg E = \lim \deg \bar{E}^k = \lim \deg \bar{F}^k = \deg F. \qquad \square$$

Sard, Weierstrass, and a technical but not at all profound limiting argument permit us to reprove the homotopy invariance theorem for continuous functions.

That the definition of degree is well defined follows using an argument similar to the proof of Theorem 22.4.2. Take two sequences converging to F,

$F^k \longrightarrow F$ and $\bar{F}^k \longrightarrow F$, where the F^k and $\bar{F}^k$ are polynomials and regular. Also, call $\bar{F} = F$, so that $\bar{F}^k \longrightarrow \bar{F}$. The "artificial" homotopy

$$H(x, t) = (1 - t)F(x) + t\bar{F}(x) = F(x)$$

is boundary-free because, by assumption, F^{-1} is not in the boundary of D.
Define

$$H^k(x, t) = (1 - t)F^k(x) + t\bar{F}^k(x) - e^k.$$

Using Sard, we may select $e^k \longrightarrow 0$, so that H^k is regular. Theorem 3.4.3 ensures that

$$\deg H^k(\cdot, 0) = \deg H^k(\cdot, 1).$$

Also, since H^k is regular and since for k sufficiently large $e^k \longrightarrow 0$, it is an easy exercise (Exercise 6) to show that

$$\deg H^k(\cdot, 0) = \deg F^k$$
$$\deg H^k(\cdot, 1) = \deg \bar{F}^k.$$

Hence

$$\lim \deg F^k = \lim \deg \bar{F}^k.$$

Thus even if we take two different subsequences converging to F, the same limit results. Degree is well-defined.

A most useful extension of these ideas is the following. By definition of degree for regular twice-continuously differentiable functions, we know from Chapter 3 that if $\deg F \neq 0$, then the system $F(x) = 0$ must possess a solution. Under our new definition of degree for continuous functions, the same holds true.

LEMMA 22.4.3   Let $F: D \longrightarrow R^n$ be continuous, $D \subset R^n$ be the closure of an open bounded set, and $\partial D \cap F^{-1}$ be empty. Suppose that

$$\deg F \neq 0.$$

Then the system

$$F(x) = 0$$

has a solution in D.

Proof.   Let $F^k \longrightarrow F$, where the F^k are polynomials and regular. Suppose, for contradiction, that $F(x) = 0$ has no solution in D. Then for k sufficiently large,

$$F^k(x) = 0$$

also has no solution in D. Consequently, by definition of degree in Chapter 3, all for k large,

$$\deg F^k = 0.$$

Yet then

$$\deg F = \lim \deg F^k = 0. \qquad \square$$

STABILITY

Degree is a fundamental concept because it is stable under perturbation. In Figure 22.4.1, functions A and C are small perturbations of B, and all three functions A, B, and C all have the same degree, $+1$. Function B is not regular, but its degree by definition of degree for continuous functions is the same as the "nearby" regular and continuously differentiable functions A and C.

Examining Figure 22.4.1, also perceive that the number of solutions of functions A, B, and C is different for each. Thus the number of solutions is not stable under small perturbation.

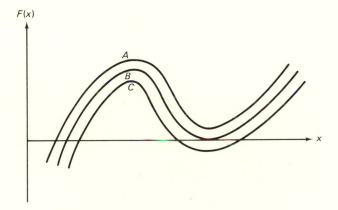

Functions A, B, and C all have degree $+1$. Function A has one solution, B has two solutions, and C has three solutions.

Figure 22.4.1

It is interesting that the number-of-solutions concept which is so important for practical purposes is not stable. (However, this number is stable in complex space. See Garcia and Li [1980].) This is perhaps one reason why it is often so difficult to calculate solutions in the reals. The more fundamental concept, degree, is easier to deal with because it is stable, yet it turns out to be less valuable from an applied perspective. This dichotomy between what is needed practically and what from a standpoint of stability is more basic needs further examination.

22.5 FIXED-POINT THEOREMS

Chapter 3 established fixed-point theorems under the assumptions that the function F be C^2 and H be regular. The fixed point also exists if F is continuous. We will rely upon Theorem 22.4.2 and, as before, take limits.

THEOREM 22.5.1   Let $D \subset R^n$ be the closure of an open bounded set and $F: D \longrightarrow R^n$ be continuous. Suppose that for $x^0 \in D^0$,

$$H(x, t) = (1 - t)(x - x^0) + t(x - F(x))$$

is boundary-free for $0 \leq t < 1$. Then F has a fixed point

$$F(x^*) = x^*$$

for $x^* \in D$.

Proof.   We want to apply Theorem 22.4.2, but that requires H boundary-free for T, which it is not. We therefore will approximate H by homotopies H^k that are boundary-free and then take limits.

Let $\tau^k \longrightarrow 0$, where all $\tau^k > 0$, and consider the homotopies for $0 \leq r \leq 1$:

$$H^k(x, r) = \left(1 - \frac{r}{1 + \tau^k}\right)(x - x^0) + \frac{r}{1 + \tau^k}(x - F(x))$$

$$= H\left(x, \frac{r}{1 + \tau^k}\right).$$

Notice that at $r = 1$, $r/(1 + \tau^k) < 1$. Thus, since H is boundary-free for $0 \leq t < 1$, H^k is boundary-free for $0 \leq r \leq 1$. Theorem 22.4.2 can then be applied to H^k.

At $r = 0$, the Jacobian of H^k is the identity, so

$$\deg H^k(\cdot, 0) = \det I = 1.$$

Using Theorem 22.4.2, since H^k is boundary-free for $0 \leq r \leq 1$,

$$\deg H^k(\cdot, 1) = \deg H^k(\cdot, 0) = 1.$$

Lemma 22.4.3 then assures us that the equation

$$H^k(x, 1) = 0$$

has a solution, call it x^k.

Let the sequence x^k have a cluster point x^*. By continuity of H and the definition of H^k, taking subsequences if necessary,

$$0 = \lim_{k \to \infty} H^k(x^k, 1) = \lim_{k \to \infty} H\left(x^k, \frac{1}{1 + \tau^k}\right) = H(x^*, 1).$$

Clearly, x^* is a fixed point. $\square$

The fact that the Leray–Schauder conditions or the Brouwer conditions implies a fixed point for F continuous follows immediately using the same reasoning as above. In particular, the general statement of the Brouwer theorem is as follows.

COROLLARY 22.5.2   Let D be the closure of an open bounded and convex set in R^n and $F: D \longrightarrow D$ be continuous. Then F has a fixed point in D.

22.6 OVERVIEW

We have now illustrated on some theorems from Chapter 3 the process of relaxing the differentiability and regularity assumptions and establishing these theorems for continuous functions. To eliminate the differentiability and regularity assumptions for other theorems, apply the same process. Specifically

1. Approximate the continuous function by a sequence of functions each of which is regular and polynomial.
2. Observe that the previous form of the theorem holds for each approximating function in the sequence.
3. Take limits and show that the theorem also holds in the limit, that is, for the continuous function.

Since the process is similar to that shown in this chapter, we will not repeat it for the other theorems. These are left as exercises (Exercises 9, 13, and 14).

THE APPROXIMATING FUNCTIONS MUST BE AT LEAST C^2

In the procedure described, we approximated the continuous function by a sequence of functions each of which is regular and polynomial. Clearly, polynomials are more than necessary since we do not need infinite differentiability, only C^2. Therefore, it is possible to relax the approximations so that they need not be polynomials.

SUMMARY

This chapter reviewed the Sard and Weierstrass theorems. These theorems permitted us to approximate a continuous function by one that is polynomial and regular. Then by taking limits, our previous theorems were reproved without the assumptions (i.e., for functions that are only continuous). The process succeeds because the properties we utilized are stable. Even if we perturb the problem slightly, the properties are undisturbed. It is the stability which permitted us to take limits and be assured that the result would still hold in the limit.

As noted, this chapter is technical and possibly superfluous. Nevertheless, it reveals that even though the theorems were established under some seemingly strong assumptions, the assumptions were not very crucial at all. If the assumptions do not hold, a slight perturbation will bring them about.

A useful extension of Sard's theorem which allows perturbation of the left-hand side is included in Appendix A of this chapter. Appendix B of the chapter considers applications to nonlinear programming.

APPENDIX A

SARD EXTENDED

We now provide a more general version of Sard's theorem that permits perturbation of the left-hand side as well as the right. As before, "almost always" means except on a set of measure zero, and a function "regular" means regular at rhs zero. Also, H regular means H' of full rank for all $(x, t) \in H^{-1}$, and H regular at t means H'_x full rank for all $x \in H^{-1}(t)$.

To begin, we must pay attention to the functions under consideration. Specify $L: B \times R^s \longrightarrow R^n$, where $B \subset R^q$, and let $w \in B$ and $e \in R^s$, so L is written as $L(w, e)$. Notice carefully that whereas before, e was fixed e is now a variable. The Jacobian L' is an $n \times (q + s)$ matrix, and L'_w is the partial Jacobian consisting of the first q columns of L'. Also,

$$L^{-1} = \{(w, e) \mid L(w, e) = 0\}.$$

Now suppose that we fix $e = \bar{e}$. Then $L(\cdot, \bar{e}): B \longrightarrow R^n$. Letting $L(\cdot, \bar{e})$ denote $e = \bar{e}$ fixed, its Jacobian $L'(\cdot, \bar{e})$ is $n \times q$ and

$$L_{\bar{e}}^{-1} = \{w \mid L(w, \bar{e}) = 0\}.$$

Observe that in L both w and e are variables, but for $L(\cdot, \bar{e})$, only w is a variable and $e = \bar{e}$ is fixed.

Recall that L is regular if its Jacobian is rank n for all $(w, e) \in L^{-1}$. Also, $L(\cdot, \bar{e})$ is regular if its Jacobian is rank n for all $w \in L_{\bar{e}}^{-1}$.

We can now state the extended Sard theorem, whose proof is in Abraham and Robbin [1967]. This theorem is also referred to as the transversality theorem.

THEOREM A.22.1 (EXTENDED SARD)   Let $L: B \times R^s \longrightarrow R^n$ for $B \subset R^q$ be C^k, where $k \geq 1 + \max\{0, q - n\}$. Also let L be regular. Now suppose that we fix $e = \bar{e}$. Then $L(\cdot, \bar{e}): B \longrightarrow R^n$ is regular for almost all $\bar{e}$.

The theorem is a bit confusing as it assumes that L is regular in order to prove that $L(\cdot, \bar{e})$ is regular. The key is understanding that for L, both w and e are variables, but for $L(\cdot, \bar{e})$, w is a variable and $\bar{e}$ is fixed. A few examples will clarify any confusion.

EXAMPLE A.22.1

Theorem A.22.1 is straightforward in the case where $B \subset R^n$ and $L: B \times R^1 \longrightarrow R^n$ is C^1 and regular.

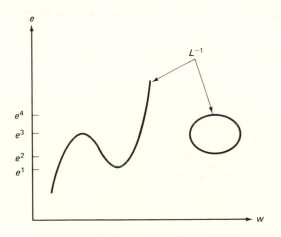

$L(\cdot, e)$ is regular except for $e = e^i, i = 1, 2, 3, 4$.

Figure A.22.1

Since L is assumed regular, the solutions (w, e) to

$$L(w, e) = 0 \qquad (A.22.1)$$

consist of paths, by the implicit function theorem. For example, examine Figure A.22.1. There L^{-1} consists of a path and a loop and $L'_w(w, e)$ is nonsingular for all $(w, e) \in L^{-1}$ provided that $e \neq e^i$, $i = 1, 2, 3, 4$. Thus $L(\cdot, e)$ is regular for all $e \neq e^i$, $i = 1, 2, 3, 4$. Hence, $L(\cdot, \bar{e}): B \longrightarrow R^n$ is regular for almost all $\bar{e}$.

EXAMPLE A.22.2

Let $F: R^n \longrightarrow R^n$ be C^1. Now define $L: R^n \times R^n \longrightarrow R^n$ by

$$L = F - e,$$

where $e \in R^n$. First examine the $n \times (n + n)$ Jacobian

$$L' = (F', -I),$$

where I is the $n \times n$ identity matrix. This follows because here e is variable. Note that because of I, the Jacobian is always of rank n. Thus L is automatically regular.

Notice that $L(\cdot, \bar{e}) = F - \bar{e}$, where $e = \bar{e}$ is now fixed. Theorem A.22.1 then applies and $L(\cdot, \bar{e})$ is regular for almost all $e = \bar{e}$. Rephrased,

$$F(x) = \bar{e}$$

is regular for rhs $\bar{e}$ for almost all $\bar{e}$. We thus obtain our previous Sard result!

EXAMPLE A.22.3

The identical argument holds for proving the regularity of $H: R^{n+1} \longrightarrow R^n$, $H \in C^2$. Define $L: R^{n+1} \times R^n \longrightarrow R^n$ by $L = H - e$. Then

$$L' = (H', -I),$$

so L is regular always. Theorem A.22.1 then ensures that

$$H(y) = e$$

is regular for rhs e for almost all e. Again our earlier conclusion holds.

EXAMPLE A.22.4

Our previous examples show how Theorem A.22.1 can be used to perturb the right-hand side. Especially with some homotopies, it is beneficial to perturb the left-hand side.

Let $H: R^{n+1} \longrightarrow R^n$, $H \in C^2$ be regular at $t = 0$ and $t = 1$. Define $L: R^{n+1} \times R^n \longrightarrow R^n$ by

$$L_i = H_i + t(1 - t)e_i, \qquad i = 1, \ldots, n,$$

so

$$L = H + t(1 - t)e.$$

The Jacobian is

$$L' = (H', t(1 - t)I).$$

Observe the $t(1 - t)$ factor in front of the $n \times n$ identity matrix I.

If $t \neq 0$ or $t \neq 1$, then the $t(1 - t)I$ part of L' ensures that L' has rank n. At $t = 0$ or $t = 1$, H is regular, so there H'_x has rank n. But H'_x is the first n columns of L', so at $t = 0$ and $t = 1$, L' has rank n. Consequently, for all t, L' has rank n, that is, L is always regular.

Theorem A.22.1 ensures that for almost all e fixed,

$$\tilde{H}(x, t) \equiv L(x, t) = H(x, t) + t(1 - t)e = 0$$

is regular. Moreover,

$$\tilde{H}(x, 0) = H(x, 0) \quad \text{and} \quad \tilde{H}(x, 1) = H(x, 1).$$

We have obtained a homotopy $\tilde{H}$ that has the same solutions at $t = 0$ and $t = 1$ as H. $\tilde{H}$ is regular at $t = 0$ and $t = 1$. Also, we can select e to ensure that $\tilde{H}$ is regular. Thus we can select e arbitrarily small so that $\tilde{H}$ is regular and regular at $t = 0$ and $t = 1$.

This result is very useful. Suppose that H is regular at $t = 0$ and $t = 1$ but we do not know if H is regular. Then we can perturb it into a homotopy $\tilde{H}$ that is regular at $t = 0$ and $t = 1$ and in addition, is regular. Most important, the solutions of the two homotopies H and $\tilde{H}$ are the same at $t = 0$ and $t = 1$.

Note that this could not be done by our earlier version of Sard because that requires a rhs perturbation. The solutions at $t = 0$ and $t = 1$ would then be different.

Indeed, this particular perturbation $t(1 - t)e$ is extremely powerful and can be used to simplify many results.

APPENDIX B

REGULARITY IN NONLINEAR PROGRAMMING

Consider the nonlinear programming problem:

$$\min f(x)$$

$$\text{subject to}$$

$$h(x) = 0,$$

where $f: R^n \longrightarrow R^1$ and $h: R^n \longrightarrow R^s$.

The Kuhn–Tucker conditions are necessary conditions for the nonlinear programming problems and are the basis for much of nonlinear programming.

THEOREM (KUHN–TUCKER)

Let the functions f and h be differentiable and suppose x^* solves the nonlinear programming problem. Then under a *constraint qualification* there exists a μ^* such that (x^*, μ^*) solves the following Kuhn–Tucker conditions:

$$\nabla f(x^*)^\tau + \sum_{j=1}^{s} \mu_j^* \nabla h_j(x^*)^\tau = 0$$

$$h(x^*) = 0.$$

The proof is easily available (see Avriel [1976], Mangasarian [1969], or Zangwill [1969]). However, our main concern here is the constraint qualification. Although there are many kinds of constraint qualifications, the one used in Chapter 4 is

(CQ):   At the minimum x^*, the derivative vectors
$\nabla h_1(x^*), \ldots, \nabla h_s(x^*)$ are linearly independent.

We now use Sard's theorem to show that the constraint qualification almost always holds. In other words, unlike much previous literature in nonlinear programming we need not split hairs over the CQ assumption. The CQ holds almost always, and when it does not hold, a slight perturbation of the constraints would make it hold.

The function h is from R^n to R^s, and suppose that h has continuous derivatives up to order $1 + \max\{0, n - s\}$. Then since at x^*

$$h(x^*) = 0,$$

Sard's theorem guarantees that $\nabla h(x^*)$ has rank s almost always. This means that the columns $\nabla h_j(x^*), j = 1, \ldots, s$, are then linearly independent. Explicitly, for almost all e

$$\min f(x)$$
$$h(x) = e$$

satisfies the CQ assumption.

Two remarks are in order. First, note that our results hold even when inequality constraints

$$h_j(x) \leq 0$$

are present. For example, we may simply rewrite an inequality constraint as

$$h_j(x) + y_j^2 = 0,$$

where y_j is a real variable.

Second, in practice, we still need to be wary of difficulties. For example, there may be a feasible set

$$\{x \mid h(x) = 0\}.$$

Yet perturbing to $h(x) = e$ could conceivably cause the set

$$\{x \mid h(x) = e\}$$

to be empty (Exercise 11). Thus we must be careful that slight perturbations do not greatly alter the feasible set.

1. Prove that if a continuously differentiable $L: B \subset R^q \longrightarrow R^n$ is such that $L'(\bar{x})$ is of rank n, then $L'(x)$ is of rank n for all x in a neighborhood of $\bar{x}$.

2. Let $B \subset R^2$ be the set

    $$B = \{(x_1, x_2) \in R^2 \,|\, 0 \le x_i \le 1, \quad i = 1, 2, \quad x_1 \text{ a rational number}\}.$$

    Is B a set of measure zero in R^2? Explain.

3. Let $L: R^q \longrightarrow R^n$ be C^1, where $q \le n - 1$. Prove via Sard's theorem that $L(R^q)$ is of measure zero. (This shows that a system with more equations than unknowns will have no solution almost always.)

4. Let $f^k: R^1 \longrightarrow R^1$ be defined by

    $$f^k(x) = \begin{cases} 1 - (kx - 1)^2, & 0 \le x \le 2/k \\ 0, & x \le 0 \text{ or } x \ge 2/k \end{cases} \qquad k = 1, 2, \ldots.$$

    We say f^k converges to a function f *pointwise* if $f^k(x) \longrightarrow f(x)$ for every x fixed. Show that f^k converges to the zero function pointwise but that $f^k \longrightarrow$ zero function (in function norm) is false.

5. Prove that if $F^k \longrightarrow F$, then F^k is pointwise convergent to F.

6. Let $F: D \subset R^n \longrightarrow R^n$ be C^1, where D is the closure of an open bounded set and $F^{-1} \cap \partial D = \varnothing$. Show that for any e sufficiently small, $\deg (F - e) = \deg F$.

7. Let $F: D \subset R^n \longrightarrow R^n$ be C^1, where D is the closure of an open bounded set and $F^{-1} \cap \partial D = \varnothing$. Show that $\deg F$ is a finite integer.

8. Consider the fixed-point homotopy

    $$H(x, t, x^0) = (1 - t)(x - x^0) + t(x - F(x)) = 0, \qquad H \in C^2.$$

    a. Prove by using the extended Sard theorem that if $(I - F')$ is nonsingular at each fixed point of F, then $H(\cdot, \cdot, x^0)$ is regular for almost all x^0.

    b. Suppose that H is not regular. Discuss which procedure is more desirable in computations: perturbing the right-hand-side zero or perturbing the initial point x^0.

9. Prove Theorem 3.3.1 where the regularity and differentiability assumptions are replaced only by continuity.

10. Suppose that $F: R^n \longrightarrow R^n$ is the linear function

    $$F(x) = Ax + b = 0, \qquad A \text{ is } n \times n \text{ and singular}.$$

    a. What happens to F^{-1} if the right-hand side is perturbed? Explain.

    b. Show how the extended Sard theorem is utilized so that F^{-1} does not become empty under a perturbation.

11. Give an example of a continuous h such that the feasible set

    $$\{x \,|\, h(x) = e\}$$

    is empty for all $e \ne 0$, and nonempty for $e = 0$.

12. Let

$$H_1(x_1, x_2, t) = (x_1)^2 + t^2 - 1$$
$$H_2(x_1, x_2, t) = (x_2)^2 + t^2 - 1.$$

   a. Show that 0 is not a regular value of H.
   b. Find all values of e close to zero for which e is a regular value of H.

13. In Theorem 5.6.2 suppose that we replace the regularity and differentiability assumptions by only continuity. Prove that an equilibrium point exists.

14. Prove that a polynomial system in n equations in n complex variables has at most
   $Q = \prod_{i=1}^{n} q_i$ isolated solutions, where q_i is the power of the highest-ordered term in the ith equation.

NOTES

For a reference on the Sard theorem, see Guillemin and Pollack [1974], Milnor [1965] or Sternberg [1964]. The Extended Sard Theorem may be found in Abraham and Robbin [1967]. Ortega and Rheinboldt [1970] contains a proof of the Weierstrass approximation theorem.

Appendix A

Derivatives
and
Differential Equations

A.1 FUNCTIONS

A FUNCTION

A function $F: R^n \longrightarrow R^m$ indicates that F takes a point $x \in R^n$ into a point $F(x) \in R^m$. The function $F(x)$ is itself a column vector

$$F(x) = \begin{pmatrix} F_1(x) \\ \cdot \\ \cdot \\ \cdot \\ F_m(x) \end{pmatrix}. \qquad (1.\text{A}.1)$$

In other words, F is a vector-valued function.

CONTINUITY

A function $F: R^n \longrightarrow R^m$ is said to be *continuous* at x if $x^k \longrightarrow x$ implies that $F(x^k) \longrightarrow F(x)$.

An equivalent definition is: Given $\epsilon > 0$, there exists a $\delta > 0$ such that

$$\|y - x\| < \delta$$

implies that

$$\| F(y) - F(x) \| < \epsilon.$$

[Here $\| x \|^2 = \sum_1^n (x_i)^2$, and $\| F(x) \|^2 = \sum_{i=1}^m (F_i(x))^2$, the Euclidean norm.]
A function is continuous if it is continuous at all points x for which it is defined.

PARTIAL DERIVATIVES

Let $F: R^n \longrightarrow R^m$ be given. If each component function $F_i(x)$ has first partial derivatives that are continuous, we write $F \in C^1$. In general, if the component functions have continuous partial derivatives of order k, we write $F \in C^k$.

GRADIENT

Consider $f: R^n \longrightarrow R^1$; that is, f is a real-valued function on R^n. If $f \in C^1$, the *gradient* of f is defined to be the vector

$$\nabla f(x) = \left(\frac{\partial f(x)}{\partial x_1}, \frac{\partial f(x)}{\partial x_2}, \ldots, \frac{\partial f(x)}{\partial x_n} \right). \tag{1.A.2}$$

The gradient is considered to be a row vector.

HESSIAN

If $f \in C^2$, the *Hessian* of f at x is the $n \times n$ matrix

$$\nabla^2 f(x) = \left(\frac{\partial^2 f(x)}{\partial x_i \partial x_j} \right). \tag{1.A.3}$$

EXAMPLE 1.A.1.

If $f(x) = c^{\tau} x + \frac{1}{2} x^{\tau} Q x$ for Q symmetric $n \times n$, then

$$\nabla f(x)^{\tau} = c + Q x$$
$$\nabla^2 f(x) = Q.$$

DERIVATIVES OF VECTOR-VALUED FUNCTIONS

Given a vector-valued $F: R^n \longrightarrow R^m$, differentiation is similar to (1.A.2). If $F \in C^1$, the first derivative of F is the $m \times n$ matrix, called the *Jacobian*,

$$F'(x) = \left(\frac{\partial F_i(x)}{\partial x_j} \right). \tag{1.A.4}$$

For example, let $F: R^3 \longrightarrow R^2$ be defined by

$$F(x) = \begin{pmatrix} x_1 + (x_2)^2 + x_1 x_3 \\ x_2 + (x_3)^2 + 7 \end{pmatrix}.$$

Then

$$F'(x) = \begin{pmatrix} 1 + x_3 & 2x_2 & x_1 \\ 0 & 1 & 2x_3 \end{pmatrix}.$$

If $F \in C^2$, it is possible to define the m Hessians $F_1''(x), \ldots, F_m''(x)$ corresponding to the m component functions $F_i(x)$. Note that the second derivative of the entire function, $F''(x)$, is a three-dimensional matrix

$$\left(\frac{\partial^2 F_i}{\partial x_j \partial x_k} \right).$$

A.2   MEAN VALUE THEOREMS

The most frequently used results about derivatives fall under the general heading of *mean value theorems* or *Taylor's theorem*.

REAL-VALUED FUNCTIONS

Suppose that $f: R^n \longrightarrow R^1, f \in C^1$ is given. Then for any two points x and y in R^n, there is a $\theta, 0 \leq \theta \leq 1$ such that

$$f(y) = f(x) + \nabla f(x + \theta(y - x))(y - x). \tag{2.A.1}$$

Moreover, if $f \in C^2$, there is a $\theta, 0 \leq \theta \leq 1$, such that

$$f(y) = f(x) + \nabla f(x)(y - x)$$
$$+ \tfrac{1}{2}(y - x)^\tau \nabla^2 f(x + \theta(y - x))(y - x). \tag{2.A.2}$$

VECTOR-VALUED FUNCTIONS

It is important to note that (2.A.1) does *not* in general hold for mappings $F: R^n \longrightarrow R^m, m > 1$. A slightly different theorem does hold, however. If $F \in C^1$, there exists $\theta = (\theta_1, \ldots, \theta_m), 0 \leq \theta_i \leq 1$, all i, such that

$$F(y) = F(x) + \begin{pmatrix} F_1'(x + \theta_1(y - x)) \\ \cdot \\ \cdot \\ \cdot \\ F_m'(x + \theta_m(y - x)) \end{pmatrix} (y - x). \tag{2.A.3}$$

Note that usually the θ_i will all be different. That is, the F'_i will be evaluated at different intermediate points.

A.3 THE IMPLICIT FUNCTION THEOREM

FORMULATION

Suppose that given $H: R^{n+k} \longrightarrow R^n$, we are to solve

$$H(y) = 0, \qquad y \in R^{n+k}. \tag{3.A.1}$$

There are n equations and $n + k$ variables in (3.A.1). The implicit function theorem tells us when we can write the solution to (3.A.1) as a function of only k variables. In detail, let $y = (x, t)$, where $x \in R^n$, $t \in R^k$. Here $x = (x_1, \ldots, x_n)$ denotes the first n variables, $y_i, i = 1, \ldots, n$, and $t = (t_1, \ldots, t_k)$ the remaining k variables, $y_i, i = n + 1, \ldots, n + k$. We wish to determine if there is a function

$$x(t) \tag{3.A.2}$$

such that

$$H(x(t), t) = 0, \qquad t \in R^k. \tag{3.A.3}$$

Observe that the solution (3.A.2) is dependent only on the variables $t \in R^k$. Given t, the remaining n values $x(t)$ are at once determined.

EXAMPLE 3.A.1.

The simplest example of a system (3.A.1) is a linear system

$$Ay = 0, \qquad A \text{ is } n \times (n + k).$$

Let $y = (x, t)$, where $x \in R^n$, $t \in R^k$. Is there a function $x(t)$ of t such that

$$A(x(t), t) = 0? \tag{3.A.4}$$

The answer for this linear system is easily obtained. Let $A = [B, C]$, where B is the $n \times n$ matrix corresponding to the variables x and C, the $n \times k$ matrix corresponding to the variables t. Then $Ay = 0$ can be written as

$$Bx + Ct = 0.$$

If B is nonsingular, then the function $x(t)$ is clearly determined to be

$$x(t) = -B^{-1}Ct, \qquad t \in R^k. \tag{3.A.5}$$

In other words, the function $x(t)$ defined by (3.A.5) completely characterizes the solution to $Ay = 0$.

THE THEOREM

The implicit function theorem is a generalization of the linear case in Example 3.A.1. It characterizes the solution to $H(x, t) = 0$ in terms of the variables t, at least in a neighborhood of a given solution $(\bar{x}, \bar{t})$.

IMPLICIT FUNCTION THEOREM 3.A.1 Given $H \in C^1$, $H: R^{n+k} \longrightarrow R^n$, let $\bar{y} = (\bar{x}, \bar{t}), \bar{x} \in R^n, \bar{t} \in R^k$, be a point in R^{n+k} satisfying $H(\bar{y}) = 0$. Also suppose that the $n \times n$ Jacobian matrix

$$H'_x(\bar{y}) = \begin{pmatrix} \dfrac{\partial H_1(\bar{y})}{\partial x_1} & \cdots & \dfrac{\partial H_1(\bar{y})}{\partial x_n} \\ \cdot & & \cdot \\ \cdot & & \cdot \\ \cdot & & \cdot \\ \dfrac{\partial H_n(\bar{y})}{\partial x_1} & \cdots & \dfrac{\partial H_n(\bar{y})}{\partial x_n} \end{pmatrix} \tag{3.A.6}$$

is nonsingular.

Then there are open neighborhoods $N_1 \subset R^k$ and $N_2 \subset R^n$ of $\bar{t}$ and $\bar{x}$, respectively, and a function

$$x(t): \bar{N}_1 \longrightarrow \bar{N}_2 \tag{3.A.7}$$

such that

1. $x(\cdot) \in C^1$.
2. $\bar{x} = x(\bar{t})$.
3. For any $t \in \bar{N}_1$, the equation $H(x, t) = 0$ has a unique solution $x(t) \in \bar{N}_2$.

Observe that the implicit function theorem states that in a neighborhood of $(\bar{x}, \bar{t})$, the only solutions to $H(x, t) = 0$ are $H(x(t), t) = 0$.

EXAMPLE 3.A.2

Consider the equation

$$x^2 + t = 0. \tag{3.A.8}$$

A solution is $x = t = 0$. However, in a neighborhood of $(0, 0)$ there is no function $x(t)$ such that

$$x^2(t) + t = 0.$$

For (3.A.8) the Jacobian $H' = (2x, 1)$, and $H'_x(0, 0)$ is singular. Thus the theorem does not hold at $x = 0$, $t = 0$.

EXAMPLE 3.A.3

Suppose that the equation of interest is

$$x + t^2 = 0. \tag{3.A.9}$$

Clearly, (3.A.9) is just (3.A.8) but with the roles of x and t reversed. The implicit function theorem now holds at $x = t = 0$. In fact,

$$x(t) = -t^2$$

is the function we are seeking.

EXAMPLE 3.A.4

Suppose that $H: R^{n+1} \rightarrow R^n$, $H \in C^1$, is given. Let $\bar{y} = (\bar{x}, \bar{t})$, $\bar{x} \in R^n$, $\bar{t} \in R^1$ satisfy (3.A.1). If $H'_x(\bar{x}, \bar{t})$ is nonsingular, then the solution to (3.A.1) in a neighborhood of $(\bar{x}, \bar{t})$ is a unique path $(x(t), t)$. At $t = \bar{t}$, $x(\bar{t}) = \bar{x}$. Moreover, $x(\cdot) \in C^1$.

A.4 EXISTENCE AND UNIQUENESS OF SOLUTIONS FOR ORDINARY DIFFERENTIAL EQUATIONS

STANDARD FORM OF A SYSTEM OF ORDINARY DIFFERENTIAL EQUATIONS

Consider a system of m first-order ordinary differential equations in standard form:

$$\frac{dy_1}{dp} = L_1(y_1, \ldots, y_m, p)$$

$$\vdots \qquad \vdots$$

$$\frac{dy_m}{dp} = L_m(y_1, \ldots, y_m, p). \tag{4.A.1}$$

The L_i are given functions of $m + 1$ real variables $y_1, \ldots, y_m, p$. We wish to find solutions of (4.A.1), that is, sets of m functions $y_1(p), \ldots, y_m(p)$ of class C^1 which satisfy (4.A.1).

EXAMPLE 4.A.1

Let $H: R^{n+1} \rightarrow R^n$ be given, and suppose that we are to solve

$$H(x, t) = 0, \qquad x \in R^n, t \in R^1. \tag{4.A.2}$$

Calling $y = (x, t)$, the BDE of Chapter 2 are

$$\dot{y}_i = \frac{dy_i}{dp} = (-1)^i \det H'_{-i}(y), \qquad i = 1, \ldots, n + 1. \tag{4.A.3}$$

The system (4.A.3) is an example of a system of ordinary differential equations (4.A.1). In (4.A.3), $m = n + 1$ and $L_i(y, p) = (-1)^i \det H'_{-i}(y)$. Observe that in this example, L is independent of the parameter p. Such systems are called *autonomous*.

EXISTENCE AND UNIQUENESS

The basic existence and uniqueness theorem for ordinary differential equations is the following:

THEOREM 4.A.1 Consider the system (4.A.1), where $L \in C^1$. Given an initial point (y^0, p^0) in R^{m+1}, there is a solution $y(p)$ of (4.A.1), and an interval (a, b), where $a < p^0 < b$, such that:

1. $y(p^0) = y^0$.
2. If $\bar{y}(p)$ is any other solution of (4.A.1) satisfying $\bar{y}(p^0) = y^0$, then its interval of definition is contained in (a, b) and $\bar{y}(p) = y(p)$ for all p in its interval of definition.

The theorem assures that the solution to (4.A.1) is unique. The interval (a, b) is called the *maximum interval of existence* corresponding to the initial values (y^0, p^0). It is possible that $a = -\infty$ and $b = +\infty$.

If L is continuous but not C^1, existence is still guaranteed but uniqueness is not. For example, see Exercise 4 of Chapter 2.

Appendix B

Convexity

B.1 CONVEX SETS

Convexity is a fundamental concept in the theory of optimization, and it is essential to have a good understanding of its properties.

A CONVEX SET

Consider any two points $x^1, x^2 \in R^n$. The *line segment* between them is described by the point

$$w = \theta x^1 + (1 - \theta)x^2 \qquad (1.\text{B}.1)$$

as θ varies between 0 and 1. Convex sets have the property that the line segment connecting any two points in the set is also in the set. Mathematically, a set $C \subset R^n$ is *convex* if

$$x^1, x^2 \in C$$

implies that

$$w = \theta x^1 + (1 - \theta)x^2 \in C \qquad \text{for any } \theta, \quad 0 \le \theta \le 1.$$

This is illustrated in Figure 1.B.1.

Some familiar properties of convex sets are as follows.

450

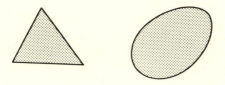
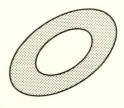

Convex sets

Nonconvex sets
Figure 1.B.1

THEOREM 1.B.1 If C is a convex set and α is a real number, the set
$$\alpha C = \{x \mid x = \alpha c, \quad c \in C\}$$
is convex.

THEOREM 1.B.2 If C and D are convex sets, the set
$$C + D = \{x \mid x = c + d, \quad c \in C, \quad d \in D\}$$
is convex.

THEOREM 1.B.3 Let C_i, $i = 1, \ldots, l$ be convex sets. Then the set
$$C = \bigcap_{i=1}^{l} C_i$$
is convex.

The proofs of these three theorems are immediate and are left as exercises. The theorems are illustrated in Figure 1.B.2.

B.2 CONVEX AND CONCAVE FUNCTIONS

Given a convex set $C \subset R^n$, a function f is *concave* on the set C if
$$x^1, x^2 \in C$$

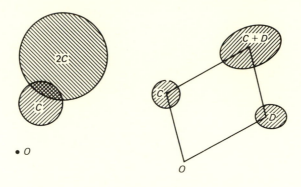

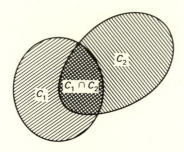

Properties of convex sets.

Figure 1.B.2

implies that

$$f(\theta x^1 + (1 - \theta)x^2) \geq \theta f(x^1) + (1 - \theta)f(x^2) \qquad (2.B.1)$$

for any θ, $0 \leq \theta \leq 1$. A function f is *convex* if $-f$ is concave. Examples of concave and convex functions are given in Figure 2.B.1.

CHARACTERIZATION OF CONCAVITY

Concavity can be characterized by the first and second derivatives of the function.

THEOREM 2.B.1 Let f be differentiable on R^n. Then f is concave if and only if

$$f(y) \leq f(x) + \nabla f(x)(y - x) \qquad \text{for any } x \text{ and } y. \qquad (2.B.2)$$

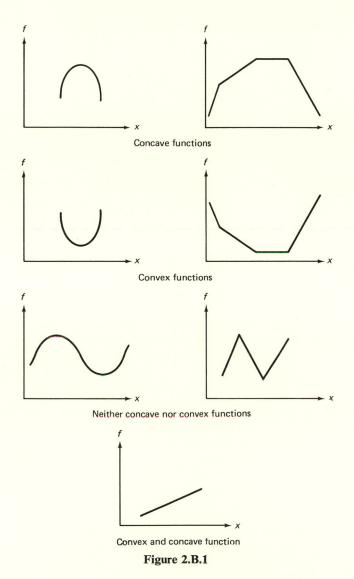

Concave functions

Convex functions

Neither concave nor convex functions

Convex and concave function

Figure 2.B.1

Essentially, the theorem states that a linear approximation to the concave function overestimates the function.

If f has second partial derivatives, there is another way to characterize concavity. If f is concave, then the Hessian $\nabla^2 f$ is negative semidefinite. Recall that a square matrix A is negative semidefinite if for any x

$$x^{\tau} A x \leq 0.$$

THEOREM 2.B.2   Let $f \in C^2$. If f is concave, then the Hessian matrix is negative semidefinite at any point x.

Let us now give the flavor of how concavity is used in optimization:

THEOREM 2.B.3   Let f be a differentiable concave function in R^n. Then
$$\nabla f(\overset{*}{x}) = 0$$
if and only if $\overset{*}{x}$ maximizes f over R^n.

Proof.   Suppose that $\nabla f(\overset{*}{x}) = 0$. For any point y, by concavity
$$f(y) \leq f(\overset{*}{x}) + \nabla f(\overset{*}{x})(y - \overset{*}{x}) = f(\overset{*}{x}).$$

Thus $\overset{*}{x}$ maximizes f over R^n. The necessity of the theorem is an elementary result in calculus and is left as an exercise. $\square$

Appendix C

The Kuhn-Tucker Conditions

C.1 THE NONLINEAR PROGRAMMING PROBLEM

Let us consider a nonlinear programming problem (NLP) of the form

$$\max f(x)$$
$$g_j(x) \geq 0, \qquad j = 1, \ldots, r$$
$$h_j(x) = 0, \qquad j = 1, \ldots, s \qquad (1.C.1)$$

where f, g_j, $j = 1, \ldots, r$ and h_j, $j = 1, \ldots, s$, are continuous functions defined in R^n. The *constraints* are $g(x) \geq 0$, $h(x) = 0$. A point $x \in R^n$ satisfying all the constraints is said to be *feasible*. An inequality constraint $g_j(x) \geq 0$ is said to be *active* at a feasible point $\bar{x}$ if $g_j(\bar{x}) = 0$, and *inactive* if $g_j(\bar{x}) > 0$. The *optimal solution* of the NLP (1.C.1) is a feasible point $\overset{*}{x}$ such that

$$f(\overset{*}{x}) \geq f(x)$$

for all feasible points x. A *local maximum point* of the NLP (1.C.1) is a feasible point $\overset{*}{x}$ such that

$$f(\overset{*}{x}) \geq f(x) \qquad \text{for all feasible points } x \text{ in a neighborhood of } \overset{*}{x}.$$

455

A *strict local maximum point* of the NLP (1.C.1) is a feasible point $\overset{*}{x}$ such that

$$f(\overset{*}{x}) > f(x) \qquad \text{for all feasible points } x \neq \overset{*}{x}, \quad x$$
$$\text{in a neighborhood of } \overset{*}{x}.$$

C.2 THE KUHN–TUCKER NECESSARY CONDITIONS

The Kuhn–Tucker (K–T) conditions under reasonable assumptions are necessarily satisfied if $\overset{*}{x}$ is a local maximum point of the NLP problem.

The K–T conditions extend Theorem 2.B.3 to constrained problems. Intuitively, they state that if one moves away from $\overset{*}{x}$ in any direction, as long as one remains in the feasible region, the objective function f cannot increase. In order that the K–T hold at $\overset{*}{x}$, a *constraint qualification* (CQ) is required. For it, let $A = \{j \mid g_j(\overset{*}{x}) = 0\}$ indicate the active constraints at $\overset{*}{x}$. Then the CQ is said to hold at $\overset{*}{x}$ if the vectors

$$\nabla g_j(\overset{*}{x}), \quad j \in A, \qquad \nabla h_j(\overset{*}{x}), \quad j = 1, \ldots, s \qquad \text{(2.C.1)}$$

are linearly independent.

KUHN–TUCKER CONDITIONS

Let $\overset{*}{x}$ be a local maximum point for the NLP (1.C.1), where $f, g_j, j = 1, \ldots, r$, and $h_j, j = 1, \ldots, s$, are differentiable. If the CQ (2.C.1) holds at $\overset{*}{x}$, then there is a vector $\lambda \in R^r$ with

$$\lambda \geq 0 \qquad \text{(2.C.2)}$$

and a vector $\mu \in R^s$ such that

$$\nabla f(\overset{*}{x})^{\tau} + \sum_{j=1}^{r} \lambda_j \nabla g_j(\overset{*}{x})^{\tau} + \sum_{j=1}^{s} \mu_j \nabla h_j(\overset{*}{x})^{\tau} = 0 \qquad \text{(2.C.3)}$$

$$\lambda_j g_j(\overset{*}{x}) = 0, \qquad j = 1, \ldots, r. \qquad \text{(2.C.4)}$$

Observe that since $\overset{*}{x}$ is a local maximum point, $\overset{*}{x}$ also satisfies

$$g(\overset{*}{x}) \geq 0 \qquad \text{(2.C.5)}$$

$$h(\overset{*}{x}) = 0. \qquad \text{(2.C.6)}$$

The conditions (2.C.2) to (2.C.6) are collectively known as the K–T conditions.

EXAMPLE 2.C.1

Consider the problem

$$\max -x_1^2 - 2x_2^2 - 2x_1x_2 + 10x_1 + 10x_2 + 4$$

subject to

$$x_1 + 3x_2 \leq 6$$
$$x_1^2 + x_2^2 \leq 5.$$

The K–T conditions for this problem are

$$-2x_1 - 2x_2 + 10 - \lambda_1 - 2\lambda_2 x_1 = 0$$
$$-4x_2 - 2x_1 + 10 - 3\lambda_1 - 2\lambda_2 x_2 = 0$$
$$\lambda_1 \geq 0 \qquad \lambda_2 \geq 0$$
$$x_1 + 3x_2 \leq 6 \qquad x_1^2 + x_2^2 \leq 5$$
$$\lambda_1(6 - x_1 - 3x_2) = 0 \qquad \lambda_2(5 - x_1^2 - x_2^2) = 0.$$

To find a solution to the problem, we can try various combinations of active constraints. Setting the first constraint inactive and the second constraint active yields

$$-2x_1 - 2x_2 + 10 - 2\lambda_2 x_1 = 0$$
$$-4x_2 - 2x_1 + 10 - 2\lambda_2 x_2 = 0$$
$$x_1^2 + x_2^2 = 5,$$

which has solution

$$x_1 = 2, \qquad x_2 = 1, \qquad \lambda_2 = 1$$

Since $x_1 + 3x_2 = 5$, $\lambda_1 = 0$, $\lambda_2 > 0$ at this solution, we conclude that the K–T conditions hold at this solution.

C.3   SUFFICIENCY OF THE K–T CONDITIONS

INSUFFICIENCY OF THE K–T CONDITIONS

The K–T conditions, by themselves, are necessary but not sufficient for a point to be optimal for the NLP. For example, consider the problem

$$\max f(x) = x^2$$

subject to

$$g_1(x) = 2 - x \geq 0$$
$$g_2(x) = 1 + x \geq 0$$

The optimal point is obviously $\overset{*}{x} = 2$, yet the K–T conditions hold at $x = -1$, as we now demonstrate.

First, $x = -1$ is clearly feasible. Now, consider the conditions (2.C.2) to (2.C.4) at $x = -1$, which become

$$0 = \lambda_1 g_1(x) = \lambda_1 \cdot 3$$
$$0 = \lambda_2 g_2(x) = \lambda_2 \cdot 0$$

and

$$0 = \nabla f(x)^\tau + \lambda_1 \nabla g_1(x)^\tau + \lambda_2 \nabla g_2(x)^\tau = -2 - \lambda_1 + \lambda_2.$$

Solving these, $\lambda_1 = 0$, $\lambda_2 = 2$. Thus at $x = -1$, the K–T conditions hold, but $x = -1$ is not optimal. The K–T conditions, therefore, may not be sufficient.

SUFFICIENCY THEOREM

The K–T conditions are sufficient if f and $g_j, j = 1, \ldots, r$, are concave and $h_j, j = 1, \ldots, s$, are linear.

THEOREM 3.C.1   In the NLP (1.C.1), suppose that f and $g_j, j = 1, \ldots, r$, are differentiable and concave, and let $h_j, j = 1, \ldots, s$, be linear. Suppose that the K–T conditions hold at $\overset{*}{x}$. Then $\overset{*}{x}$ is optimal for the NLP (1.C.1).

We can now combine the result in Section C.2 with Theorem 3.C.1 to yield a necessary and sufficient condition for optimality.

THEOREM 3.C.2   Given the NLP (1.C.1), suppose that f and $g_j, j = 1, \ldots, r$, are differentiable and concave and that $h_j, j = 1, \ldots, s$, are linear. Also suppose that the CQ (2.C.1) holds at $\overset{*}{x}$. Then $\overset{*}{x}$ is the optimal for the NLP (1.C.1) if and only if the K–T conditions hold at $\overset{*}{x}$.

Bibliography

ABRAHAM, R., AND J. ROBBIN, *Transversal Mappings and Flows.* New York: W. A. Benjamin, 1967.

AGANACIC, M., AND R. W. COTTLE, "A Note on Q-Matrices," *Mathematical Programming* 16, [1979], 374–77.

ALEXANDER, J. C., "The Topological Theory of an Embedding Method," in H. Wacker, ed., *Continuation Methods.* New York: Academic Press, 1978, pp. 37–68.

ALEXANDER, J. C., AND J. A. YORKE, "The Homotopy Continuation Method: Numerically Implementable Topological Procedures," *Transactions of the American Mathematical Society* 242 [1978], 271–84.

ALLGOWER, E. L., AND K. GEORG, "Simplicial and Continuation Methods for Approximating Fixed Points," *SIAM Review* 22 [1980], 28–85.

ALLGOWER, E. L., AND M. M. JEPPSON, "The Approximation of Solutions of Nonlinear Elliptic Boundary Value Problems with Several Solutions," *Springer Lecture Notes in Mathematics* 333 [1973], 1–20.

ALLGOWER, E. L., AND C. L. KELLER, "A Search for a Sperner Simplex," *Computing* 8 [1971], 157–65.

ARROW, K. J., AND F. H. HAHN, *General Competitive Analysis.* Edinburgh: Oliver and Boyd; San Francisco: Holden-Day, 1971.

ASMUTH, R. L., "Traffic Network Equilibria," *Technical Report SOL* 78–2, 1978.

AUMANN, R. J., "Existence of Competitive Equilibria in Markets with a Continuum of Traders," *Econometrica* 34 [1966], 1–17.

AVRIEL, M., *Nonlinear Programming: Analysis and Methods.* Englewood Cliffs, NJ: Prentice-Hall, Inc., 1976.

BANACH, S., "Sur les opérations dans les ensembles abstraits et leur applications aux équations integrales," *Fund. Math.* 3 [1922], 133–81.

BAZARAA, M. S., J. J. GOODE, AND C. M. SHETTY, "Constraint Qualifications Revisited," *Management Science* 18 [1972], 567–73.

BECKMAN, M. J., C. B. MCGUIRE, AND C. B. WINSTON, *Studies in the Economics of Transportation.* New Haven: Yale University Press, 1956.

BEDNAREK, A. R., AND L. CESARI, *Dynamical Systems.* Proceedings of a University of Florida International Symposium. New York: Academic Press, 1977.

BELLUCE, L., AND W. KIRK, "Fixed Point Theorems for Certain Classes of Nonexpansive Mappings," *Proceedings of the American Mathematical Society* 20 [1969], 141–46.

BERGE, C., *Topological Spaces.* Edinburgh: Oliver and Boyd, 1963.

BIRKHOFF, G., AND G. C. ROTA, *Ordinary Differential Equations.* Boston: Ginn & Co., 1962.

BOHL, P., "Über die Bewegung eines Mechanischen Systems in der Nähe einer Gleichgewichtslage," *J. Reine Angew. Math.* 127 [1904], 179–276.

BOSARGE, W., "Infinite Dimensional Iterative Methods and Applications," IBM Houston Science Report #320.2347, 1968.

BRANIN, F. J. JR., "Widely Convergent Method for Finding Multiple Solutions of Simultaneous Nonlinear Equations," *IBM Journal of Research Development* 16 [1972], 504–22.

BRAUER, F., AND J. A. NOHEL, *Ordinary Differential Equations.* New York: W. A. Benjamin, 1967.

BROCKER, T., AND L. LANDER, *Differentiable Germs and Catastrophes.* Cambridge: Cambridge University Press, 1975.

BROUWER, L. E., "Über Abbildungen von Mannigfaltigkeiten," *Mathematische Annalen* 71 [1912], 97–115.

BROWDER, F. E., "On Continuity of Fixed Points under Deformations of Continuous Mappings," *Summa Brasiliensis Mathematicae* 4 [1960], 183–91.

———, "Nonexpansive Nonlinear Operators in a Banach Space," *Proceedings of the National Academy of Science* 54 [1965], 1041–44.

BROWDER, F. E., AND W. PETRYSHYN, "Construction of Fixed Points of Nonlinear Mappings in Hilbert Space," *Journal of Mathematical Analysis and Applications* 20 [1967], 197–228.

CHANDRASEKARAN, R., "A Special Case of the Complementary Pivot Problem," *Operations Research* 7 [1970], 263–68.

CHARNES, A., C. B. GARCIA, AND C. E. LEMKE, "Constructive Proofs of Theorems Relating to $F(x) = y$, with Applications," *Mathematical Programming* 12 [1977], 328–43.

CHOW, S. N., J. MALLET-PARET, AND J. A. YORKE, "Finding Zeros of Maps: Homotopy Methods that are Constructive with Probability One," *Mathematics of Computation* 32 [1978], 887–99.

———, "A Homotopy Method for Locating Zeros of a System of Polynomials," University of Maryland Technical Report, 1978.

COHEN, D. I. A., "On the Sperner Lemma," *Journal of Combinatorial Theory* 2 [1967], 585–87.

——, "On the Kakutani Fixed Point Theorem," in W. Forster, ed., *Numerical Solution of Highly Nonlinear Problems*. Amsterdam: North Holland, 1980, pp. 239–40.

COLLATZ, L., *Functional Analysis and Numerical Mathematics* (translated by H. Oser), New York: Academic Press, 1964.

COTTLE, R. W., "Nonlinear Programs with Positively Bounded Jacobians," *SIAM Journal of Applied Mathematics* 14 [1966], 147–58.

——, "On a Problem in Linear Inequalities," *Journal of the London Mathematical Society* 43 [1968], 378–84.

——, "Monotone Solutions of the Parametric Linear Complementarity Problem," *Mathematical Programming* 3 [1972], 210–24.

——, "Completely Q-Matrices," *Mathematical Programming* 19, [1980] 347–51.

COTTLE, R. W., AND G. B. DANTZIG, "Positive (semi) Definite Programming," in *Nonlinear Programming*, ed. J. Abadie. Amsterdam: North-Holland, 1967, pp. 55–73.

——, "Complementary Pivot Theory of Mathematical Programming," *Linear Algebra and Applications* 1 [1968], 103–25.

——, "A Generalization of the Linear Complementarity Problem," *Journal of Combinatorial Theory* 8 [1970], 79–90.

COTTLE, R. W., G. J. HABETLER, AND C. E. LEMKE, "Quadratic Forms Semi-Definite over Convex Cones," *Proceedings of the International Symposium on Mathematical Programming* [1967], 551–65.

COTTLE, R. W., AND R. VON RANDOW, "On Q-Matrices, Centroids, and Simplotopes," *Stanford Technical Report* 79–10, 1979.

CRONIN, J., *Fixed Points and Topological Degree in Nonlinear Analysis*. Providence: American Mathematics Society, 1964.

DANTZIG, G. B., "A Proof of the Equivalence of the Programming Problem and the Game Problem," in *Activity Analyses of Production and Allocation*, ed. T. C. Koopmans. New York: John Wiley & Sons, Inc., 1951.

——, *Linear Programming and Extensions*. Princeton: Princeton University Press, 1963.

DANTZIG, G. B., B. C. EAVES, AND D. GALE, "An Algorithm for a Piecewise Linear Model of Trade and Production with Negative Prices and Bankruptcy," *Mathematical Programming* 16 [1979], 210–27.

DAVIDENKO, D., "On a New Method of Numerically Integrating a System of Nonlinear Equations," *Doklady Akademie Nauk SSSR* 88 [1953a], 601–604.

——, "On the Approximate Solution of a System of Nonlinear Equations," *Ukraine Mat Zurnal* 5 [1953b], 196–206.

——, "An Application of the Method of Variation of Parameters to the Construction of Iterative Formulas of Higher Accuracy for the Determination of the Elements of the Inverse Matrix," *Doklady Akademie Nauk SSSR* 162 [1965a], 743–46; *Soviet Math Doklady* 6, 738–42.

————, "An Application of the Method of Variation of Parameters to the Construction of Iterative Formulas of Increased Accuracy for Numerical Solutions of Nonlinear Integral Equations," *Doklady Akademie Nauk SSSR* 162 [1965b], 499–502; *Soviet Math Doklady* 6, 702–706.

DAVIS, J., "The Solution of Nonlinear Operator Equations with Critical Points," Ph.D. dissertation, Oregon State University, Corvallis, 1966.

DEBREU, G., "A Social Equilibrium Theorem," *Proceedings of the National Academy of Science* 38 [1952], 886–93.

————, *Theory of Value*. New York: John Wiley & Sons, Inc., 1959.

————, "Economies with a Finite Set of Equilibria," *Econometrica* 38 [1970], 387–92.

————, "Four Aspects of the Mathematical Theory of Economic Equilibrium," Working Papers No. IP-211, University of California, Berkeley, 1975.

DEIST, F., AND L. SEFOR, "On Newton's Method and Nonlinear Equations by Parameter Variation," *Computer Journal* 10 [1967], 78–82.

DIEUDONNÉ, J., *Foundations of Modern Analysis*. New York: Academic Press, 1960.

DOVERSPIKE, R. D., AND C. E. LEMKE, "A Partial Characterization of a Class of Matrices Defined by Solutions to the Linear Complementarity Problem," Rensselaer Polytechnic Institute Technical Report, 1979.

DREXLER, F. J., "Eine methode zur Berechnung sämtlicher lösungen von polynomgleichungessystemen," *Numerische Mathematic* 29 [1977], 45–58.

————, "A Homotopy Method for the Calculation of all Zeros of Zero-Dimensional Polynomial Ideals," in *Continuation Methods*, ed. H. Wacker. New York: Academic Press, 1978.

DUGUNDJI, J., *Topology*. Boston: Allyn and Bacon, 1966.

EAVES, B. C., "An Odd Theorem," *Proceedings of the American Mathematical Society* 26 [1970], 509–13.

————, "Computing Kakutani Fixed Points," *SIAM Journal of Applied Mathematics* 21 [1971a], 236–44.

————, "On the Basic Theorem of Complementarity," *Mathematical Programming* 1 [1971b], 68–75.

————, "The Linear Complementarity Problem," *Management Science* 17 [1971c], 612–34.

————, "Homotopies for the Computation of Fixed Points," *Mathematical Programming* 3 [1972], 1–22.

————, "Solving Piecewise Linear Convex Equations," *Mathematical Programming Study* 1 [1974], 96–119.

EAVES, B. C., AND R. SAIGAL, "Homotopies for the Computation of Fixed Points on Unbounded Regions," *Mathematical Programming* 3 [1972], 225–37.

EAVES, B. C., AND H. SCARF, "The Solution of Systems of Piecewise Linear Equations," *Mathematics of Operations Research* 1 [1976], 1–27.

EDGEWORTH, F. Y., *Mathematical Psychics*. London: Kegan Paul, 1881.

FICKEN, F., "The Continuation Method for Functional Equations," *Communications on Pure and Applied Mathematics* 4 [1951], 435–56.

FISHER, M. L., AND F. J. GOULD, "A Simplicial Algorithm for the Nonlinear Complementarity Problem," *Mathematical Programming* 6 [1974], 281–300.

FISHER, M. L., F. J. GOULD, AND J. W. TOLLE, "A Simplicial Approximation Algorithm for Solving Systems of Nonlinear Equations," *Instituto Nazionale di Alta Matematica: Symposia Matematica* 19 [1976], 73–90.

————, "A New Simplicial Approximation Algorithm with Restarts: Relations between Convergence and Labelling," in *Fixed Points: Algorithms and Applications*, eds. S. Karamardian and C. B. Garcia. New York: Academic Press, 1977, pp. 41–58.

FISHER, M. L., AND J. W. TOLLE, "The Nonlinear Complementarity Problem: Existence and Determination of Solutions," *SIAM Journal of Control and Optimization* 15 [1977], 612–23.

FLORIAN, M. A., ED., *Traffic Equilibrium Methods*. Heidelberg; New York: Springer-Verlag, 1976.

FOSTER, W., ED., *Numerical Solution of Nonlinear Problems*. Amsterdam: North-Holland, 1980.

FREIDENFELDS, J., "Fixed Point Algorithms and Almost Complementary Sets," TR71-17, Operations Research House, Stanford University, 1971.

FREUDENSTEIN, F., AND B. ROTH, "Numerical Solution of Systems of Nonlinear Equations," *Journal of the Association of Computing Machinery* 10 [1963], 550–56.

FREUDENTHAL, H., "Simplizialzerlegungen von Beschränkter Flachheit," *Annals of Mathematics* 43 [1942], 580–82.

————, *L. E. J. Brouwer: Collected Works. Vol. 2: Geometry, Analysis, Topology, and Mechanics*. Amsterdam: North-Holland, 1976.

FUJISAWA, T., AND E. S. KUH, "Piecewise-Linear Theory of Nonlinear Networks," *SIAM Journal of Applied Mathematics* 22 [1972], 307–28.

GALE, D., "The Game of Hex and the Brouwer Fixed-Point Theorem," *American Mathematics Monthly* 86 [1979], 818–27.

GALE, D., H. W. KUHN, AND A. W. TUCKER, "Linear Programming and the Theory of Games," in *Activity Analysis of Production and Allocation*, ed. T. C. Koopmans. New York: John Wiley & Sons, Inc., 1951.

GALE, D., AND A. MAS-COLELL, "An Equilibrium Existence Theorem for a General Model without Order Preferences," *Journal of Mathematical Economics* 2 [1975], 9–16.

GALE, D., AND H. NIKAIDO, "The Jacobian Matrix and Global Univalence of Mappings," *Mathematische Annalen* 159 [1965], 81–93.

GARCIA, C. B., "Some Classes of Matrices in Linear Complementarity Theory," *Mathematical Programming* 5 [1973], 299–310.

————, "A Global Existence Theorem for the Equation $F(x) = y$," Center for Mathematical Studies in Business and Economics, University of Chicago, 1975.

————, "A Hybrid Algorithm for the Computation of Fixed Points," *Management Science* 22 [1976], 606–13.

————, "Continuation Methods for Simplicial Mappings," in *Fixed Points: Algorithms and Applications*, eds. S. Karamardian and C. B. Garcia. New York: Academic Press, 1977a, pp. 149–64.

———, "Computation of Solutions to Nonlinear Equations under Homotopy Invariance," *Mathematics of Operations Research* 2 [1977b], 25–29.

GARCIA, C. B., AND F. J. GOULD, "A Theorem on Homotopy Paths," *Mathematics of Operations Research* 3 [1978], 282–89.

———, "Scalar Labelings for Homotopy Paths," *Mathematical Programming* 17 [1979], 184–97.

———, "Relations between Several Path Following Algorithms and Local and Global Newton Methods," *SIAM Review* 22 [1980a], 263–74.

———, "Studies in Linear Complementarity Theory," University of Chicago Center Report, 1980b.

GARCIA, C. B., C. E. LEMKE, AND H. LÜTHI, "Simplicial Approximation of an Euilibrium Point for Non-Cooperative N-person Games," in *Mathematical Programming*, eds. T. C. Hu and S. Robinson. New York: Academic Press, 1973.

GARCIA, C. B., AND T. Y. LI, "On a Path Following Method for Systems of Equations," University of Wisconsin MRC Technical Report No. 1983, 1979.

———, "On the Number of Solutions to Polynomial Systems of Equations," *SIAM Journal of Numerical Analysis* 17 [1980], 540–46.

GARCIA, C. B., AND W. I. ZANGWILL, "Determining All Solutions to Certain Systems of Nonlinear Equations," *Mathematics of Operations Research* 4 [1979a], 1–14.

———, "Finding All Solutions to Polynomial Systems and Other Systems of Equations," *Mathematical Programming* 16 [1979b], 159–76.

———, "On Univalence and P-Matrices," *Linear Algebra and its Applications* 24 [1979c], 239–50.

———, "An Approach to Homotopy and Degree Theory," *Mathematics of Operations Research* 4 [1979d], 390–405.

———, "Global Calculation of Catastrophes," University of Chicago Center Report, 1979e.

———, "A Flex Simplicial Algorithm," in *Numerical Solution of Highly Nonlinear Problems*, ed. W. Forster. Amsterdam: North-Holland, 1980a.

———, "Global Calculation Methods for Finding All Solutions to Polynomial Systems of Equations in N Variables," in *Extremal Methods and Systems Analysis*. Heidelberg; New York: Springer-Verlag, 1980b.

GOULD, F. J., "Recent and Past Developments in the Simplicial Approximation Approach to Solving Nonlinear Equations—A Subjective View," in *Extremal Methods and System Analysis*. Heidelberg; New York: Springer-Verlag, 1980.

GOULD, F. J. AND C. P. SCHMIDT, "An Existence Result for the Global Newton Method," in *Variational Inequalities and Complementarity Problems: Theory and Applications*. New York: John Wiley and Sons, 1980.

GOULD, F. J., AND J. W. TOLLE, "Geometry of Optimality Conditions and Constraint Qualifications," *Mathematical Programming* 2 [1972], 1–18.

———, "A Unified Approach to Complementarity in Optimization," *Discrete Mathematics* 7 [1974], 225–71.

———, "An Existence Theorem for Solutions to $f(x) = 0$," *Mathematical Programming* 11 [1976], 252–62.

GRAVES, R. L., "A Principal Pivoting Simplex Algorithm for Linear and Quadratic Programming," *Operations Research* 15 [1967], 482–94.

GUILLEMIN, V., AND A. POLLACK, *Differential Topology*. Englewood Cliffs, N.J.: Prentice-Hall, 1974.

HABETLER, G. J., AND M. M. KOSTREVA, "On a Direct Algorithm for Nonlinear Complementarity Problems," *SIAM Journal of Control and Optimization* 16 [1978], 504–11.

———, "Sets of Generalized Complementarity Problems and P-Matrices," *Mathematics of Operations Research* 5 [1980], 280–84.

HABETLER, G. J., AND A. L. PRICE, "Existence Theory for Generalized Nonlinear Complementarity Problems," *Journal of Optimization Theory and Applications* 7 [1971], 223–39.

HANSEN, T. H., "On the Approximation of Nash Equilibrium Points in an N-person Noncooperative Game," *SIAM Journal of Applied Mathematics* 26 [1974], 622–37.

HANSEN, T. H., AND H. SCARF, "On the Applications of a Recent Combinatorial Algorithm," Cowles Foundation Discussion Paper No. 272, Yale University, 1969.

HARSANYI, J. C., "Oddness of the Number of Equilibrium Points: a New Proof," *International Journal of Game Theory* 2 [1973], 235–50.

———, "The Tracing Procedure: a Bayesian Approach to Defining a Solution for n-Person Noncooperative Games," *International Journal of Game Theory* 4 [1975], 61–94.

HASELGROVE, C., "Solution of Nonlinear Equations and of Differential Equations with Two-Point Boundary Conditions," *Computer Journal* 4 [1961], 255–59.

HEINZ, E., "An Elementary Theory of the Degree of a Mapping in N-Dimensional Space," *Journal of Mathematical Mechanics* 8 [1959], 231–47.

HILDENBRAND, W., AND A. P. KIRMAN, *Introduction to Equilibrium Analysis*. Amsterdam: North-Holland; New York: American Elsevier, 1976.

HIRSCH, M. W., "A Proof of the Nonretractability of a Cell Onto Its Boundary," *Proceedings of the American Mathematical Society* 14 [1963], 364–65.

HIRSCH, M. W., AND S. SMALE, *Differential Equations, Dynamical Systems, and Linear Algebra*. New York: Academic Press, 1974.

———, "On Algorithms for Solving $f(x) = 0$," *Communication on Pure and Applied Mathematics* 32 [1979], 281–312.

HOWE, R., "Linear Complementarity and the Degree of Mappings," Cowles Foundation Discussion Paper No. 542, Yale University, 1980.

INGLETON, A. W., "A Problem in Linear Inequalities," *Proceedings of the London Mathematical Society* 16 [1966], 519–36.

INTRILIGATOR, M. D., *Mathematical Optimization and Economic Theory*. Englewood Cliffs, N.J.: Prentice-Hall, 1971.

JAKOVLEV, M., "The Solution of Sytems of Nonlinear Equations by a Method of Differentiation with Respect to a Parameter," *Z. Vycisl. Mat. i. Mat. Fiz.* 4 [1964], 146–49.

JEPPSON, M. M., "A Search for the Fixed Points of a Continuous Mapping," in *Mathematical Topics in Economics Theory and Computation*, eds. R. H. Day and S. M. Robinson. New York: Academic Press, 1972, pp. 122–29.

KAKUTANI, S., "A Generalization of Brouwer's Fixed Point Theorem," *Duke Mathematical Journal* 8 [1941], 457–59.

KANEKO, I., "The Number of Solutions of a Class of Linear Complementarity Problems," *Mathematical Programming* 17 [1979], 104–105.

KANTOROVICH, L., AND G. AKILOV, *Functional Analysis in Normed Spaces* (translated by D. Brown and A. Robertson, 1964). Oxford: Pergamon Press, 1959.

KARAMARDIAN, S., "The Nonlinear Complementarity Problem with Applications, Parts I and II," *Journal of Optimization Theory and Applications* 4 [1969], 87–98 and 167–81.

———, "The Complementarity Problem," *Mathematical Programming* 2 [1972], 107–29.

KARUSH, W., "Minima of Functions of Several Variables with Inequalities as Side Conditions," M.S. thesis, Department of Mathematics, University of Chicago, 1939.

KATZENELSON, J., "An Algorithm for Solving Nonlinear Resistor Networks," *Bell Telephone Technical Journal* 44 [1965], 1605–20.

KELLER, H., *Numerical Methods for Two-Point Boundary Value Problems*. Boston: Ginn & Co., 1968.

———, "Global Homotopies and Newton Methods," in *Numerical Analysis*. New York: Academic Press, 1978.

KELLOGG, R. B., T. Y. LI, AND J. YORKE, "A Constructive Proof of the Brouwer Fixed-Point Theorem and Computational Results," *SIAM Journal of Numerical Analysis* 13 [1976], 473–83.

KELLY, L. M., AND L. T. WATSON, "Q-Matrices and Spherical Geometry," *Linear Algebra and Its Applications* 25 [1979], 175–89.

KIRK, W., "A Fixed Point Theorem for Mappings Which Do Not Increase Distance," *American Mathematical Monthly* 72 [1965], 1004–1006.

KNASTER, B., C. KURATOWSKI, AND S. MAZURKIEWICZ, "Ein Beweis des Fixpunktsatzes für n-Dimensionale Simplexe," *Fund. Math.* 14 [1929], 132–37.

KOJIMA, M., "Computational Methods for Solving the Nonlinear Complementarity Problem," *Keio Engineering Reports* 27 [1974], 1–41.

———, "On the Homotopic Approach to Systems of Equations with Separable Mappings," in *Mathematical Programming Study* 7, eds. M. L. Balinski and R. W. Cottle. Amsterdam: North-Holland, 1978a, pp. 170–84.

———, "A Modification of Todd's Triangulation J_3," *Mathematical Programming* 15 [1978b], 223–27.

———, "A Complementarity Pivoting Approach to Parametric Nonlinear Programming," *Mathematics of Operations Research* 4 [1979], 464–77.

KOJIMA, M., H. NISHINO, AND T. SEKINE, "An Extension of Lemke's Method to the Piecewise Linear Complementarity Problem," *SIAM Journal of Applied Mathematics* 31 [1976], 600–613.

KOJIMA, M., AND R. SAIGAL, "On the Number of Solutions to a Class of Linear Complementarity Problems," *Mathematical Programming* 17 [1979], 136–39.

KOSTREVA, M. M., "Direct Algorithms for Complementarity Problems," Ph.D. dissertation, Rensselaer Polytechnic Institute, 1976.

KRASNOSELSKII, M. A., *Topological Methods in the Theory of Nonlinear Integral Equations.* (Translation by A. Armstrong, Pergamon Press, 1964.) Moscow: Gostekhteoretizdat, 1956.

KRONECKER, L., "Über Systeme von Funktionen Mehrerer Variablen," *Monatsb. Deutsch Akad. Wiss. Berlin*, 1869, pp. 159–93 and 688–98.

KUHN, H. W., "Some Combinatorial Lemmas in Topology," *IBM Journal of Research and Development* 4 [1960], 518–24.

———, "Simplicial Approximation of Fixed Points," *Proceedings of the National Academy of Sciences* 61 [1968], 1238–42.

———, "Approximate Search for Fixed Points," in *Computing Methods in Optimization Problems*, 2. New York: Academic Press, 1969.

———, "A New Proof of the Fundamental Theorem of Algebra," *Mathematical Programming Study* 1 [1976], 148–58.

———, "Finding Roots by Pivoting," in *Fixed Points: Algorithms and Applications*, eds. S. Karamardian and C. B. Garcia. New York: Academic Press, 1977.

KUHN, H. W., AND J. G. MACKINNON, "Sandwich Method for Finding Fixed Points," *Journal of Optimization Theory and Applications* 17 [1975], 189–204.

KUHN, H. W., AND A. W. TUCKER, "Nonlinear Programming," in *Proceedings of the Second Berkeley Symposium on Mathematical Statistics and Probability*, ed. J. Neyman. Berkeley: University of California Press, 1951.

LAHAYE, E., "Solution of Systems of Transcendental Equations," *Acad. Roy. Belg. Bull. Cl. Sci.* 5 [1948], 805–22.

LEMKE, C. E., "Bimatrix Equilibrium Points and Mathematical Programming," *Management Science* 11 [1965], 681–89.

———, "On Complementary Pivot Theory," in *Mathematics of Decision Sciences*, eds. G. B. Dantzig and A. F. Vienott, Jr. Providence: American Mathematical Society, 1968.

———, "Recent Results on Complementarity Problems," in *Nonlinear Programming*, eds. O. L. Mangasarian and K. Ritter. New York: Academic Press, 1970, pp. 349–84.

LEMKE, C. E., AND S. J. GROTZINGER, "On Generalizing Shapley's Index Theory to Labelled Pseudomanifolds," *Mathematical Programming* 10 [1976], 245–62.

LEMKE, C. E., AND J. T. HOWSON, "Equilibrium Points of Bimatrix Games," *SIAM Review* 12 [1964], 413–23.

LERAY, J., AND J. SCHAUDER, "Topologie et équations fonctionelles," *Ann. Sci. Ecole. Norm. Sup.* 51 [1934], 45–78.

LÜTHI, H. J., *Komplementaritäts- und Fixpunktalgorithmen in der mathematischen Programmierung Spietheorie und Ökonomie.* Berlin: Springer Verlag 129, 1976a.

———, "A Simplicial Approximation of a Solution for the Nonlinear Complementarity Problem," *Mathematical Programming* 9 [1976b], 278–93.

McCAMMON, S. R., "Complementary Pivoting," Ph.D. Thesis, Rensselaer Polytechnic Institute, 1970.

McKENZIE, L., "Why Compute Economic Equilibria?" *Computing Equilibria: How and Why.* Amsterdam: North-Holland, 1976.

MACKINNON, J. G., "Solving Economic General Equilibrium Models by the Sandwich Method," in *Fixed Points: Algorithms and Applications*, eds. S. Karamardian and C. B. Garcia. New York: Academic Press, 1977.

MAGNANTI, T. L., AND B. L. GOLDEN, "Transportation Planning: Network Models and Their Implementation," in *Studies in Operations Management*, ed. A. C. Hax. New York: North-Holland, 1978, pp. 465–518.

MAIER, G., "A Matrix Structural Theory of Piecewise-Linear Elastoplasticity with Interacting Yield-Planes," *Meccanica* 5 [1970], 54–66.

MAJTHAY, A., "On Complementary Pivot Theory," *Studia Scientiarum Mathematicarum Hungarica* 4 [1969], 213–24.

MANGASARIAN, O. L., *Nonlinear Programming.* New York: McGraw-Hill, 1969.

——, "Equivalence of the Complementarity Problem to a System of Nonlinear Equations," University of Wisconsin Technical Report No. 227, 1974.

——, "Linear Complementarity Problems Solvable by a Single Linear Program," *Mathematical Programming* 10 [1976], 263–70.

MAS-COLELL, A., "A Note on a Theorem of F. Browder," *Mathematical Programming* 6 [1974a], 229–33.

——, "An Equilibrium Existence Theorem without Complete or Transitive Preferences," *Journal of Mathematical Economics* 1 [1974b], 237–46.

MEGIDDO, N., AND M. KOJIMA, "On the Existence and Uniqueness of Solutions in Nonlinear Complementarity Theory," *Mathematical Programming* 12 [1977], 110–30.

MENZEL, R., AND H. SCHWETLICK, "Zur Lösung Parameterabhängiger Nichtlinearer Gleichungen Mit Singulären Jacobi-Matrizen," *Numerische Mathematic* 30 [1978], 65–79.

MERRILL, O. H., "Applications and Extensions of an Algorithm that Computes Fixed Points of Certain Non-Empty Convex Upper Semi-Continuous Point to Set Mappings," Technical Report 71–7, University of Michigan, 1971.

MEYER, G., "On Solving Nonlinear Equations with a One-Parameter Operator Imbedding," *SIAM Journal of Numerical Analysis* 5 [1968], 739–52.

MILNOR, J., *Topology from the Differentiable Viewpoint.* Charlottesville: University Press of Virginia, 1965.

MORÉ, J., "Classes of Functions and Feasibility Conditions in Nonlinear Complementarity Problems," *Mathematical Programming* 6 [1974], 327–38.

MURCHLAND, J. D., "Braess Paradox of Traffic Flow," *Transportation Research* 4 [1970], 391–94.

MURTY, K. G., "On a Characterization of P-Matrices," *SIAM Journal of Applied Mathematics* 20 [1971], 378–84.

——, "On the Number of Solutions to the Complementarity Problems and Spanning Properties of Complementarity Cones," *Linear Algebra and Applications* 5 [1972], 65–108.

NAGUMO, M., "A Theory of Degree of Mappings Based on Infinitesimal Analysis," *American Journal of Mathematics* 73 [1951], 485–96.

NASH, J. F., "Equilibrium Points in n-Person Games," *Proceedings of the National Academy of Sciences* 36 [1950a], 48–49.

———, "The Bargaining Problem," *Econometrica* 18 [1950b], 155–62.

NEUMANN, J. VON, "Zur Theorie des Gesellshaftsspiele," *Mathematische Annalen* 100 [1928], 295–320.

———, "Über ein Ökonomisches Gleichungssystem und eine Verallgemeinerung des Brouwerschen Fixpunktsatzes," *Ergebnisse eines Mathematischen Kolloquiums* 8 [1937], 73–83.

NEUMANN, J. VON, AND O. MORGENSTERN, *Theory of Games and Economic Behavior.* Princeton, N.J.: Princeton Press, 1947.

ODEN, J. T., *Finite Elements of Nonlinear Continua.* New York: McGraw-Hill, 1972.

ORTEGA, J. M., AND W. C. RHEINBOLDT, *Iterative Solution of Nonlinear Equations in Several Variables.* New York: Academic Press, 1970.

OSTROWSKI, A., *Solution of Equations and Systems of Equations.* New York: Academic Press, 1960.

PANG, J. S., "On Q-Matrices," *Mathematical Programming* 17 [1979], 243–47.

PANG, J. S., I. KANEKO, AND W. P. HALLMAN, "On the Solution of Some (Parametric) Linear Complementarity Problems with Applications to Portfolio Selection, Structural Engineering, and Actuarial Graduation," *Mathematical Programming* 16 [1979], 325–47.

PANNE, C. VAN DE, "A Complementary Variant of Lemke's Method for the Linear Complementarity Problem," *Mathematical Programming* 7 [1974], 283–310.

PEITGEN, H. O., *Approximation of Fixed Points and Functional Differential Equations.* New York: Springer-Verlag, 1979.

PEITGEN, H. O., AND M. PRÜFER, "The Leray-Schauder Continuation Method Is a Constructive Element in the Numerical Study of Nonlinear Eigenvalue and Bifurcation Problems," Report No. 122, Universität Bremen, 1979.

POINCARE, H., "Sur les courbes défines par une équation differentielle I," Oeuvres I, Paris: Gauthier-Villars, 1881.

PONTRYAGIN, L. S., *Foundations of Combinatorial Topology.* Rochester: Graylock Press, 1952.

PRÜFER, M., AND H. W. SIEGBERG, "Complementary Pivoting and the Hopf Degree Theorem," Universitat Bremen Report No. 20, 1980.

RABINOWITZ, P. H., "Some Global Results for Nonlinear Eigenvalue Problems," *Journal of Functional Analysis* 7 [1971], 487–513.

———, ED., *Applications of Bifurcation Theory.* New York: Academic Press, 1977.

RHEINBOLDT, W. C., "Numerical Continuation Methods for Finite Element Applications," in *Formulations and Computational Algorithms in Finite Element Analysis,* ed. J. Bathe. Cambridge: MIT Press, 1977, pp. 599–631.

———, "Numerical Methods for a Class of Finite Dimensional Bifurcation Problems," *SIAM Journal of Numerical Analysis* 15 [1978], 1–11.

———, "Solution Fields of Nonlinear Equations and Continuation Methods," *SIAM Journal of Numerical Analysis* 17 [1980], 221–37.

ROBERTS, S. M., AND J. S. SHIPMAN, "Continuation in Shooting Methods for Two-Point Boundary Value Problems," *Journal of Mathematical Analysis and Applications* 21 [1967], 23–30.

———, *Two-Point Boundary Value Problems: Shooting Methods*. New York: Elsevier, 1972.

ROCKAFELLAR, R. T., *Convex Analysis*. Princeton, N.J.: Princeton Press, 1970.

SAIGAL, R., "On the Class of Complementary Cones and Lemke's Algorithm," *SIAM Journal of Applied Mathematics* 23 [1972], 46–60.

———, "Extension of the Generalized Complementarity Problem," *Mathematics of Operations Research* 1 [1976a], 260–66.

———, "On Paths Generated by Fixed Point Algorithms," *Mathematics of Operations Research* 1 [1976b], 359–80.

———, "On the Convergence Rate of Algorithms for Solving Equations that Are Based on Methods of Complementary Pivoting," *Mathematics of Operations Research* 2 [1977], 108–24.

———, "On Piecewise Linear Approximations to Smooth Mappings," *Mathematics of Operations Research* 4 [1979], 153–61.

SAIGAL, R. AND C. SIMON, "Generic Properties of the Complementarity Problem," *Mathematical Programming* 4 [1973], 324–35.

SAMELSON, H., R. M. THRALL, AND O. WESLER, "A Partition Theorem for Euclidean N-Space," *Proceedings of the American Mathematical Society* 9 [1958], 805–807.

SARD, A., "The Measure of the Critical Points of Differentiable Maps," *Bulletin of the American Mathematics Society* 48 [1942], 883–90.

SCARF, H. E., "The Approximation of Fixed Points of Continuous Mappings," *SIAM Journal of Applied Mathematics* 15 [1967], 1328–43.

SCARF, H. E. AND T. HANSEN, *Computation of Economic Equilibria*. New Haven: Yale University Press, 1973.

SCHMIDT, C. P., "Homotopies and the Existence of Solutions of Systems of Nonlinear Equations," Ph.D. Thesis, University of Chicago, 1979.

SCHRAMM, R., "On the Number of Solutions of Piecewise Linear Equations," in *Numerical Solution of Highly Nonlinear Problems*, ed. W. Forster. Amsterdam: North-Holland, 1980, pp. 239–40.

SCHWARTZ, J. T., *Nonlinear Functional Analysis*. New York: Gordon and Breach, 1969.

SHAFER, W., AND H. SONNENSCHEIN, "Equilibrium in Abstract Economies without Ordered Preferences," *Journal of Mathematical Economics* 2 [1975], 345–48.

SHAMPINE, L. F., AND M. K. GORDON, *Computer Solution of Ordinary Differential Equations: The Initial Value Problem*. San Francisco: Freeman Press, 1975.

SHAPLEY, L. S., "On Balanced Games Without Side Payments," in *Mathematical Programming*, eds. T. C. Hu and S. M. Robinson. New York: Academic Press, 1973.

———, "A Note on the Lemke-Howson Algorithm," *Mathematical Programming Study* 1 [1974], 175–89.

SHAPLEY, L. S., AND H. E. SCARF, "On Cores and Indivisibility," *Journal of Mathematical Economics* 1 [1974], 23–38.

SHOVEN, J. B., "Applying Fixed Points Algorithms to the Analysis of Tax Policies," in *Fixed Points: Algorithms and Applications*, eds. S. Karamardian and C. B. Garcia. New York: Academic Press, 1977, pp. 403–34.

SIDLOVSKAYA, N., "Application of the Method of Differentiation with Respect to a Parameter to the Solution of Nonlinear Equations in Banach Spaces" (Russian), *Leningrad Gos. Univ. Učen. Zap. Ser. Mat. Nauk* 33 [1958], 3–17.

SIEGBERG, H. W., *Abbildungsgrade in Analysis und Topologie*. Diplomarbeit, Universitat Bonn, 1976.

SLATER, M., "Lagrange Multipliers Revisited: A Contribution to Nonlinear Programming," Cowles Discussion Paper No. 403, 1950.

SMALE, S., "A Convergent Process of Price Adjustment and Global Newton Methods," *Journal of Mathematical Economics* 3 [1976], 107–20.

SPERNER, E., "Neur Beweis für die Invarianz der Dimensionszahl und des Gebietes," *Abh. a.d. Math. Sem. d. Univ. Hamburg* 6 [1928], 265–72.

SPINGARN, J. E., AND R. T. ROCKAFELLAR, "The Generic Nature of Optimal Conditions in Nonlinear Programs," *Mathematics of Operations Research* 4 [1979], 425–30.

STEINBERG, R. AND W. I. ZANGWILL, "The Prevalence of Braess' Paradox," Graduate School of Business Columbia University Report, 1981.

STERNBERG, S., *Lectures on Differential Geometry*. Englewood Cliffs, N.J.: Prentice-Hall, 1964.

THOM, R., *Stabilité Structurelle et Morphogénèse*. Boston: W. A. Benjamin, 1972.

TODD, M. J., "A Generalized Complementary Pivoting Algorithm," *Mathematical Programming* 6 [1974], 243–63.

————, *The Computation of Fixed Points and Applications*. Heidelberg; New York: Springer-Verlag, 1976a.

————, "Orientation in Complementarity Pivoting," *Mathematics of Operations Research* 1 [1976b], 54–66.

————, "On Triangulations for Computing Fixed Points," *Mathematical Programming* 10 [1976c], 322–46.

————, "Union Jack Triangulations," in *Fixed Points: Algorithms and Applications*, eds. S. Karamardian and C. B. Garcia. New York: Academic Press, 1977, pp. 315–36.

————, "Improving the Convergence of Fixed Point Algorithms," in *Mathematical Programming Study* 7, eds. M. L. Balinski and R. W. Cottle. Amsterdam: North-Holland, 1978, pp. 151–69.

————, "Traversing Large Pieces of Linearity in Algorithms that Solve Equations by Following Piecewise-Linear Paths," *Mathematics of Operations Research* 5 [1980a], 242–57.

————, "Exploiting Structure in Piecewise-Linear Homotopy Algorithms for Solving Equations," *Mathematical Programming* 18 [1980b], 233–47.

TUY, H., "Pivotal Methods for Computing Equilibrium Points: Unified Approach and New Restart Algorithm," *Mathematical Programming* 16 [1979], 210–27.

VAINBERG, M., *Variational Methods for the Study of Nonlinear Operators*, Gostekhteoretizdat, Moscow. (Translation by A. Feinstein, 1964.) New York: Holden-Day, 1956.

VAN DER HEYDEN, L., "A Variable Dimension Algorithm for the Linear Complementarity Problem," *Mathematical Programming* 19 [1980], 328–46.

VAN DER LAAN, G. AND A. J. J. TALMAN, "A Restart Algorithm for Computing Fixed Points without an Extra Dimension," *Mathematical Programming* 17 [1979], 74–84.

———, "An Improvement of Fixed Point Algorithms by Using a Good Triangulation," *Mathematical Programming* 18 [1980], 274–85.

VAN DER WAERDEN, B. L., *Modern Algebra*, I and II. New York: F. Ungar Publishing, 1931.

VARGA, R., *Matrix Iterative Analysis*. Englewood Cliffs, N.J.: Prentice-Hall, 1962.

VERTGEIM, B. A., "On an Approximate Determination of the Fixed Points of Continuous Mappings," *Soviet Math Doklady* 11 [1970], 295–98.

VILLE, J., "Sur la théorie générale des jeux où intervient l' habilité des joueurs," *Traité du calcul des probabilités et de ses applications*, ed. E. Borel and collaborators, 2. Paris: Gauthier-Villars et. Cie, 1938, pp. 105–13.

WACKER, H., *Continuation Methods*. New York: Academic Press, 1977.

WALRAS, L., *Elements d'économie politique pure*. Lausanne: L. Corbaz, 1874.

WARDROP, J. G., "Some Theoretical Aspects of Road Traffic Research," *Proceedings of the Institute of Civil Engineers, Part II* 1 [1952], 325–78.

WASSERSTROM, E., "Numerical Solutions by the Continuation Method," *SIAM Review* 15 [1973], 89–119.

WATSON, L. T., "A Variational Approach to the Linear Complementarity Problem," Ph.D. thesis, University of Michigan, 1974.

WEYL, H., "Elementary Proof of a Minimax Theorem Due to Von Neumann," in *Contributions to the Theory of Games*, eds. H. W. Kuhn and A. W. Tucker. Princeton, N.J.: Princeton Press, 1950.

WHALLEY, J., "Fiscal Harmonization in the EEC; Some Preliminary Findings of Fixed Point Calculations," in *Fixed Points: Algorithms and Applications*, eds. S. Karamardian and C. B. Garcia. New York: Academic Press, 1977, pp. 435–72.

WHITNEY, H., *Geometric Integration Theory*. Princeton, N.J.: Princeton Press, 1957.

WILMUTH, R. J., "The Computations of Fixed Points," Ph.D. Thesis, Stanford University, 1973.

YAMAMURO, S., "Some Fixed Points Theorems in Locally Convex Linear Spaces," *Yokohama Math Journal* 11 [1963], 5–12.

ZANGWILL, W. I., *Nonlinear Programming: A Unified Approach*. Englewood Cliffs, N.J.: Prentice-Hall, 1969.

———, "An Eccentric Barycentric Fixed Point Algorithm," *Mathematics of Operations Research* 2 [1977], 343–59.

ZANGWILL, W. I., AND C. B. GARCIA, "Equilibrium Programming: The Path-Following Approach and Dynamics," to be published, *Mathematical Programming* [1981].

ZEEMAN, E. C., *Catastrophe Theory: Selected Papers 1972–1977*. Reading, Mass.: Addison-Wesley, 1977.

Author Index

Index